# Nanocomposite Design for Energy-Related Applications

# Nanocomposite Design for Energy-Related Applications

Guest Editors

**Hanfeng Liang**
**Qiu Jiang**
**Gang Huang**
**Yi-Zhou Zhang**

MDPI

Basel • Beijing • Wuhan • Barcelona • Belgrade • Novi Sad • Cluj • Manchester

*Guest Editors*

Hanfeng Liang
College of Chemistry and Chemical Engineering
Xiamen University
Xiamen
China

Qiu Jiang
School of Materials and Energy
University of Electronic Science and Technology of China
Chengdu
China

Gang Huang
State Key Laboratory of Rare Earth Resource Utilization
Changchun Institute of Applied Chemistry, CAS
Changchun
China

Yi-Zhou Zhang
School of Chemistry and Materials Science
Nanjing University of Information Science and Technology
Nanjing
China

*Editorial Office*
MDPI AG
Grosspeteranlage 5
4052 Basel, Switzerland

This is a reprint of the Special Issue, published open access by the journal *Nanomaterials* (ISSN 2079-4991), freely accessible at: https://www.mdpi.com/journal/nanomaterials/special_issues/nano_composite_energy.

For citation purposes, cite each article independently as indicated on the article page online and as indicated below:

Lastname, A.A.; Lastname, B.B. Article Title. *Journal Name* **Year**, *Volume Number*, Page Range.

**ISBN 978-3-7258-5487-5 (Hbk)**
**ISBN 978-3-7258-5488-2 (PDF)**
**https://doi.org/10.3390/books978-3-7258-5488-2**

# Contents

# About the Editors

**Hanfeng Liang**

Hanfeng Liang is an associate professor of Chemical Engineering at the College of Chemistry and Chemical Engineering, Xiamen University, China. His research focuses on nanomaterials and functional coatings as well as their applications in energy, medical, and optical devices. He is a Clarivate Highly Cited Researcher and was included in the "World's Top 2% Scientists – Career-long Impact list". Currently he serves as the subject editor of International Journal of Hydrogen Energy, and is on the (youth) editorial board of several journals.

**Qiu Jiang**

Qiu Jiang is an associate professor at the School of Materials and Energy at the University of Electronic Science and Technology of China (UESTC). He obtained his Ph.D. at King Abdullah University of Science and Technology in 2019, followed by a postdoc at the University of Cambridge. His research interests focus on developing methods for the controlled synthesis of two-dimensional materials, understanding their fundamental properties, and the surface modifications of those materials for energy storage and conversion applications.

**Gang Huang**

Gang Huang is a full professor at the State Key Laboratory of Rare Earth Resources and Utilization, Changchun Institute of Applied Chemistry, Chinese Academy of Sciences. He was named among the "World's Top 2% Scientists" by Stanford University. His work primarily focuses on fundamental research and technological development in the field of advanced battery materials and devices, particularly key materials for metal-air batteries.

**Yi-Zhou Zhang**

Yi-Zhou Zhang is a professor of Materials Science and Engineering at Nanjing University of Information Science and Technology (NUIST) and head of the institute of Advanced Materials and Flexible electronics in NUIST. Prof. Zhang mainly works on functional materials for printed and flexible energy storage and electronics. Due to his contributions to this field, he has been recognized as a Highly Cited Researcher by Clarivate in 2023 and 2024.

nanomaterials

MDPI

Editorial

# Nanocomposite Design for Energy-Related Applications

Qiu Jiang [1,2,*], Hanfeng Liang [3], Yizhou Zhang [4] and Gang Huang [5]

1 Yangtze Delta Region Institute (Huzhou), University of Electronic Science and Technology of China, Huzhou 313000, China
2 School of Materials and Energy, University of Electronic Science and Technology of China, Chengdu 611731, China
3 State Key Laboratory of Physical Chemistry of Solid Surfaces, Tan Kah Kee Innovation, Laboratory (IKKEM), College of Chemistry and Chemical Engineering, Xiamen University, Xiamen 361005, China; hfliang@xmu.edu.cn
4 Institute of Advanced Materials and Flexible Electronics (IAMFE), School of Chemistry and Materials Science, Nanjing University of Information Science and Technology, Nanjing 210044, China; yizhou.zhang@nuist.edu.cn
5 State Key Laboratory of Rare Earth Resource Utilization, Changchun Institute of Applied Chemistry, Chinese Academy of Sciences, Changchun 130022, China; ghuang@ciac.ac.cn
* Correspondence: jiangqiu@uestc.edu.cn

**Abstract:** Nanocomposites, which combine various nanomaterials, offer immense potential in the design of advanced materials for energy-related applications. These materials, engineered at the nanoscale, exhibit enhanced properties compared to their bulk counterparts, such as improved electrical conductivity, mechanical strength, and thermal stability. Nanocomposites have emerged as promising candidates for use in energy storage systems, including batteries and supercapacitors, by improving energy density, cycle life, and charge–discharge rates. In renewable energy technologies such as fuel cells, nanocomposites play a crucial role in enhancing efficiency and stability, which are vital for reducing costs and promoting the adoption of clean energy solutions. The unique properties of nanocomposites, such as high surface area and tunable composition, allow for the integration of multiple functionalities, making them ideal for multifunctional catalysts in energy conversion and environmental remediation. Additionally, nanocomposites enable the development of energy harvesting systems with improved performance and durability. These materials can be tailored by adjusting the composition of the nanomaterials, opening new opportunities for energy applications. The increasing research into nanocomposites continues to drive innovation in energy-related technologies, positioning them as a key enabler for sustainable energy solutions and future advancements in renewable energy systems.

**Keywords:** nanocomposites; energy; lithium-ion battery; thermochromic; SOFC; binder

Nanocomposites, materials that combine multiple nanomaterial components, have emerged as a significant area of research due to their unique properties and potential applications across a wide range of industries. When engineered at the nanoscale, materials often exhibit distinct electronic, physical, and mechanical characteristics that are drastically different from their bulk counterparts. These unique properties are particularly advantageous in the design of advanced technologies in energy systems, such as batteries, supercapacitors, solar cells, fuel cells, and multifunctional catalysts. Nanocomposites take advantage of the synergistic combination of the properties of different nanomaterials, allowing for the creation of materials with enhanced or even entirely new functionalities that cannot be achieved through the use of individual components alone. This synergy offers nanocomposites greater flexibility in terms of chemical composition, morphology, and other properties, making them suitable for highly specialized applications that require optimized performance. For example, in energy storage systems such as batteries and supercapacitors, nanocomposites provide enhanced energy density, longer cycle life, and faster charge–discharge capabilities compared to conventional materials. They offer improved efficiency and stability in solar cells and fuel cells, crucial for reducing costs and increasing the feasibility of renewable energy technologies. Additionally, nanocomposites play an essential role in multifunctional catalysts, where their unique surface properties can accelerate chemical reactions, increasing their utility in sustainable energy production and environmental remediation. The flexibility of nanocomposites allows researchers to tailor their properties by adjusting the composition of the components, opening new avenues for energy harvesting, conversion, and storage technologies. Their potential to improve existing materials and enable the development of new energy systems is transformative, positioning nanocomposites as a cornerstone for future advances in renewable energy and sustainability. As research on nanocomposites continues to evolve, their applications in novel energy-related fields are expected to expand, driving innovation in the quest for more efficient and sustainable solutions to global energy challenges. This Special Issue features a selection of six original research articles and two in-depth reviews, focusing on the latest developments, ongoing challenges, and future prospects in this evolving field.

One often-overlooked aspect of the energy storage field is its environmental impact. A prime example is the widespread use of polyvinylidene fluoride (PVDF) as a binder. These substances are notorious for their persistent and bioaccumulating pollution, particularly during production processes, leading to significant environmental concerns. To address this, Nirschl et al. investigated the potential of using silica as a substitute for PVDF as a binder for carbon black [1]. They synthesized colloidal silica particles via the Stoeber process and coated them onto carbon black using a spray flame method. The study revealed that larger silica particles, due to their higher mass, tended to deflect upon contact with carbon black aggregates, while smaller particles, especially those with a primary particle size of 10 nm, adhered more readily and sintered to the carbon black. The findings suggest that the optimal conditions for the hetero-aggregation of carbon black and silica occur when the mass ratios of silica to carbon black are equal to or below one and when the silica particles are small in size.

Regarding materials engineering for energy storage, Wang and colleagues introduced a novel three-dimensional porous nanosheet antimony/carbon (3DPNS-Sb/C) anode [2]. This composite demonstrates a reversible capacity of 511.5 mAh $g^{-1}$ when cycled at 0.5 A $g^{-1}$ over 100 cycles, and maintains 289.5 mAh $g^{-1}$ at an elevated rate of 10 A $g^{-1}$. The synthesis protocol entails a controlled hydrothermal reaction followed by thermal annealing, during which Sb nanoparticles are homogeneously generated in situ within a self-supporting organic carbon matrix. The uniform dispersion of Sb is ascribed to the facile reduction of sodium antimonate and the strong interfacial affinity afforded by the carbon network. Consequently, the resulting 3DPNS-Sb/C composite leverages both the structural advantages of its polymer-derived framework and the enhanced electrical conductivity imparted by the carbon scaffold. Similarly, Ma et al. synthesized a series of Sn-doped $Li_{1.3}Al_{0.3}Sn_xTi_{1.7-x}(PO_4)_3$ (LATP-xSn) ceramic electrolytes via a conventional solid-state route, systematically varying the $Sn^{4+}$ content to probe its effects on crystal structure and ion-transport characteristics [3]. To mitigate interfacial resistance between the ceramic and polymer phases, they fabricated a sandwich-type composite electrolyte by casting a PEO-based polymer precursor onto both faces of the modified LATP pellet, rather than blending the two constituents homogeneously. This bilayer architecture was shown to markedly lower solid–solid interface impedance, thereby enhancing overall electrochemical performance.

Thermochromic materials—which reversibly alter their optical or electrical properties in response to temperature changes—hold significant promise for energy-efficient technologies, adaptive optics, and environmental sensing. By dynamically modulating solar transmittance or reflectance, thermochromic coatings can reduce building energy consumption for heating and cooling without relying on external power, thereby lowering greenhouse gas emissions and operational costs. In its monoclinic form, vanadium dioxide ($VO_2$) undergoes a reversible phase transition to the rutile structure upon heating, a transformation accompanied by a sharp drop in near-infrared transmittance that makes it an attractive candidate for smart window coatings. Zhao et.al reported a straightforward one-step annealing procedure to synthesize $VO_2$ powders [4]. Following synthesis, the particles were rapidly quenched in either deionized water or ethanol to introduce controlled surface lattice distortions. The resulting composite films exhibited pronounced localized surface plasmon resonance (LSPR) effects, leading to a significant enhancement in solar-energy modulation performance compared to undistorted $VO_2$ coatings.

Molecular oxygen, upon activation by visible light, generates radicals with substantial oxidation capabilities, offering significant potential for environmental remediation. In a noteworthy study, Zhang et al. demonstrated that coupling two-dimensional (2D) $CdIn_2S_4$ with 2D graphitic carbon nitride (g-$C_3N_4$) represents an efficient strategy to enhance the photocatalytic activity of $CdIn_2S_4$ while simultaneously reducing the reliance on cadmium species [5]. Their research reported the successful construction of 2D/2D $CdIn_2S_4$/g-$C_3N_4$ nanocomposites with varying weight ratios, achieved through a straightforward and cost-effective electrostatic self-assembly technique. This approach not only improves the photocatalytic performance but also offers a sustainable method for reducing the environmental impact associated with Cd-based materials.

Transition metal nitrides have emerged as materials of considerable research interest due to their hybrid bonding character—combining metallic, ionic, and covalent interactions—which underpins their remarkable mechanical and electronic performance. In a recent first-principles investigation, Zhang and co-workers explored the impact of manganese substitution on the cubic phase of molybdenum nitride (d-MoN:Mn) [6]. Through detailed electronic structure calculations and comparative analysis of multiple spin arrangements, they unraveled the mechanisms through which Mn dopants induce magnetic behavior and mediate exchange interactions. Their results reveal how the introduction of Mn modifies the host's band structure and fosters long-range magnetic coupling among dopant sites.

In addition to the recent experimental advancements in the field, two comprehensive reviews have made substantial contributions by providing in-depth analyses of specific materials relevant to energy applications. One of these reviews focuses on the two prominent families of oxygen-conducting electrolytes: doped lanthanum aluminates ($LaAlO_3$) and lanthanum gallates ($LaGaO_3$) [7]. These materials are of particular interest due to their crucial role in solid oxide fuel cells (SOFCs), where high ionic conductivity and chemical stability at elevated temperatures are essential. The review meticulously examines the preparation methods of these electrolytes, their chemical stability, thermal behavior, and transport properties, all of which are deeply intertwined with the performance characteristics required for efficient SOFC operation. The second review delves into the promising field of inks based on two-dimensional (2D) MXene sheets, which have garnered significant attention due to their unique properties and versatile applications [8]. This review comprehensively explores the recent developments in the synthesis of MXene-based inks, highlighting their potential for use in various applications such as flexible electronics, sensors, energy storage devices, and as conductive materials in printed electronics. The properties, advantages, and limitations of MXene inks are thoroughly discussed, with particular attention to their future potential in the rapidly evolving field of additive manufacturing and printed electronics.

**Author Contributions:** All the authors wrote and reviewed this Editorial Letter. All authors have read and agreed to the published version of the manuscript.

**Funding:** This research was funded by the "Pioneer" and "Leading Goose" R&D Program of Zhejiang (No. 2023C03017), the National Natural Science Foundation of China (22405035), the Natural Science Foundation of Sichuan Province (No. 2024NSFSC1104), and the Huzhou Science and Technology Bureau (No. 2023GZ02).

**Conflicts of Interest:** The authors declare no conflicts of interest.

## References

1. Buchheiser, S.; Kistner, F.; Rhein, F.; Nirschl, H. Spray Flame Synthesis and Multiscale Characterization of Carbon Black–Silica Hetero-Aggregates. *Nanomaterials* **2023**, *13*, 1893. [CrossRef] [PubMed]
2. Wang, Z.; Zeng, F.; Zhang, D.; Shen, Y.; Wang, S.; Cheng, Y.; Li, C.; Wang, L. Antimony Nanoparticles Encapsulated in Self Supported Organic Carbon with a Polymer Network for High-Performance Lithium-Ion Batteries Anode. *Nanomaterials* **2022**, *12* 2322. [CrossRef] [PubMed]
3. Xu, A.; Wang, R.; Yao, M.; Cao, J.; Li, M.; Yang, C.; Liu, F.; Ma, J. Electrochemical Properties of an Sn-Doped LATP Ceramic Electrolyte and Its Derived Sandwich-Structured Composite Solid Electrolyte. *Nanomaterials* **2022**, *12*, 2082. [CrossRef] [PubMed]
4. Wu, S.; Zhou, L.; Li, B.; Tian, S.; Zhao, X. Enhanced Thermochromic Performance of $VO_2$ Nanoparticles by Quenching Process *Nanomaterials* **2023**, *13*, 2252. [CrossRef] [PubMed]
5. Yin, H.; Yuan, C.; Lv, H.; He, X.; Liao, C.; Liu, X.; Zhang, Y. Construction of Electrostatic Self-Assembled 2D/2D CdIn2S4/g-C3N4 Heterojunctions for Efficient Visible-Light-Responsive Molecular Oxygen Activation. *Nanomaterials* **2021**, *11*, 2342. [CrossRef] [PubMed]
6. Wang, K.; Yu, J.; Chi, C.; Zhang, G. Magnetic Properties in Mn-Doped δ-MoN: A Systematic Density Functional Theory Study *Nanomaterials* **2022**, *12*, 747. [CrossRef] [PubMed]
7. Filonova, E.; Medvedev, D. Recent Progress in the Design, Characterisation and Application of $LaAlO_3$- and $LaGaO_3$-Based Solid Oxide Fuel Cell Electrolytes. *Nanomaterials* **2022**, *12*, 1991. [CrossRef] [PubMed]
8. Aghayar, Z.; Malaki, M.; Zhang, Y. MXene-Based Ink Design for Printed Applications. *Nanomaterials* **2022**, *12*, 4346. [CrossRef] [PubMed]

*nanomaterials*

*Article*

# Spray Flame Synthesis and Multiscale Characterization of Carbon Black–Silica Hetero-Aggregates

**Simon Buchheiser *, Ferdinand Kistner, Frank Rhein and Hermann Nirschl ***

Process Machines, Institute of Mechanical Process Engineering and Mechanics, Karlsruhe Institute of Technology, 76131 Karlsruhe, Germany; ukxmk@student.kit.edu (F.K.); frank.rhein@kit.edu (F.R.)

* Correspondence: simon.buchheiser@kit.edu (S.B.); hermann.nirschl@kit.edu (H.N.)

**Abstract:** The increasing demand for lithium-ion batteries requires constant improvements in the areas of production and recycling to reduce their environmental impact. In this context, this work presents a method for structuring carbon black aggregates by adding colloidal silica via a spray flame with the goal of opening up more choices for polymeric binders. The main focus of this research lies in the multiscale characterization of the aggregate properties via small-angle X-ray scattering, analytical disc centrifugation and electron microscopy. The results show successful formation of sinter-bridges between silica and carbon black leading to an increase in hydrodynamic aggregate diameter from 201 nm to up to 357 nm, with no significant changes in primary particle properties. However, segregation and coalescence of silica particles was identified for higher mass ratios of silica to carbon black, resulting in a reduction in the homogeneity of the hetero-aggregates. This effect was particularly evident for silica particles with larger diameters of 60 nm. Consequently, optimal conditions for hetero-aggregation were identified at mass ratios below 1 and particle sizes around 10 nm, at which homogenous distributions of silica within the carbon black structure were achieved. The results emphasise the general applicability of hetero-aggregation via spray flames with possible applications as battery materials.

**Keywords:** hetero-aggregation; small-angle X-ray scattering; carbon black; silica; spray flame; nanoparticle characterization

## 1. Introduction

The urgent need for a reduction in the use of fossil fuels in order to reduce $CO_2$ emissions has led to the emergence of electro mobility, e.g., in the form of electric cars. Currently, lithium-ion batteries are the most frequently used component for energy storage. The batteries consist of multiple components, including active materials, polymer binders and carbon black. Therein, the active materials are responsible for energy storage, and the carbon black increases the conductivity via the creation of electrical pathways between the active material particles. At the cathode side, a frequently used binder is polyvinylidene fluoride (PVDF), which ensures the uniformity and binding of the active materials, correct slurry rheology [1] and the dispersion of the carbon black [2]. However, PVDF is part of the group of per- and polyfluoroalkyl substances (PFASs), which are responsible for the persistent and bioaccumulating pollution of the environment during production [3] and recycling [4]. Thus, the European Union is evaluating a possible ban on the production, use, sale and import of PFASs until September 2023, which is why the development and evaluation of alternatives is of the utmost importance. However, PVDF has the advantages of a high mechanical strength, corrosion resistance and chemical stability, making it difficult to replace [5]. One option is to improve the properties of the carbon black network with another additive to reduce the technical requirements on the binder. In polymers, the addition of silica to carbon black is used to adjust mechanical properties. One main application is the reinforcement of tires [6], in which the miscibility of both components is linked to durability [7]. Although silica has insulating properties, it is able to improve the generation of electrical pathways via carbon black, which may lead to increased conductivities [8]. Recently investigated carbon–silica structures for possible applications in Li-ion batteries as anode materials include hollow spheres [9], aerogels [10], amorphous powders [11] and carbon-coated silica nanoparticles [12]. To overcome the limitations of volume expansion due to lithiation at the anode, the silica is nanostructured [13,14]. Moreover, on the cathode side, nanostructured silica has been shown to improve the electrochemical cycling stability [15]. Regarding the desired structure of carbon black, an open, high-surface-area structure [16] that forms a robust carbon black–binder network is important for the electrochemical performance [17,18]. Therein, the aggregate size of the carbon black has a significant impact on the carbon black–binder network, with smaller particles (<250 nm) forming a more cross-linked network [19]. The current methods for the synthesis of carbon black–$SiO_2$ composite materials include complex manufacturing methods that are not easily scalable for industrial production. Therefore, this work introduces a hetero-aggregation process carried out via the sintering of silica nanoparticles directly onto the carbon black aggregates in a spray flame. Spray flames are scalable [20] and allow for the dispersion of both particles due to the active mixing zone above the nozzle [21]. The feasibility of the spray flame synthesis of hetero-aggregates has been shown frequently for catalysts in which the sintered contact between the catalytic material and promoter is able to improve the catalytic activity [22,23]. For hetero-aggregates of carbon black and silica, the sintered hetero-contact may lead to improvements in mechanical strength and dispersibility. In theory, this leads to a reduction in the demand on the multi-functionality of the polymeric binder, opening up alternatives to PVDF. Another possible improvement gained via hetero-aggregation is an increase in aggregate stability, leading to easier processing during dry mixing, during which conductivity due to the deformation and breakage of carbon black is possible [24].

In order to achieve the desired macroscopic properties, the hetero-aggregation process and its influencing factors must be understood on a microscopic level. However, the resulting hetero-aggregates are highly fractal nanoscale structures that consist of two different amorphous materials and are therefore difficult to reliably characterize. Hence, a multiscale characterization ranging from primary particle properties to aggregate properties is needed. Common characterization methods for hetero-aggregates are laser light diffraction [25,26], X-Ray diffraction [26] and electron microscopy [27,28]. In order to obtain comprehensive information about primary particles, as well as aggregate and mixing properties, a combination of ultra-small-angle X-ray scattering (USAXS), high-angle annular dark-field scanning transmission electron microscopy (HAADF-STEM) with energy-dispersive X-ray spectroscopy (EDXS) and analytical disc centrifugation (ADC) is used. USAXS provides integral information about the aggregates and is routinely used for aggregates of carbon and silica, as well as products obtained via spray flame synthesis [29–32]. The results include fractal dimensions of the aggregates as well as the mean primary particle sizes of both materials. ADC allows for the determination of changes in the hydrodynamic aggregate size for varying flame parameters. The combination of HAADF-STEM and EDXS yields qualitative information about aggregate shapes, primary particle size distributions and the homogeneity of the dispersion. In order to understand the behaviour of hetero-aggregation in a two-material system, the aggregation mechanisms of pure carbon black and pure silica particles (one-material system) are first investigated. Afterwards, the obtained findings are compared with the experimental results of the hetero-aggregation process in order to determine the influence of the primary silica particle size on the hetero-aggregate size, primary particle properties, fractal dimensions and the homogeneity of the hetero-aggregation process. As a result, optimized experimental conditions are presented.

## 2. Materials and Methods

### 2.1. Synthesis of Silica Particles

The colloidal silica particles were synthesised via the Stoeber process [33], using ammonia (25% ammonia solution for analysis, Merck, Darmstadt, Germany) as a catalyst and tetraethyl orthosilicate (TEOS) (98% tetraethoxysilane, Alfa Aesar, Karlsruhe, Germany) as a precursor. The cosolvent was ethanol (VWR Chemicals, Darmstadt, Germany). Temperature control and mixing were achieved using a magnetic stirrer with a heating plate. The exact experimental conditions are provided in the Supplementary Information. Table 1 lists the names of the samples with the corresponding mean particle diameter, $d_{50,0}$ +/− one standard deviation of the number size distribution averaged over a triple measurement, as determined via dynamic light scattering (Zetasizer nano ZS, Malvern Panalytical, Malvern, UK).

**Table 1.** Overview of the colloidal silica particles produced and their mean diameters.

| Name | $d_{50,0}$ in nm |
|---|---|
| 60 nm | 63.6 ± 14.9 |
| 40 nm | 39.2 ± 11.9 |
| 30 nm | 32.0 ± 8.1 |
| 10 nm | 11.6 ± 3.4 |

### 2.2. Production of the Hetero-Aggregates

For the production of the hetero-aggregates, a carbon black of the type TIMCAL Super C65 (Nanografi Nano technology, Çankaya/Ankara, Turkey) was suspended in ethanol with a mass concentration $x_{CB}$ of 0.5 weight-%. The suspension of the colloidal silica particles was added to the stock solution in varying mass ratios to the concentration of carbon black that ranged from 5 $\frac{x_{CB}}{x_{SiO_2}}$ to 0.2 $\frac{x_{CB}}{x_{SiO_2}}$. The particles were then dispersed with ultrasonic waves (Branson Sonifier 450, Branson Ultrasonics, Danbury, CT, USA) to break up the existing agglomerates. For clarification, in this publication, loosely bonded particles, e.g., particles bonded by Van der Waals forces, are referred to as agglomerates, whereas aggregates are considered particle networks bonded by strong interactions, e.g., solid sinter bridges. The suspension was then fed into the spray flame using a SpraySyn burner (University of Duisburg-Essen, Duisburg, Germany). The gas flows were controlled via mass flow controllers by Bronkhorst (Ruurlo, The Netherlands ). A laminar pilot flame was ignited with 2 standard litres per minute (slm) of $CH_4$ and 12 slm of $O_2$. To stabilise the flame, a sheath gas flow of 120 slm of pressurized air was added. The spray flame was created in the centre of the burner, where a canula with an annular gap was located. A syringe pump (Harvard Instruments, Holliston, MA, USA) conveyed the prepared suspension through the canula. The gas flow of 10 slm of $O_2$ through the annular gap atomized the suspension. The resulting spray was then continuously ignited by the pilot flame, which created the spray flame. A scheme of the burner is presented in the Supplementary Material with an overview of the burner parameters. Further general data regarding the burner are published in [34]. At 12 cm above the burner surface, a hole-in-tube probe collected the hetero-aggregates on a nanoporous track-etched membrane (Whatman Nuclepore Track-Etched Membranes, Cytiva, Amersham, UK) with a pore diameter of 200 nm.

### 2.3. Aggregate Characterization

The hetero-aggregates produced were characterized using small-angle X-ray scattering (Xeuss 2.0 Q-Xoom, Xenocs SA, Sassenage, France) in order to obtain information about the multiple structural characteristics ranging from the primary particle diameters to the fractal dimension of the aggregate. Beam generation was achieved with the X-ray micro focus source Genix3D Cu ULC (Ultra Low divergence), which emits Cu-K$\alpha$ radiation with an energy of 8.04 keV and a wavelength of 1.5406 Å. For an extended measuring range, a Bonse-Hart-module [35] for USAXS measurements was installed. The powder samples were prepared for measurement on adhesive polyimide foil. The chosen exposure time was 30 min, with a distance of 2500 mm from the sample to the detector. A comparative image analysis of the HAADF-STEM (FEI Tecnai Osiris, FEI company, Hillsboro, OR, USA) images, carried out via ImageJ [36], yielded a number-based size distribution of the primary particles. The hydrodynamic aggregate size distribution was measured with an analytical disc centrifuge (CPS instruments, Prairieville, LA, USA). Before the measurement, the particles are suspended in deionised water (0.01 mass-%) and dispersed with ultrasonic waves (Branson Sonifier 450, Branson Ultrasonics, Danbury, CT USA) for the breakage of the agglomerates.

Small-Angle X-ray Scattering

The scattering of incoming X-Rays on nanoparticles is characteristic to their nanoscale structures, such as particle size and morphology. The resulting scattering curve is a double logarithmic plot of an intensity I over the scattering vector q in $^{-1}$. The scattering vector q in Equation (1) describes the scattering angle 2θ independent of the wavelength λ of the primary beam and is provided by:

$$q = \frac{4\pi}{\lambda}\sin(\theta). \quad (1)$$

Guinier's law (Equation (2)) yields information about particle or aggregate size in the form of the radius of gyration $R_g$ with pre-factor G at small angles ($qR_g < 1$):

$$I(q) = G\,\exp\left(-\frac{q^2R_g^2}{3}\right) \quad (2)$$

The slope of the scattering curve is provided by a local power law fit with prefactor B according to:

$$I(q) = B\,q^{-P}. \quad (3)$$

For fractal aggregates, the exponent P is equal to the fractal dimension of the mass of the aggregate $D_{FM}$ [37]. For surface fractals, the exponent P is also proportional to the fractal dimension of the surface $D_{FS}$.

$$P = 6 - D_{FS}, \quad (4)$$

resulting in values for P between 3 and 4, making it possible to distinguish mass and surface fractals. If the exponent P equals 4, the particles exhibit an ideally smooth surface with a sharp density transition, fulfilling Porod's law [38]. Further evaluation of the scattering data was performed via the unified fit model (Irena Package 2.71 [39], IgorPro, Wavemetrics, Portland, OR, USA), according to Beaucage [40]. The results of the unified fit yield a log-normal size distribution with the assumption of spherical primary particles. The geometric standard deviation $\sigma_g$ is calculated from the parameters of Guinier's law and the power law fit with the polydispersity index (PDI) [41]:

$$\sigma_g = \exp\left(\sqrt{\frac{\ln(PDI)}{12}}\right);\ PDI = \frac{B\,R_g^4}{1.62\,G}. \quad (5)$$

The mean diameter of the distribution is then calculated as follows [32]:

$$d_{SAXS} = 2\sqrt{\frac{5}{3}}R_g\exp(-13\frac{\ln(PDI)}{24}). \quad (6)$$

Due to the possible convolution of scattering information from the silica and carbon black in the hetero-aggregates, the above-mentioned methodology was only applied to pure materials.

## 3. Results

### 3.1. Pure Carbon Black in the Spray Flame

Figure 1 shows the changes in the primary particle and aggregate properties of the pure carbon black in a spray flame using ethanol as dispersion liquid. Before the spray flame, the carbon black had an open fractal structure with a fractal dimension of mass of 2.2, which was derived from the scattering curve with a unified fit, as depicted in Figure 1a. The open-branched structure is also observable in the HAADF-STEM image, Figure 1d. The radius of gyration of the primary particles was $R_g = 46.4$ nm and exhibited a sharp density transition with $P = 4$, suggesting a smooth particle surface. From the fit parameters $B = 2.06$ and $G = 8.2 \cdot 10^9$, a number-weighted primary particle size distribution with a median diameter of 38 nm and a geometric standard deviation of $\sigma_g = 0.4$ was calculated using Equation 5. This particle size is in good agreement with the primary particle size reported by Spahr et al. [42]. The ADC yielded a hydrodynamic aggregate size distribution with a mode at 159 nm. After the spray flame, the hydrodynamic aggregate size of the carbon black increased to $d_{mod} = 201$ nm, with a reduction in the fractal dimension of mass to 1.8, suggesting further aggregation in the spray flame. In the TEM image (Figure 1e), the fractal structure of the carbon black is still observable. The analysis of the primary particles via TEM imaging resulted in a mean particle diameter of 39.6 nm with a geometric standard deviation from $\sigma_g = 0.28$. The retrieved fit parameters of the unified fit of the scattering data after the spray flame were $R_g = 33.3$ nm, $G = 1.1 \cdot 10^{10}$ and $B = 6.7$. With $P = 4$, the primary particle surface properties are unchanged. The calculated size distribution shows a decrease in polydispersity in the geometric standard deviation from $\sigma_g = 0.4$ to 0.35 and a decreased primary particle diameter of 35.5 nm. The differences in absolute values are the results of the resolution limit for TEM and potential deviations from the assumptions drawn from the SAXS data.

One major advantage of the holistic approach is the combined evaluation of multiple particle and aggregate properties. Therein, only negligible changes in the primary particle properties after the spray flame process are revealed, whereas the hydrodynamic aggregate size increased due to further aggregation. Because of the newly formed inter-aggregate connections, the structure became more open, which led to a decrease in the fractal dimension of mass.

### 3.2. Pure Colloidal Silica in the Spray Flame

In order to evaluate the changes in the particle properties of the pure colloidal silica in the spray flame using ethanol as dispersion liquid, a comparison of the SAXS data and the TEM analysis was conducted. Figure 2 shows exemplary results for the 60 nm colloidal silica particles before and after the spray flame. In Figure 2a, the scattering curve of the sample after the spray flame exhibits multiple structure levels, which are also visible in the accompanying HAADF STEM image, Figure 2b. For scattering vectors in the range of 0.01 Å$^{-1}$ and 0.1 Å$^{-1}$, the properties of the primary particles (marked in red) were retrieved. The local Guinier fit yielded a radius of gyration $R_g = 25.3$ nm which corresponds to a sphere equivalent diameter of 65.3 nm. A comparison with the TEM data yielded a mean diameter of 61.1 nm and shows good agreement. Furthermore, oscillations characteristic for monodisperse spheres are observed in the scattering data. With a fractal dimension of surface $D_{FS} = 2$, the particles exhibited a smooth surface with a spherical shape, which is identical to the properties prior to the spray flame. The silica particles visibly aggregated and formed sinter bridges in the flame, which were both observable via TEM and SAXS (marked in blue). A local power law fit results in a fractal dimension of mass $D_{FM} = 1.7$. In contrast, the scattering curve of the sample before the spray flame levels off, indicating the characteristic spherical shape. In consequence, no significant aggregation or other large structures were present before the spray flame. The TEM images of the sample after the spray flame revealed that most silica particles are directly connected to one to three other particles, resulting in a mean coordination number of $2.1 \pm 0.9$. The contact length between two aggregated silica primary particles was roughly half of the particle diameter. However, an infrequent coalescence of the silica particles, leading to the formation of large structures in the micron size range, was detected (marked with green). These structures were also observed in the USAXS data with a radius of gyration $R_g = 367.4$ nm. The local power law fit yields an exponent $P = 2.8$, showing an overlap between the fractal dimension of surface of the structure (P between 3 and 4) and the fractal dimension of mass of the silica aggregate, which mainly consists of the smaller primary particles ($P = 1.7$).

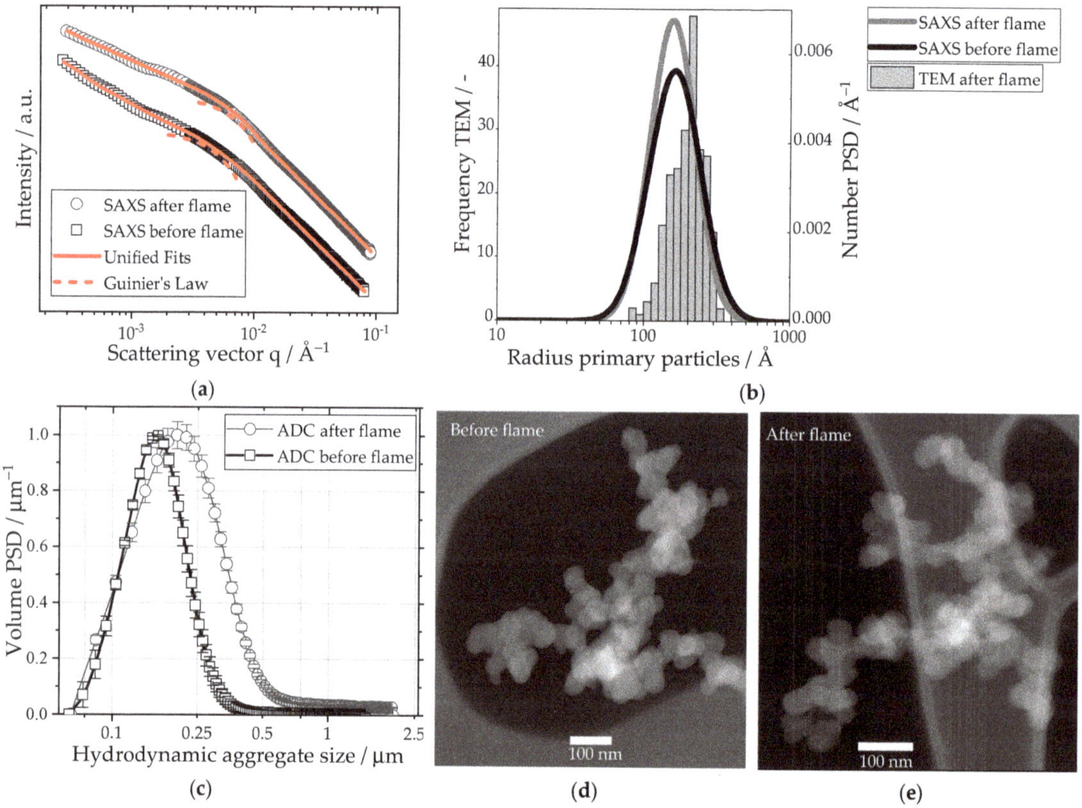


**Figure 1.** (**a**) SAXS data of carbon black before and after the spray flame. The original scattering data has been fitted with a unified fit. The local Guinier fit of the primary particles is highlighted with a dashed line. (**b**) Size distributions calculated from the fit parameters of the unified fit. Additional TEM data analysis of 224 counted primary particles of the carbon black after the spray flame yields a comparative size distribution. (**c**) Normalized volume PSDs of the carbon black aggregates before and after the spray flame, obtained via ACD, yield a hydrodynamic diameter. (**d**) HAADF-STEM image of the carbon black before the spray flame and (**e**) after the spray flame.

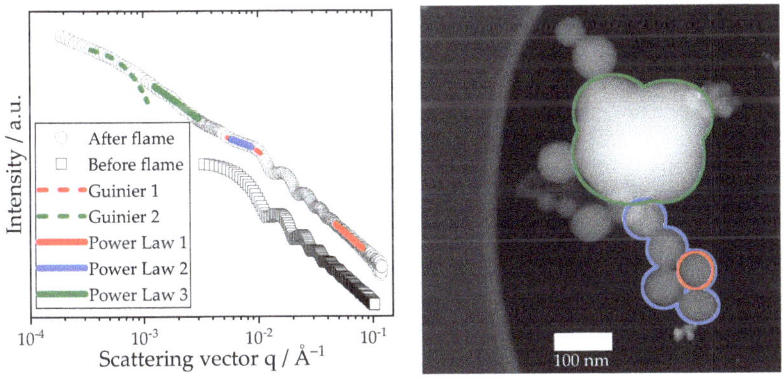


(a) (b)

**Figure 2.** (**a**) SAXS data of the investigated colloidal silica particles before and after the spray flame. The scattering curve of the particles was fitted with local power law and Guinier fits. The colours of the local fits correspond to the colours of the structures formed, which are marked in the HAADF-STEM image of the sample after the spray flame in (**b**). The sample before the flame was measured as a dilute suspension.

In conclusion, the colloidal silica particles aggregated heavily in the spray flame. Although coalescence was observed, most primary particles preserved their spherical shape. Their infrequent coalescence into larger structures may be attributed to the residence time distribution of the particles in the turbulent spray flame.

### 3.3. Characterization of the Hetero-Aggregates

The findings obtained for the spray flame experiments of the pure materials were also observed for the hetero-aggregates of carbon black and silica. In Figure 3a, a HAADF-STEM image of a hetero-aggregate with visible hetero-contact is shown. Closeups of the hetero-contact between the carbon black and silica are depicted in Figure 3b–d. Due to sintering, a large contact area between the carbon black and silica was formed. A clearly defined border between the two materials is detected within the resolution limit.

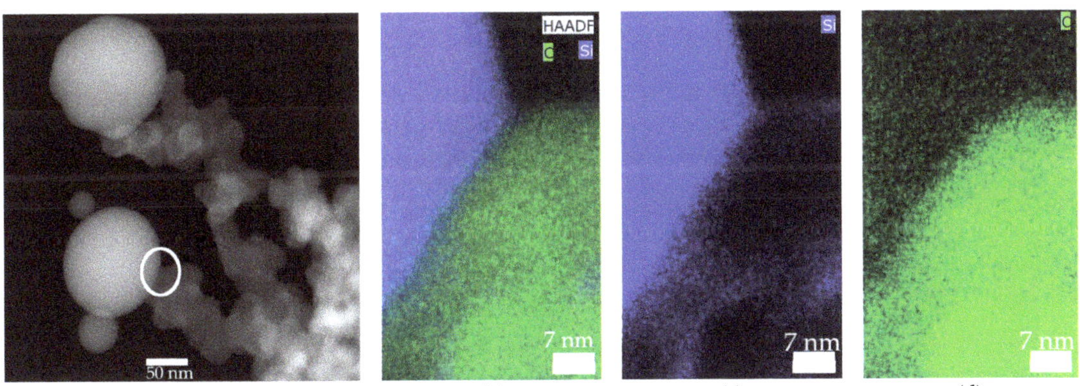


(a) (b) (c) (d)

**Figure 3.** (**a**) HAADF-STEM image of a hetero-aggregate with visible hetero-contact (white circle) (**b–d**) Closeup HAADF-STEM image with EDXS of the hetero-contact, with carbon (symbol C) coloured in green and silicon (symbol Si) coloured in blue. Experimental conditions were a mass ratio of silica to carbon black of 3 and original colloidal silica primary particle size of 30 nm.

As shown in Section 3.2, silica–silica homo-contacts and coalescence may lead to the formation of homo-aggregates consisting of pure silica. The formation of these homo-aggregates is visible in the SAXS data shown in Figure 4a. The slope at ultra-small angles ($q < 10^{-3}$Å$^{-1}$) was fitted with a power law fit (Equation (3)), and the exponent P was evaluated for different original particle sizes of silica and for different mass ratios of silica to carbon black in Figure 4b. Similar to the data shown in Figure 2a the exponent P is influenced both by the fractal dimension of mass of the aggregate ($P \sim 1.8 - 2.1$ for carbon black) as well as the fractal dimension of surface of the homo-aggregate ($P = 4$ for colloidal silica). Consequently, a higher exponent corresponds with a higher relative degree of homo-aggregation. In the scattering data, two major trends were observed: on one hand, the exponent P is influenced by the particle size of the silica. For the hetero-aggregates produced with 10 nm and 40 nm silica particles with mass ratios of one, the exponent P was close to the fractal dimension of mass of the pure carbon black aggregates, with values typically between 2.2 and 2.3 (see Section 3.1). For hetero-aggregates produced at the same mass ratio but with silica particles 60 nm in size, the exponent P is increased to 2.8. On the other hand, the mass concentration of silica in the hetero-aggregates significantly influences homo-aggregation. For a silica particle size of 40 nm, the exponent increased from 1.8 for a mass ratio of 0.33 to 2.5 for a mass ratio of 3. The HAADF-STEM image in Figure 4c shows the formation of an irregular structure due to coalescence accompanied by homo-aggregated 30 nm silica primary particles for a mass ratio of 3 and thus supports the findings acquired via the SAXS analysis. For this particle size, an increase from 1.9 (mass ratio 0.33) up to 2.5 emphasizes the importance of choosing lower mass ratios in order to avoid homo-aggregation and possible coalescence.

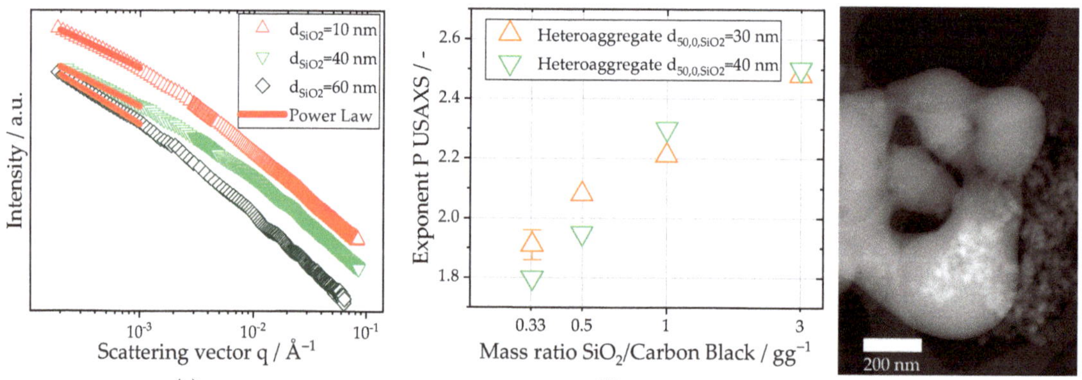


(a) (b) (c)

**Figure 4.** (**a**) SAXS data of hetero-aggregates produced at a mass ratio of silica to carbon black of 1 with varying sizes of the colloidal silica particles ranging from 10 nm to 60 nm. The evaluated power law fit in the USAXS area is highlighted in red. (**b**) Derived exponents of the local power law fit in the USAXS area for hetero-aggregates with different mass ratios of silica to carbon black ranging from 0.3 to 3. Two different primary particle size ranges of the colloidal silica were investigated: 30 nm and 40 nm. In order to evaluate the experimental spread, three individual experiments were evaluated for a mass ratio of 0.33. (**c**) Observed homo-aggregation and coalescence of silica particles for a mass ratio of $SiO_2$ to carbon black of 3 and 30 nm silica particles in HAADF-STEM images.

The trends observed in the USAXS measurements are further supported by the ADC measurements depicted in Figure 5. For mass ratios above or equal to three, the PSD is bimodal (Figure 5a). The first mode is close to the observed hydrodynamic aggregate size of pure carbon black, while the second mode corresponds to a structure in the micron size range, possibly the homo-aggregates. The relative weight of the second mode increases with a decrease in the silica particle size. Even though the formation of homo-aggregates is observed, the carbon black aggregates were incorporated into the formed silica structures for smaller particle sizes. This was especially evident for the 10 nm silica particles at a mass ratio of 3 for which only a single mode in the micron size range is measured. Therefore, both materials show better miscibility for smaller silica particle sizes. For lower mass ratios, the PSD resembled the PSD of pure carbon black after the spray flame with one mode (Figure 5b). Figure 6b shows the dependency of the first mode of particle size on the mass ratio of the silica particles. Due to the formed hetero-contact, by sintering silica particles onto the carbon black aggregate structure, an increase in the mode hydrodynamic aggregate diameter is expected, as the sedimentation speed is influenced by the aggregate density, which may be increased by the attached silica ($\rho_{CB} = 1.86\ \frac{g}{cm^3}$ [42]; $\rho_{SiO2} = 2.1\ \frac{g}{cm^3}$ depending on porosity [43]). Therefore, the hydrodynamic aggregate size is a measure for the effectiveness of the incorporation of silica into existing CB structures. For all particle sizes, the largest mode value is observed for an equal mass ratio of silica to carbon black, showing the optimal incorporation of the silica in the carbon black structure. For the hetero-aggregates produced with a silica particle size of 60 nm, the least pronounced change in the hydrodynamic aggregate size is observed with a spread between 218 ± 19 nm and 255 ± 8 nm. Furthermore, in Figure 6c, a TEM image of a hetero-aggregate produced with 60 nm silica particles is depicted. Therein, a carbon black aggregate exhibits only one visible silica hetero-contact despite a mass ratio of silica to carbon black of 3, supporting the drawn hypothesis of increasing the segregation of the two materials for larger particle sizes of silica. Furthermore, the change in the hydrodynamic aggregate size for varying mass ratios of silica was the most pronounced for hetero-aggregates produced with 10 nm silica particles. Here, the maximum measured hydrodynamic aggregate size was 357 ± 10 nm. Figure 6a underlines the influence of silica particle size on the size of the hydrodynamic aggregate. In the EDXS scan, a homogenous distribution of silica on the carbon black structure is apparent, meaning that both materials were successfully hetero-aggregated. The decreasing mode aggregate diameter for mass ratios larger than 1 independent of silica particle size indicates segregation due to the homo-aggregation of silica. However, in all cases, the mode of the hetero-aggregates was larger than for the pure carbon black in the spray flame. This means in all cases, some degree of hetero-aggregation was achieved.

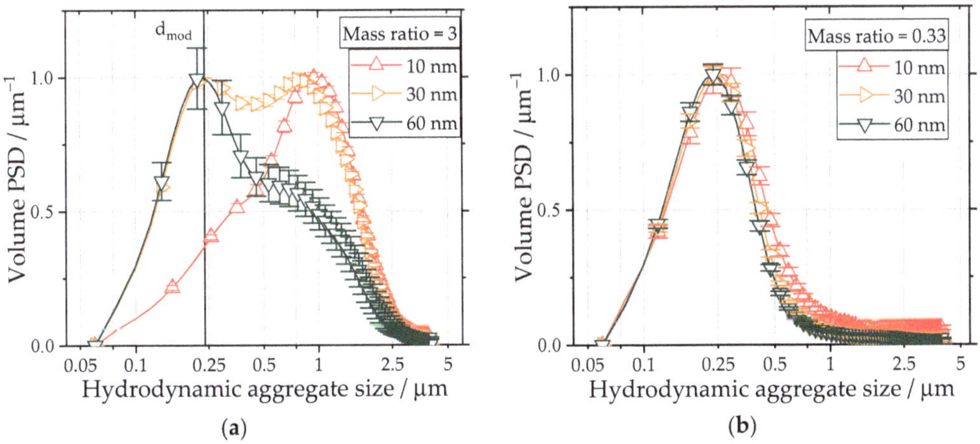


**Figure 5.** (**a**,**b**) Normalized volume PSDs of hetero-aggregates, as measured via ADC. For better visual clarity, only every fifth data point is shown. The investigated hetero-aggregates were produced with mass ratios of silica to carbon black of 3 (**a**) and 0.33 (**b**) for 10 nm, 30 nm and 60 nm colloidal silica particles, respectively.

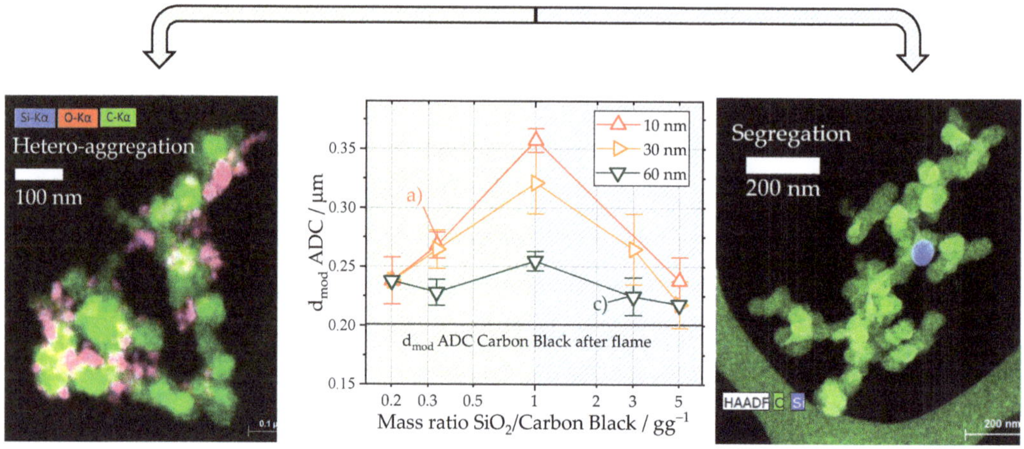


(**a**) (**b**) (**c**)

**Figure 6.** (**a**,**c**) HAADF-STEM images with EDXS of a hetero-aggregate produced with (**a**) 10 nm silica particles at a mass ratio of 0.33 and (**c**) 60 nm silica particles at a mass ratio of 3. Carbon is coloured in green (symbol C), silicon in blue (symbol Si) and oxygen in red (symbol O). Therefore, the silica ($SiO_2$) particles in (**a**) show a pinkish colour. (**b**) First mode (highlighted in Figure 5a with a straight line) of the PSD of hetero-aggregates for mass ratios of silica to carbon black ranging from 0.2 to 5 for three silica particle sizes. The data points each represent two experiments with three respective ADC measurements. The highlighted data points (**a**,**c**) correspond to the TEM images depicted in (**a**,**c**).

In conclusion, the data retrieved via both ADC and SAXS suggest that optimal hetero-aggregation is achieved for mass ratios of silica to carbon black equal to or below 1 and particle sizes of silica in the size range of 10 nm. This assumption is supported by the USAXS data, which reveal a fractal dimension of mass of 2.2, which is equal to the aggregates of pure carbon black (Figure 3a). The increased hydrodynamic aggregate size of 258 nm in comparison to pure carbon black after the spray flame with 201 nm shows successful hetero-aggregation. For higher mass ratios, increased segregation due to the homo-aggregation of silica is observed.

As an extra finding, a layer of carbon was formed on the surface of the silica particles in the spray flame, as revealed in Figure 7. Although this was not observed for all surveyed aggregates, the carbon layer is expected to increase the overall conductivity of the aggregate and is therefore desirable.

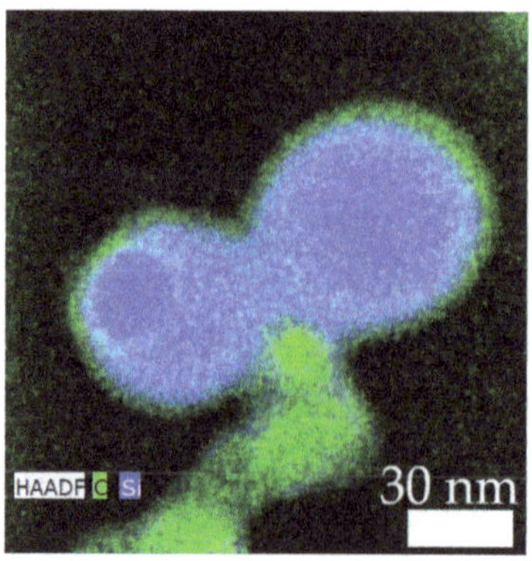


**Figure 7.** Carbon layer on the surface of the silica particles (diameter of silica = 60 nm; mass ratio = 3). Carbon is coloured in green (symbol C) and silicon in blue (symbol Si).

## 4. Discussion

Carbon black is a widely used additive for to improve conductivity in batteries. Therein, binders like PVDF ensure the binding of the active material as well as the stabilization of the electric pathways of carbon black. However, the environmental hazards of PVDF motivate the search for alternatives. A novel hetero-aggregation process is presented with the goal of improving the inherent mechanical strength and dispersibility of carbon black by incorporating colloidal silica into the carbon black structure via sintering in a spray flame. This work focuses on a comprehensive description of hetero-aggregate properties via multi-scale characterization through the use of SAXS, HAADF-STEM with EDXS and ADC. Therein, the particle size and mass ratio of the silica particles were identified as main factors in successful hetero-aggregation, which is achieved when a sinter contact is formed without significant changes in primary particle properties. In the evaluated TEM images, the majority of silica particles preserved their spherical shape and original diameter while forming aggregates connected by sinter bridges. Therefore, it is suggested that the surface of the silica is partially melted upon particle–particle contact which, in turn, leads to plastic deformation, forming a contact area. The negligible changes in the primary particle properties can be attributed to the short flame residence times of ≈0.7 ms [44]. However, previous researchers suggested the existence of a toroidal vortex within the spray flame of the SpraySyn burner [45]. Recirculation within this vortex may explain the infrequent observed coalescence of silica particles. The coalescence of silica particles has been reported for temperatures ranging from 1300 to 1700 K [46,47], which are similar to the gas phase temperature of the SpraySyn burner (around 1500 K in the spray flame [34]) used in this study. The rate of coalescence is directly influenced by the particle size, leading to increased coalescence rates for smaller particles [48]. However, the observed segregation of silica and carbon black for 60 nm silica particles resulted in an increased number of silica homo-contacts which, in turn, facilitated the creation of homo-aggregates. A similar trend was observed for mass ratios of silica to carbon black above 1. The increased concentration led to an increased particle collision frequency which, in turn, increased the coalescence rate [49] and the formation of homo-aggregates. The results of the different measurement systems are consistent with each other. In the USAXS data, the formation of homo-aggregates is identified by an increased exponent in the power law fit, leading to values of up to 2.8 in comparison to the original carbon black used (1.8–2.2). Additionally, in the ADC measurements, the homo-aggregates appear as a second mode in the micron size range.

In conclusion, larger silica particles may deflect upon contact with the carbon black aggregates due to their higher mass, whereas smaller silica particles tend to adhere and sinter to the carbon black. Therefore, the results suggest that the optimal experimental conditions for the hetero-aggregation of carbon black and silica are mass ratios equal to or below one and low silica primary particle sizes (10 nm). Smaller silica particle sizes are only achievable via a modification of the Stoeber method, e.g., through the addition of Triton X-100, which would lead to impurities in the spray flame [50]. TEM images of the hetero-aggregates produced under these experimental conditions show a homogeneous distribution of the silica on the carbon black structure. Furthermore, a fractal dimension of mass close to the original value for carbon black (2.2) and an increased hydrodynamic aggregate diameter of 258 nm were measured, which further supports the assumption of high-quality hetero-aggregation. Additionally, individual silica particles coated with carbon layers were identified. They are expected to exhibit advantageous properties for Li-ion battery applications [12,51] and should be investigated in further studies. In summary, the obtained results emphasize the general applicability of spray flames for the production of novel hetero-aggregate materials. Future studies will focus on the determination of functional properties like conductivity, dispersibility and the application of these hetero-aggregates in batteries.

**Supplementary Materials:** The following supporting information can be downloaded at: https://www.mdpi.com/article/10.3390/nano13121893/s1, Figure S1: Schematic drawing of the used SpraySyn burner; Table S1: Experimental conditions of the stoeber process; Table S2: Flame parameters of the SpraySyn burner.

**Author Contributions:** Conceptualization, S.B.; methodology, S.B.; formal analysis, S.B. and F.K.; investigation, S.B. and F.K.; data curation, S.B. and F.R.; writing—original draft preparation, S.B.; writing—review and editing, S.B. and F.R.; visualization, S.B. and F.R.; supervision, H.N.; project administration, H.N.; funding acquisition, H.N. All authors have read and agreed to the published version of the manuscript.

**Funding:** This research was funded by the German Research Foundation (DFG) within the priority program "SPP 2289: Creation of synergies in tailor-made mixtures of heterogeneous powders: Hetero aggregations of particulate systems and their properties", project number 462264068 and the grant number NI 414/41-1.

**Institutional Review Board Statement:** Not applicable.

**Informed Consent Statement:** Not applicable.

**Data Availability Statement:** Data will be made available upon request.

**Acknowledgments:** We would like to thank H. Störmer (Laboratory for Electron Microscopy, Karlsruhe Institute of Technology) for her help with the electron microscopy. We acknowledge support from the KIT-Publication Fund of the Karlsruhe Institute of Technology.

**Conflicts of Interest:** The authors declare no conflict of interest.

## References


1. Kwon, Y.I.; Kim, J.D.; Song, Y.S. Agitation Effect on the Rheological Behavior of Lithium-Ion Battery Slurries. *J. Electron. Mate* **2015**, *44*, 475–481. [CrossRef]
2. Sung, S.H.; Kim, S.; Park, J.H.; Park, J.D.; Ahn, K.H. Role of PVDF in Rheology and Microstructure of NCM Cathode Slurries fo Lithium-Ion Battery. *Materials* **2020**, *13*, 4544. [CrossRef]
3. Song, X.; Vestergren, R.; Shi, Y.; Huang, J.; Cai, Y. Emissions, Transport, and Fate of Emerging Per- and Polyfluoroalkyl Substance from One of the Major Fluoropolymer Manufacturing Facilities in China. *Environ. Sci. Technol.* **2018**, *52*, 9694–9703. [CrossRef]
4. Wang, M.; Liu, K.; Yu, J.; Zhang, Q.; Zhang, Y.; Valix, M.; Tsang, D.C.W. Challenges in Recycling Spent Lithium-Ion Batteries Spotlight on Polyvinylidene Fluoride Removal. *Glob. Chall.* **2023**, *7*, 2200237. [CrossRef]
5. Wang, N.; NuLi, Y.; Su, S.; Yang, J.; Wang, J. Effects of binders on the electrochemical performance of rechargeable magnesiur batteries. *J. Power Sources* **2017**, *341*, 219–229. [CrossRef]
6. Lechtenboehmer, A.; Moneypenny, H.G.; Mersch, F. A Review of Polymer Interfaces in Tyre Technology. *Brit. Poly. J.* **1990**, *22* 265–301. [CrossRef]
7. Mora-Barrantes, I.; Ibarra, L.; Rodríguez, A.; González, L.; Valentín, J.L. Elastomer composites based on improved fumed silic and carbon black. Advantages of mixed reinforcing systems. *J. Mater. Chem.* **2011**, *21*, 17526. [CrossRef]
8. Zhang, W.; Blackburn, R.S.; Dehghani-Sanij, A.A. Effect of silica concentration on electrical conductivity of epoxy resin–carbor black–silica nanocomposites. *Scr. Mater.* **2007**, *56*, 581–584. [CrossRef]
9. An, W.; Fu, J.; Su, J.; Wang, L.; Peng, X.; Wu, K.; Chen, Q.; Bi, Y.; Gao, B.; Zhang, X. Mesoporous hollow nanospheres consisting o carbon coated silica nanoparticles for robust lithium-ion battery anodes. *J. Power Sources* **2017**, *345*, 227–236. [CrossRef]
10. Meng, J.; Cao, Y.; Suo, Y.; Liu, Y.; Zhang, J.; Zheng, X. Facile Fabrication of 3D SiO2@Graphene Aerogel Composites as Anode Material for Lithium Ion Batteries. *Electrochim. Acta* **2015**, *176*, 1001–1009. [CrossRef]
11. Lv, P.; Zhao, H.; Wang, J.; Liu, X.; Zhang, T.; Xia, Q. Facile preparation and electrochemical properties of amorphous SiO2/C composite as anode material for lithium ion batteries. *J. Power Sources* **2013**, *237*, 291–294. [CrossRef]
12. Yao, Y.; Zhang, J.; Xue, L.; Huang, T.; Yu, A. Carbon-coated $SiO_2$ nanoparticles as anode material for lithium ion batteries. *J. Powe Sources* **2011**, *196*, 10240–10243. [CrossRef]
13. Li, H.-H.; Wu, X.-L.; Sun, H.-Z.; Wang, K.; Fan, C.-Y.; Zhang, L.-L.; Yang, F.-M.; Zhang, J.-P. Dual-Porosity $SiO_2$/C Nanocomposite with Enhanced Lithium Storage Performance. *J. Phys. Chem. C* **2015**, *119*, 3495–3501. [CrossRef]
14. Yan, N.; Wang, F.; Zhong, H.; Li, Y.; Wang, Y.; Hu, L.; Chen, Q. Hollow porous $SiO_2$ nanocubes towards high-performance anodes for lithium-ion batteries. *Sci. Rep.* **2013**, *3*, 1568. [CrossRef]
15. Fan, Y.; Wang, J.; Tang, Z.; He, W.; Zhang, J. Effects of the nanostructured $SiO_2$ coating on the performance of LiNi0.5Mn1.5O4 cathode materials for high-voltage Li-ion batteries. *Electrochim. Acta* **2007**, *52*, 3870–3875. [CrossRef]
16. Fransson, L.; Eriksson, T.; Edström, K.; Gustafsson, T.; Thomas, J. Influence of carbon black and binder on Li-ion batteries. *J. Power Sources* **2001**, *101*, 1–9. [CrossRef]

17. Liu, G.; Zheng, H.; Song, X.; Battaglia, V.S. Particles and Polymer Binder Interaction: A Controlling Factor in Lithium-Ion Electrode Performance. *J. Electrochem. Soc.* **2012**, *159*, A214–A221. [CrossRef]
18. Zheng, H.; Yang, R.; Liu, G.; Song, X.; Battaglia, V.S. Cooperation between Active Material, Polymeric Binder and Conductive Carbon Additive in Lithium Ion Battery Cathode. *J. Phys. Chem. C* **2012**, *116*, 4875–4882. [CrossRef]
19. Mayer, J.K.; Almar, L.; Asylbekov, E.; Haselrieder, W.; Kwade, A.; Weber, A.; Nirschl, H. Influence of the Carbon Black Dispersing Process on the Microstructure and Performance of Li-Ion Battery Cathodes. *Energy Technol.* **2020**, *8*, 1900161. [CrossRef]
20. Gröhn, A.J.; Pratsinis, S.E.; Sánchez-Ferrer, A.; Mezzenga, R.; Wegner, K. Scale-up of Nanoparticle Synthesis by Flame Spray Pyrolysis: The High-Temperature Particle Residence Time. *Ind. Eng. Chem. Res.* **2014**, *53*, 10734–10742. [CrossRef]
21. Teoh, W.Y.; Amal, R.; Mädler, L. Flame spray pyrolysis: An enabling technology for nanoparticles design and fabrication. *Nanoscale* **2010**, *2*, 1324–1347. [CrossRef] [PubMed]
22. Stahl, J.; Ilsemann, J.; Pokhrel, S.; Schowalter, M.; Tessarek, C.; Rosenauer, A.; Eickhoff, M.; Bäumer, M.; Mädler, L. Comparing Co-catalytic Effects of ZrOx SmOx and Pt on COx Methanation over Co-based Catalysts Prepared by Double Flame Spray Pyrolysis. *ChemCatChem* **2021**, *13*, 2815–2831. [CrossRef]
23. Height, M.J.; Pratsinis, S.E.; Mekasuwandumrong, O.; Praserthdam, P. Ag-ZnO catalysts for UV-photodegradation of methylene blue. *Appl. Catal. B Environ.* **2006**, *63*, 305–312. [CrossRef]
24. Bauer, W.; Nötzel, D.; Wenzel, V.; Nirschl, H. Influence of dry mixing and distribution of conductive additives in cathodes for lithium ion batteries. *J. Power Sources* **2015**, *288*, 359–367. [CrossRef]
25. Men, J.; Kolan, S.R.; Massomi, A.; Hoffmann, T.; Schmidt, J.; Tsotsas, E.; Bück, A. Formulation of Nanostructured Heteroaggregates by Fluidization Technologies. *Chem. Ing. Tech.* **2023**, *95*, 107–113. [CrossRef]
26. Gäßler, M.; Stahl, J.; Schowalter, M.; Pokhrel, S.; Rosenauer, A.; Mädler, L.; Güttel, R. The Impact of Support Material of Cobalt-Based Catalysts Prepared by Double Flame Spray Pyrolysis on $CO_2$ Methanation Dynamics. *ChemCatChem* **2022**, *14*, e202200286. [CrossRef]
27. Strobel, R. Flame spray synthesis of $Pd/Al_2O_3$ catalysts and their behavior in enantioselective hydrogenation. *J. Catal.* **2004**, *222*, 307–314. [CrossRef]
28. Gerken, B.; Mahr, C.; Stahl, J.; Grieb, T.; Schowalter, M.; Krause, F.F.; Mehrtens, T.; Mädler, L.; Rosenauer, A. Material Discrimination in Nanoparticle Hetero-Aggregates by Analysis of Scanning Transmission Electron Microscopy Images. *Part Part Syst. Charact.* **2023**, 2300048. [CrossRef]
29. Hyeon-Lee, J.; Beaucage, G.; Pratsinis, S.E.; Vemury, S. Fractal Analysis of Flame-Synthesized Nanostructured Silica and Titania Powders Using Small-Angle X-ray Scattering. *Langmuir* **1998**, *14*, 5751–5756. [CrossRef]
30. Simmler, M.; Meier, M.; Nirschl, H. Characterization of Fractal Structures by Spray Flame Synthesis Using X-ray Scattering. *Materials* **2022**, *15*, 2124. [CrossRef]
31. Rieker, T.P.; Hindermann-Bischoff, M.; Ehrburger-Dolle, F. Small-Angle X-ray Scattering Study of the Morphology of Carbon Black Mass Fractal Aggregates in Polymeric Composites. *Langmuir* **2000**, *16*, 5588–5592. [CrossRef]
32. Sztucki, M.; Narayanan, T.; Beaucage, G. In situ study of aggregation of soot particles in an acetylene flame by small-angle X-ray scattering. *J. Appl. Phys.* **2007**, *101*, 114304. [CrossRef]
33. Stöber, W.; Fink, A.; Bohn, E. Controlled growth of monodisperse silica spheres in the micron size range. *J. Colloid Interface Sci.* **1968**, *26*, 62–69. [CrossRef]
34. Schneider, F.; Suleiman, S.; Menser, J.; Borukhovich, E.; Wlokas, I.; Kempf, A.; Wiggers, H.; Schulz, C. SpraySyn-A standardized burner configuration for nanoparticle synthesis in spray flames. *Rev. Sci. Instrum.* **2019**, *90*, 85108. [CrossRef]
35. Bonse, U.; Hart, M. Small angle X-ray scattering by spherical particles of Polystyrene and Polyvinyltoluene. *Z. Phys.* **1966**, *189*, 151–162. [CrossRef]
36. Schneider, C.A.; Rasband, W.S.; Eliceiri, K.W. NIH Image to ImageJ: 25 years of image analysis. *Nat. Methods* **2012**, *9*, 671–675. [CrossRef]
37. Schmidt, P.W. Small-angle scattering studies of disordered, porous and fractal systems. *J. Appl. Cryst.* **1991**, *24*, 414–435. [CrossRef]
38. Porod, G. Die Röntgenkleinwinkelstreuung von dichtgepackten kolloiden Systemen. *Kolloid Z.* **1951**, *124*, 83–114. [CrossRef]
39. Ilavsky, J.; Jemian, P.R. Irena tool suite for modeling and analysis of small-angle scattering. *J. Appl. Cryst.* **2009**, *42*, 347–353. [CrossRef]
40. Beaucage, G. Approximations Leading to a Unified Exponential/Power-Law Approach to Small-Angle Scattering. *J. Appl. Cryst.* **1995**, *28*, 717–728. [CrossRef]
41. Beaucage, G.; Kammler, H.K.; Pratsinis, S.E. Particle size distributions from small-angle scattering using global scattering functions. *J. Appl. Cryst.* **2004**, *37*, 523–535. [CrossRef]
42. Spahr, M.E.; Goers, D.; Leone, A.; Stallone, S.; Grivei, E. Development of carbon conductive additives for advanced lithium ion batteries. *J. Power Sources* **2011**, *196*, 3404–3413. [CrossRef]
43. Bogush, G.H.; Tracy, M.A.; Zukoski, C.F. Preparation of monodisperse silica particles: Control of size and mass fraction. *J. Non-Cryst. Solids* **1988**, *104*, 95–106. [CrossRef]
44. Bieber, M.; Al-Khatib, M.; Fröde, F.; Pitsch, H.; Reddemann, M.A.; Schmid, H.-J.; Tischendorf, R.; Kneer, R. Influence of angled dispersion gas on coaxial atomization, spray and flame formation in the context of spray-flame synthesis of nanoparticles. *Exp. Fluids* **2021**, *62*, 98. [CrossRef]

45. Martins, F.J.; Kirchmann, J.; Kronenburg, A.; Beyrau, F. Experimental investigation of axisymmetric, turbulent, annular jets discharged through the nozzle of the SPP1980 SpraySyn burner under isothermal and reacting conditions. *Exp. Therm. Fluid Sci.* **2020**, *114*, 110052. [CrossRef]
46. Xiong, Y.; Pratsinis, S.E. Formation of agglomerate particles by coagulation and sintering—Part I. A two-dimensional solution of the population balance equation. *J. Aerosol Sci.* **1993**, *24*, 283–300. [CrossRef]
47. Ehrman, S.H.; Friedlander, S.K.; Zachariah, M.R. Characteristics of $SiO_2/TiO_2$ nanocomposite particles formed in a premixed flat flame. *J. Aerosol Sci.* **1998**, *29*, 687–706. [CrossRef]
48. Ehrman, S.H. Effect of Particle Size on Rate of Coalescence of Silica Nanoparticles. *J. Colloid Interface Sci.* **1999**, *213*, 258–261. [CrossRef] [PubMed]
49. Ulrich, G.D.; Milnes, B.A.; Subramanian, N.S. Particle Growth in Flames. II: Experimental Results for Silica Particles. *Combust. Sci. Technol.* **1976**, *14*, 243–249. [CrossRef]
50. Liu, X.; Lu, X.; Wen, P.; Shu, X.; Chi, F. Synthesis of ultrasmall silica nanoparticles for application as deep-ultraviolet antireflection coatings. *Appl. Surf. Sci.* **2017**, *420*, 180–185. [CrossRef]
51. Cao, L.; Huang, J.; Lin, Z.; Yu, X.; Wu, X.; Zhang, B.; Zhan, Y.; Xie, F.; Zhang, W.; Chen, J.; et al. Amorphous $SiO_2/C$ composite as anode material for lithium-ion batteries. *J. Mater. Res.* **2018**, *33*, 1219–1225. [CrossRef]

nanomaterials

Article

# Antimony Nanoparticles Encapsulated in Self-Supported Organic Carbon with a Polymer Network for High-Performance Lithium-Ion Batteries Anode

Zhaomin Wang [1,2], Fanming Zeng [1,2,*], Dongyu Zhang [3], Yabin Shen [3], Shaohua Wang [3], Yong Cheng [3,4,5], Chun Li [1,2,*] and Limin Wang [3,*]

1 School of Materials Science and Engineering, Changchun University of Science and Technology, Changchun 130022, China; zmwang@ciac.ac.cn
2 Collaborative Innovation Center of Optical Materials and Chemistry, Changchun University of Science and Technology, Changchun 130022, China
3 State Key Laboratory of Rare Earth Resource Utilization, Changchun Institute of Applied Chemistry, CAS, Changchun 130022, China; dyzhang@ciac.ac.cn (D.Z.); ybshen@ciac.ac.cn (Y.S.); shwang@ciac.ac.cn (S.W.); cyong@ciac.ac.cn (Y.C.)
4 Key Laboratory of Preparation and Applications of Environmental Friendly Materials, Jilin Normal University, Ministry of Education, Changchun 130103, China
5 State Key Laboratory of Metastable Materials Science and Technology, Yanshan University, Qinhuangdao 066004, China
* Correspondence: zengfm@126.com (F.Z.); lichun1210@163.com (C.L.); lmwang@ciac.ac.cn (L.W.)

**Abstract:** Antimony (Sb) demonstrates ascendant reactive activation with lithium ions thanks to its distinctive puckered layer structure. Compared with graphite, Sb can reach a considerable theoretical specific capacity of 660 mAh $g^{-1}$ by constituting $Li_3Sb$ safer reaction potential. Hereupon, with a self-supported organic carbon as a three-dimensional polymer network structure, Sb/carbon (3DPNS-Sb/C) composites were produced through a hydrothermal reaction channel followed by a heat disposal operation. The unique structure shows uniformitarian Sb nanoparticles wrapped in a self-supported organic carbon, alleviating the volume extension of innermost Sb alloying, and conducive to the integrality of the construction. When used as anodes for lithium-ion batteries (LIBs), 3DPNS-Sb/C exhibits a high invertible specific capacity of 511.5 mAh $g^{-1}$ at a current density of 0.5 A $g^{-1}$ after 100 cycles and a remarkable rate property of 289.5 mAh $g^{-1}$ at a current density of 10 A $g^{-1}$. As anodes, LIBs demonstrate exceptional electrochemical performance.

**Keywords:** Sb/C; anode; alloying-conversion action; lithium-ion batteries

## 1. Introduction

What accompanies the swift advancement of various intelligent mobile appliances is the enhancement of energy requirements, and LIBs have become attractive energy storage and conversion devices [1–4]. Finding electrodes with superior capacity is one of the most diffusely researched subjects in the domain of LIBs since the invertible ability is diametrically relevant to the useful life of the cell [5–9]. Sb-based anode material has received much attention and combines Li to form the $Li_3Sb$ alloy and gives rise to an excellent theoretical specific capacity (660 mAh $g^{-1}$) [10–16]. Furthermore, Sb is a member of the most prospective anode materials for LIBs, which can be alloyed with Li at a low reaction potential of approximately 0.8 V [12]. Nevertheless, Li-ion insertion/extraction procedures lead to severe volume effect, which causes the prompt exacerbation of cycle property [11,17].

Therefore, numerous strategies have been utilized to mitigate these issues of Sb-based material anodes. For example, shrinking the grain diameter can curtail the Li-ion convey way and dramatically mitigate the mechanical stress during alloy reaction and thus moderate the pulverization trouble, which is a pervasive medium to improve the property of Sb-based anodes [11,18–20]. However, the machinery unsteadiness associated with the lithium alloy reaction cannot be entirely resolved only by reducing the particle dimension. The carbon matrix may undertake a rampart to adapt the polymerization and pulverization of active granules while enhancing the conductivity, which is deemed as a member of the prospective means to ameliorate the electrochemical property [14,21–41]. Noteworthily, since antimony is a heat-shrinking, cold-expanding metal, intermetallic systems (Sb-based alloys) can possess a powerful structural relationship with Li-ion, which leads to minor volume effects during the charge/discharge process [42–46]. Although the series of preparations mentioned above effectively buffer the volume effect and mechanical tension of Sb-based materials, the synthesis of nanoporous Sb-based composites through an uncomplicated and extensible method is still essential for practical application in LIBs.

Herein, an innovative 3DPNS-Sb/C nanoparticle anode material is fabricated based on the above discussions. It involves the uniformly in situ insertion of Sb nanoparticles into self-supported organic carbon, exploiting a manageable hydrothermal synthesis reaction and annealing treatment. The generation of the evenly distributed structure could be attributed to the facile reduction of sodium antimonate ($NaSbO_3$) and the formidable binding interaction of the carbon network. The 3DPNS-Sb/C composites have the merits of the polymer network structure effect and high conductivity. In addition, the unique construction plays an appreciable role in enhancing the charge transfer kinetics and structure steadiness during the repeated insertion/deinsertion procedure of Li-ion, which ultimately exhibited excellent cyclability and rate property.

## 2. Experimental Section

### 2.1. Materials

Glucose ($C_6H_{12}O_6$, CP, 99%), sodium antimonate ($NaSbO_3$, 99.9%) and sodium polyacrylate ($(C_3H_3NaO_2)_n$, 99%) were stocked from Aladdin Reagent Co. Ltd. Shanghai, China. All the chemicals and solvents were exploited as acquired without further depuration.

### 2.2. Synthesis of the3DPNS-Sb/C Composites

The 3DPNS-Sb/C composites were fabricated utilizing a common hydrothermal synthesis reaction and annealing treatment. Typically, 1 g $NaSbO_3$ and 0.02 g $(C_3H_3NaO_2)_n$ were dispersed in a Teflon-lined autoclave with 100 mL aqueous liquor consisting of 3 g $C_6H_{12}O_6$, and the autoclave was shut and conserved at 180 °C for 12 h, followed by a return to room temperature. Next, the as-synthesized precursor was subjected to an annealing process at 450 °C for 6 h with a warming speed of 3 °C $min^{-1}$ under a perpetual high-purity Ar. After spontaneously dropping down to ambient temperature, the obtained product was denoted as 3DPNS-Sb/C-2. To assess the impact of the carbon content of the 3DPNS-Sb/C composites, two distinct $C_6H_{12}O_6$ concentrations ($m(C_6H_{12}O_6)$ = 2.5/3.5 g) were also carried out while keeping other factors unchanged, which were denoted as 3DPNS-Sb/C-1 and 3DPNS-Sb/C-3, respectively.

### 2.3. Materials Characterization

X-ray diffraction (XRD, Bruker D8 Advance diffractometer using Cu Kα radiation ($\lambda$ = 1.5418 Å)) was used to authenticate the component and crystal structure of the as-obtained 3DPNS-Sb/C composites. Field emission scanning electron microscopy (FESEM, Hitachi S-4800, Tokyo, Japan) and transmission electron microscopy (TEM, FEI Tecnai G2 S-Twin, Hillsboro, OR, America) were utilized to identify the morphology and structural characteristics. Thermogravimetric analysis (TGA) was executed utilizing a Q50 (Guangzhou, China) thermogravimetric analyzer from 25 through 800 °C at a velocity of 10 °C $min^{-1}$ in an atmosphere of air. Nitrogen desorption/adsorption isotherms were assessed by nitrogen adsorption at 77 K using a Quadrachrome Adsorption Apparatus (Beijing, China). The Raman spectrum was acquired using a Renishaw Invia Raman microscope (Beijing, China). The X-ray photoelectron spectra (XPS) were recorded on a

Thermo Scientific ESCALAB 250Xi (Shanghai, China) X-ray photoelectron spectrometer with a monochromatized Al-Kα X-ray (1486.6 eV) as the excitation source to estimate the apparent component.

### 2.4. Electrochemical Characterization

The anode electrode was prepared by mixing 80 wt.% 3DPNS-Sb/C, 10 wt.% acetylene black and 10 wt.% carboxymethyl cellulose sodium (CMC) with an appropriate amount of DI water as the solvent to produce a homogeneous phase of the slurry. The slurry was uniformly spread onto pure copper foil (of thickness 10 μm) current collector and dried at 60 °C for 6 h under vacuum conditions. Subsequently, the loaded collector was punched into a circular slice with an area of 1.13 $cm^2$. The electrodes were then pressed using a stainless-steel metal disc to enhance the contact between the material and the Cu foil. The mass of material loading on each electrode was about 0.93–1.12 mg $cm^{-2}$ (including the weight of acetylene black and the binder). The electrochemical assessments were executed using CR2025 coin-type batteries. Lithium foil was utilized as both the counter and the reference electrode, while the 3DPNS-Sb/C electrode was the working electrode. Polypropylene membrane (Celgard 2400) was employed as the separator for LIBs. The electrolyte was constituted of a solution of 1 M $LiPF_6$ dispersed in a blend of dimethyl carbonate (DMC), diethyl carbonate (DEC) and ethylene carbonate (EC) (1:1:1 vol.%) with the addition of 10 vol.% fluoroethylene carbonate (FEC) for LIBs. The electrochemical properties of all the manufactured half-batteries was assessed by cyclic voltammetry (CV) measurement using the BioLogic VMP3 instrument. The charge/discharge performance was measured at room temperature with disparate current densities under the potential window of 0.01~2 V (vs. $Li/Li^+$) employing the LAND CT2001A multichannel battery examination system.

## 3. Results and Discussion

### 3.1. Experimental Synthesis Mechanism

The prototypical design approach and synthesis path applied for manufacturing the 3DPNS-Sb/C nanoparticle materials are schematically demonstrated in Figure 1. In the first step, glucose molecules engender dihydroxyacetone, glyceraldehyde, erythrose, organic acids, aldehydes and other small molecular substances by cracking. On the other side, the glucose molecules produce anhydroglucose polymers via mutual dehydration or generate 5-hydroxymethylfurfural by self-isomerization. These molecules, of distinct sizes, are dehydrated under hydrothermal conditions and condense with each other to form the soluble polymer. [47] At the same time, the $NaSbO_3$ is heated and hydrolyzed into antimonic acid ($HSbO_3$) colloid, which is uniformly dispersed in the soluble glucose polymer under the action of $(C_3H_3NaO_2)_n$. With the temperature reaching a critical value, the soluble glucose polymer is progressively carbonized and the $HSbO_3$ is gradually decomposed into antimonic oxide ($Sb_2O_5$) and ultimately, the 3DPNS-$Sb_2O_5$/hydrochar nanoparticle composites are obtained. [47] Subsequently, the Sb nanoparticles were evenly distributed in an organic carbon skeleton during the annealing treatment, which originated from the in-situ reduction reaction of $Sb_2O_5$ and hydrochar.

### 3.2. Morphology Analysis

The morphologies of the acquired 3DPNS-Sb/C composites are shown in Figure 2. As illustrated in Figure 2a–c, the 3DPNS-Sb/C composites clearly show an interconnected 3D polymer network framework structure and individual Sb/C nanoparticles with a diameter of about 50–200 nm. Furthermore, the size of this individual Sb/C nanoparticle enlarges with increasing carbon content. As shown in Figure 2d,e, the TEM and high-resolution TEM (HRTEM) figures (3DPNS-Sb/C-2) indicate the lattice fringes with an interval of 0.22 nm, coinciding with the (110) planes of hexagonal Sb, further verifying the high crystallinity of the Sb. In addition, the tiny Sb nanoparticles are equally distributed in a thin carbon layer structure, which can supply a more favorable appearance and curtail the diffusion interval for ions to inner pores. As revealed in Figure 2f–h, the detected element mappings, such as Sb and C, display a uniform dispersion in the 3DPNS-Sb/C-2 composites.

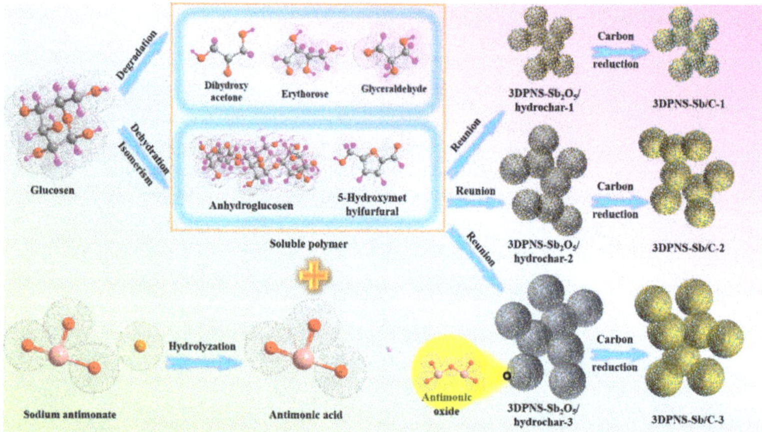


**Figure 1.** Schematic picture of the synthesis process of 3DPNS-Sb/C nanoparticle materials.

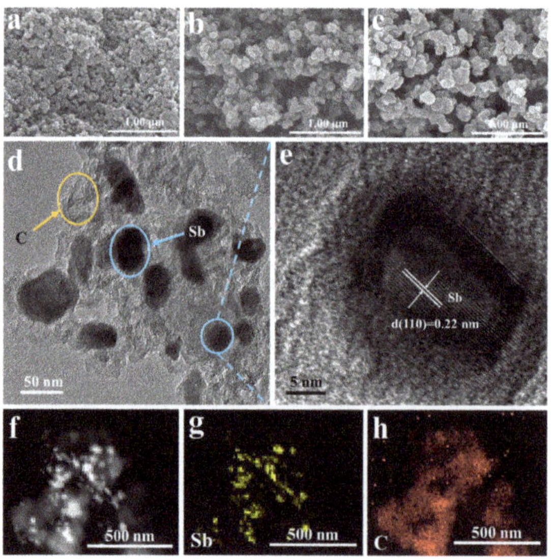


**Figure 2.** (**a–c**) SEM images of the 3DPNS-Sb/C-1, 3DPNS-Sb/C-2 and 3DPNS-Sb/C-3 composites, (**d**,**e**) TEM and HRTEM images of the 3DPNS-Sb/C-2 composites, (**f**) STEM image, (**g**) Sb and (**h**) C element mappings of the 3DPNS-Sb/C-2 composites.

### *3.3. Microstructure and Component Analysis*

The structure of 3DPNS-Sb/C composites is demonstrated by the XRD examination. As revealed in Figure 3a, the significant characteristic peak of 23.6°, 28.6°, 40°, 41.9°, 47°, 48.4°, 51.5°, 59.3°, 62.7°, 65.9°, 68.5° and 75.3° is admirably indexed to the (003), (012), (104), (110), (015), (006), (202), (024), (107), (116), (122) and (214) crystal face of elemental Sb (JCPDS 35-0732), respectively, which is consistent with the scrutinization in the HRTEM image. In addition, the quantitative component of the 3DPNS-Sb/C specimens is confirmed by TGA. In accordance with the TGA consequences in Figure 3b, the Sb content is computed to be 40.23, 36.79 and 32.51 wt.% for 3DPNS-Sb/C-1, 3DPNS-Sb/C-2 and 3DPNS-Sb/C-3 composites, respectively. The subsequent weight increase corresponds to Sb oxidation [48,49]. The specific surface acreage and the pore size distribution of 3DPNS-Sb/C samples are afterwards explained via nitrogen adsorption/desorption isotherm measuring. Figure 3c demonstrates that the isotherms present type IV features, which means they belong to mesoporous substances [48]. The BET-specific surface acreage of 3DPNS-Sb/C-1, 3DPNS-Sb/C-2 and 3DPNS-Sb/C-3 composites are 136.7, 141.5 and 147.8 $m^2\ g^{-1}$, respectively, which could be owing to the coordination of the large surface area of the 3D polymer network structure and small Sb nanoparticles. The relevant pore diameter distribution curve (inset in Figure 3c) also evidences a mesoporous constitution, and the pore dimension ratio (3.5–4.8 nm) is evidenced. For the cell electrode, the powerful specific surface area and mesoporous construction of 3DPNS-Sb/C samples can accelerate the pervasion of Li-ions and electrons. The Raman spectroscopy analysis was conducted further to inspect the structural characteristics of the 3DPNS-Sb/C-2 composites, with the acquired spectrogram revealed in Figure 3d. Two typical bands located at 107 and 139 $cm^{-1}$ are associated with the Sb phase in the nanohybrid [26,50–53]. Moreover, Raman scattering measurement analysis confirmed that the Sb nanoparticles contain traces of $Sb_2O_3$ [54]. The peak at 1346 $cm^{-1}$ indicates the disarray-induced D-band, which connects with flaws in the $sp^2$ lattice construction of carbon. The peak at 1592 $cm^{-1}$ correlates with the graphitic G-band, which insinuates the $sp^2$ lattice of carbon. The above results indicate the amorphous character of the carbon contained in the 3DPNS-Sb/C-2 composites [23,49]. The surface chemical composition of the 3DPNS-Sb/C-2 composites was also studied by XPS characterization, and the Sb and C spectrums of the 3DPNS-Sb/C-2 composites are demonstrated in Figure 3e,f, which absolutely correspond to the EDX mapping aforementioned. Among these, two peaks at 531.7 eV and 533.1 eV are from the O 1s core-level XPS spectrum, while the other peaks arise from the XPS spectrum of Sb 3d. In particular, the two peaks centered at 539.23 eV (Sb $3d_{3/2}$) and 529.85 eV (Sb $3d_{5/2}$) represent $Sb_2O_3$. Meanwhile, the peak appearing at 528.7 eV (Sb $3d_{5/2}$) represents metallic Sb (Figure 3e) [23,28,33,55]. Hence, this result confirms the partial oxidation of Sb by oxygen. Meanwhile, the high-resolution C 1s spectrum displayed in Figure 3f could be fitted into three peaks correlated with C-C (284.55 eV), C-O (286.01 eV) and C=O (288.44 eV) bonds [55–57].

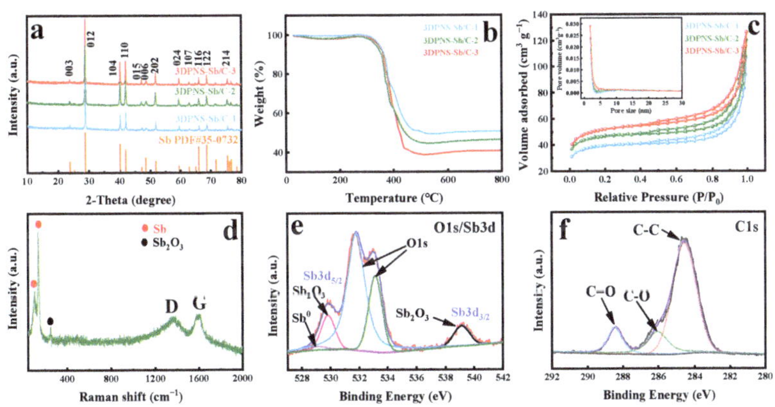


**Figure 3.** (**a**) XRD patterns of 3DPNS-Sb/C-1, 3DPNS-Sb/C-2 and 3DPNS-Sb/C-3. (**b**) TG curves of 3DPNS-Sb/C-1, 3DPNS-Sb/C-2 and 3DPNS-Sb/C-3. (**c**) Nitrogen adsorption–desorption isotherms and related pore dimension distribution curves of 3DPNS-Sb/C-1, 3DPNS-Sb/C-2 and 3DPNS-Sb/C-3. (**d**) Raman spectra of 3DPNS-Sb/C-2. XPS spectra of 3DPNS-Sb/C-2: (**e**) Sb 3d and O 1s, (**f**) C 1s.

## 4. Electrochemical Evaluation in LIBs

The cyclic voltammetry (CV) curves of the 3DPNS-Sb/C-2 electrode for the initial three cycles are demonstrated in Figure 4a. In the primary cathode scanning, the capacious summit between 0.8 and 0.6 V could be ascribed to a suite of Li-insertion reactions, containing the reaction of metallic Sb to alloyed $Li_3Sb$ and the generation of an SEI film on the cover of the electrode from electrolyte decomposition. During the initial invertible anode scanning, the anode summit at around 1.14 V conforms to the Li-extraction reaction of $Li_3Sb$ to metallic Sb [11,17,58]. In following cyclings, peaks tend to overlap, suggesting excellent electrochemical invertibility of the 3DPNS-Sb/C-2 samples in the lithiation–delithiation procedure. Subsequently, it still displays an illustrious invertible specific capacity of 511.5 mAh $g^{-1}$ at a current density of 0.5 A $g^{-1}$ after 100 cycles, with a primary charge/discharge specific capacity of 775.3/1117.8 mAh $g^{-1}$ and a first coulombic efficiency (CE) of 69.35% (Figure 4b). Figure S1 shows the SEM images of 3DPNS-Sb/C-2 after 100 repeated cycles at a current density of 0.5 A $g^{-1}$. Distinctly, the structure is nearly maintained, suggesting excellent structural stability. Moreover, with the unique network constructure of 3DPNS-Sb/C-2 composites, a pronouncedly high-rate capacity is acquired (Figure 4d). Though the current density enhances from 0.1 to 10 A $g^{-1}$, it could release a high specific capacity of 289.5 mAh $g^{-1}$. Noteworthily, the gaps between charge/discharge of the 3DPNS-Sb/C-2 composites reduce slightly with the current density increase, which signifies the weakness of polarization and mechanical effect in the cycling procedure [59,60]. Figure 4e displays the cycle property of the 3DPNS-Sb/C-1, 3DPNS-Sb/C-2 and 3DPNS-Sb/C-3 electrodes, and the reversible specific capacities are still 359.3, 440.5 and 341 mAh $g^{-1}$ at a current density of 1 A $g^{-1}$ after 250 cycles, respectively. Obviously, the 3DPNS-Sb/C-2 electrode exhibits superior cycle stability and reversible specific capacity. Furthermore, 3DPNS-Sb/C also reveals surpassing performance in comparison with the commercial LIBs anode materials (graphite, LTO) (Figure S2). In contrast with other statements on diverse Sb-C positive materials, the 3DPNS-Sb/C composites also show excellent electrochemical properties (Table 1).

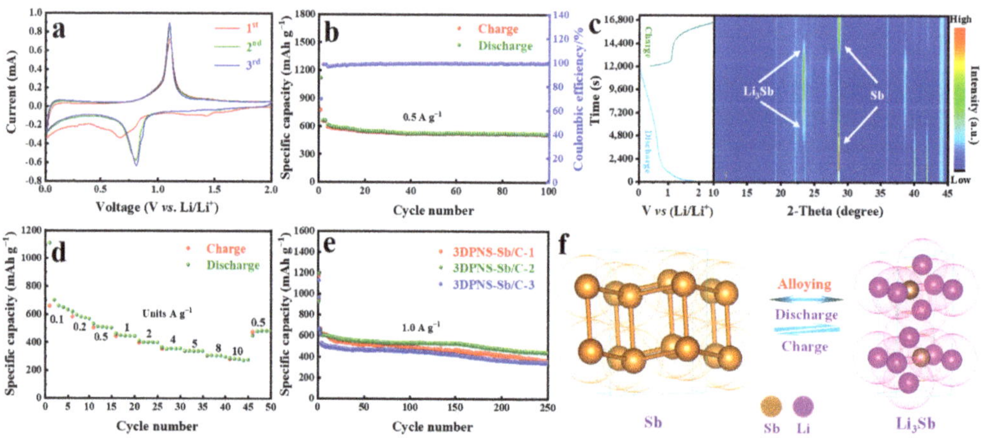


**Figure 4.** (**a**) CV curves of 3DPNS-Sb/C-2 electrodes at a scan rate of 0.1 mV $s^{-1}$, (**b**) Cycle performance of 3DPNS-Sb/C-2 electrodes at 0.5 A $g^{-1}$ for 100 cycles, (**c**) In-situ XRD of 3DPNS-Sb/C-2 electrodes, (**d**) Rate capability of 3DPNS-Sb/C-2 electrodes at current densities from 0.1 to 10 A $g^{-1}$, (**e**) Cycle performance of 3DPNS-Sb/C-1, 3DPNS-Sb/C-2 and 3DPNS-Sb/C-3 electrodes at 1 A $g^{-1}$ for 250 cycles, (**f**) The crystal structures of the active Sb in 3DPNS-Sb/C in the charge/discharge course.

**Table 1.** Contrast of the electrochemical properties of 3DPNS-Sb/C composites (this work) and various reported Sb-C as anodes for LIBs.

| Material | Reversible Capacity/mAh $g^{-1}$ | Current Density (mA $g^{-1}$) | Areal Mass Loading (mg $cm^{-2}$) | Batteries | Ref. |
|---|---|---|---|---|---|
| Hollow Sb Nanoparticles | 615/100th cycles | 120 | | Li-ion | [61] |
| Sb nanoparticles | 120/70th cycles | 120 | | Li-ion | [62] |
| Sb-carbon nanocomposite | 550/250th cycles | 230 | 1.07–1.11 | Li-ion | [51] |
| Sb/C composite fibers | 315.9/100th cycles | 100 | | Li-ion | [63] |
| Sb HNSs | 627.3/50th cycles | 100 | | Li-ion | [12] |
| Sb nanocrystals | 600/100th cycles | 660 | | Li-ion | [11] |
| Spherical Sb/C Composites | 590/80th cycles | 100 | 1 | Li-ion | [49] |
| Sb@C nanosponges | 447.1/500th cycles | 660 | 1.5 | Li-ion | [23] |
| Sb/C micro-/nanohybrid | 793/100th cycles | 66 | | Li-ion | [26] |
| Sb@C composites | 598.6/100th cycles | 100 | 1.132 | Li-ion | [21] |
| Sb/C/G nanocomposites | 413/700th cycles | 1000 | 1.0 | Li-ion | [32] |
| Sb/NPC | 556/100th cycles | 200 | 1.00 | Li-ion | [14] |
| Sb@C composites | 280/500th cycles | 100 | 1.35 | Li-ion | [30] |
| Sb@CNFs | 394.5/2000th cycles | 2000 | 0.8 | Li-ion | [28] |
| Sb2Se3/Sb/C nanofibers | 764/300th cycles | 100 | | Li-ion | [33] |
| Sb@C/EG | 486/600th cycles | 1000 | 0.5 | Li-ion | [34] |
| Ni-Co-Sb/C Nanosphere | 354/100th cycles | 100 | ~0.55 | Li-ion | [31] |
| Sb@C | 525/400th cycles | 500 | 1.2–1.5 | Li-ion | [24] |
| 3DPNS-Sb/C composites | 511.5/100th cycles | 500 | 0.93–1.12 | Li-ion | this work (586 Wh $L^{-1}$) (ICE: 69.35%) |

To further research the phase changes and the reaction mechanism, the 3DPNS-Sb/C-2 electrodes at diverse conditions of charge/discharge during the cycle were inspected using in-situ XRD. As shown in Figure 4c, the phase change of Sb could be watched from the peak intention variation on the contour map of the in-situ XRD consequence. The primary phase is at 28.7°, matched with the (012) crystal plane of Sb (JCPDS: 35-0732). As the discharge procedure continues, the diffraction peaks of Sb crystal progressively step down with alloying reaction between $Li^+$ and Sb to firstly form $Li_xSb$ ($x \leq 3$) phase located at 23.4°. In the reversible charge process, the diffraction peaks of the $Li_3Sb$ phase gradually disappear with the representation of the Sb phase, manifesting the arising of a dealloying reaction. Especially, Figure 4f interprets the alloying mechanism of the active Sb in the 3DPNS-Sb/C composites, which could be deemed as a type of alloying-typed material with a better electrochemical property.

## 5. Conclusions

In conclusion, the 3DPNS-Sb/C composites are fabricated with Sb nanoparticles uniformly embedded in the 3D polymer network structure via an uncomplicated and controllable synthetic medium. Based on the 3D polymer network structure, the 3DPNS-Sb/C composites employed as an anode display excellent electrochemical properties in LIBs. Specifically, they demonstrate a high invertible specific capacity of 511.5 mAh $g^{-1}$ at a current density of 0.5 A $g^{-1}$ after 100 cycles and a remarkable rate property of 289.5 mAh $g^{-1}$ at a current density of 10 A $g^{-1}$. This study explicitly demonstrates the promising potential of the 3DPNS-Sb/C composites as well-performaning LIBs anode.

**Supplementary Materials:** The following supporting information can be downloaded at: https://www.mdpi.com/article/10.3390/nano12142322/s1, Figure S1: SEM images of 3DPNS-Sb/C-2 after 100 cycles.; Figure S2: (a) Cycle performances of $Li_4Ti_5O_{12}$ at 0.5 A $g^{-1}$ for 50 cycles, (b) Cycle performances of graphite at 0.5 A $g^{-1}$ for 50 cycles.

**Author Contributions:** Methodology: Z.W. and Y.C.; Resources: F.Z., C.L. and L.W.; Software: D.Z., Y.S. and S.W. All authors have read and agreed to the published version of the manuscript.

**Funding:** This research were funded by [National Key R&D Program of China] grant number [2017YFE0198100], [National Natural Science Foundation of China] grant number [21975250], [Scientific and Technological Developing Project of Jilin Province] grant number [YDZJ202101ZYTS185], [Open Project Program of Key Laboratory of Preparation and Application of Environmental Friendly Materials (Jilin Normal University) Ministry of Education, China] grant number [No. 2020005], and [Open Pogram of State Key Laboratory of Metastable Materials Science and Technology (Yanshan University), China] grant number [No. 202110].

**Data Availability Statement:** Not applicable.

**Conflicts of Interest:** The authors declare no conflict of interest.

## References

1. Demirocak, D.E.; Srinivasan, S.S.; Stefanakos, E.K. A Review on Nanocomposite Materials for Rechargeable Li-ion Batteries. *Appl. Sci.* **2017**, *7*, 731. [CrossRef]
2. Deng, D. Li-ion batteries: Basics, progress, and challenges. *Energy Sci. Eng.* **2015**, *3*, 385–418. [CrossRef]
3. Shahjalal, M.; Roy, P.K.; Shams, T.; Fly, A.; Chowdhury, J.I.; Ahmed, R.; Liu, K. A review on second-life of Li-ion batteries: Prospects, challenges, and issues. *Energy* **2021**, *241*, 122881. [CrossRef]
4. Tian, Y.; Zeng, G.; Rutt, A.; Shi, T.; Kim, H.; Wang, J.; Koettgen, J.; Sun, Y.; Ouyang, B.; Chen, T.; et al. Promises and Challenges of Next-Generation "Beyond Li-ion" Batteries for Electric Vehicles and Grid Decarbonization. *Chem. Rev.* **2020**, *121*, 1623–1669. [CrossRef]
5. Nie, P.; Le, Z.; Chen, G.; Liu, D.; Liu, X.; Bin Wu, H.; Xu, P.; Li, X.; Liu, F.; Chang, L.; et al. Graphene Caging Silicon Particles for High-Performance Lithium-Ion Batteries. *Small* **2018**, *14*, e1800635. [CrossRef] [PubMed]
6. Ong, S.P.; Chevrier, V.L.; Hautier, G.; Jain, A.; Moore, C.; Kim, S.; Ma, X.; Ceder, G. Voltage, stability and diffusion barrier differences between sodium-ion and lithium-ion intercalation materials. *Energy Environ. Sci.* **2011**, *4*, 3680–3688. [CrossRef]
7. Roselin, L.S.; Juang, R.-S.; Hsieh, C.-T.; Sagadevan, S.; Umar, A.; Selvin, R.; Hegazy, H.H. Recent Advances and Perspectives of Carbon-Based Nanostructures as Anode Materials for Li-ion Batteries. *Materials* **2019**, *12*, 1229. [CrossRef]
8. Roy, P.; Srivastava, S.K. Nanostructured anode materials for lithium ion batteries. *J. Mater. Chem. A* **2014**, *3*, 2454–2484. [CrossRef]
9. Saritha, D. A concise review on the advancement of anode materials for Li-ion batteries. *Mater. Today Proc.* **2019**, *19*, 726–730. [CrossRef]
10. Fan, L.; Zhang, J.; Cui, J.; Zhu, Y.; Liang, J.; Wang, L.; Qian, Y. Electrochemical performance of rod-like Sb-C composite as anodes for Li-ion and Na-ion batteries. *J. Mater. Chem. A* **2015**, *3*, 3276–3280. [CrossRef]
11. He, M.; Kravchyk, K.; Walter, M.; Kovalenko, M.V. Monodisperse Antimony Nanocrystals for High-Rate Li-ion and Na-ion Battery Anodes: Nano versus Bulk. *Nano Lett.* **2014**, *14*, 1255–1262. [CrossRef]
12. Hou, H.; Jing, M.; Yang, Y.; Zhu, Y.; Fang, L.; Song, W.; Pan, C.; Yang, X.; Ji, X. Sodium/Lithium Storage Behavior of Antimony Hollow Nanospheres for Rechargeable Batteries. *ACS Appl. Mater. Interfaces* **2014**, *6*, 16189–16196. [CrossRef]
13. Liu, J.; Yu, L.; Wu, C.; Wen, Y.; Yin, K.; Chiang, F.-K.; Hu, R.; Liu, J.; Sun, L.; Gu, L.; et al. New Nanoconfined Galvanic Replacement Synthesis of Hollow Sb@C Yolk–Shell Spheres Constituting a Stable Anode for High-Rate Li/Na-Ion Batteries. *Nano Lett.* **2017**, *17*, 2034–2042. [CrossRef]
14. Yang, Q.; Zhou, J.; Zhang, G.; Guo, C.; Li, M.; Zhu, Y.; Qian, Y. Sb nanoparticles uniformly dispersed in 1-D N-doped porous carbon as anodes for Li-ion and Na-ion batteries. *J. Mater. Chem. A* **2017**, *5*, 12144–12148. [CrossRef]
15. Yuan, Y.; Jan, S.; Wang, Z.; Jin, X. A simple synthesis of nanoporous Sb/C with high Sb content and dispersity as an advanced anode for sodium ion batteries. *J. Mater. Chem. A* **2018**, *6*, 5555–5559. [CrossRef]
16. Wang, B.; Deng, Z.; Xia, Y.; Hu, J.; Li, H.; Wu, H.; Zhang, Q.; Zhang, Y.; Liu, H.; Dou, S. Realizing Reversible Conversion-Alloying of Sb(V) in Polyantimonic Acid for Fast and Durable Lithium- and Potassium-Ion Storage. *Adv. Energy Mater.* **2019**, *10*, 1903119. [CrossRef]
17. Darwiche, A.; Marino, C.; Sougrati, M.T.; Fraisse, B.; Stievano, L.; Monconduit, L. Better Cycling Performances of Bulk Sb in Na-Ion Batteries Compared to Li-Ion Systems: An Unexpected Electrochemical Mechanism. *J. Am. Chem. Soc.* **2012**, *134*, 20805–20811. [CrossRef]
18. Ding, Y.-L.; Wu, C.; Kopold, P.; van Aken, P.A.; Maier, J.; Yu, Y. Graphene-Protected 3D Sb-based Anodes Fabricated via Electrostatic Assembly and Confinement Replacement for Enhanced Lithium and Sodium Storage. *Small* **2015**, *11*, 6026–6035. [CrossRef]
19. Ko, Y.N.; Kang, Y.C. Electrochemical properties of ultrafine Sb nanocrystals embedded in carbon microspheres for use as Na-ion battery anode materials. *Chem. Commun.* **2014**, *50*, 12322–12324. [CrossRef]
20. Luo, W.; Li, F.; Gaumet, J.-J.; Magri, P.; Diliberto, S.; Zhou, L.; Mai, L. Bottom-Up Confined Synthesis of Nanorod-in-Nanotube Structured Sb@N-C for Durable Lithium and Sodium Storage. *Adv. Energy Mater.* **2018**, *8*, 1703237. [CrossRef]
21. Han, Q.; Zhang, X.; Li, X.; Li, Y.; Zhang, W.; Sheng, Y. Temperature-Dependent Nanopolyhedron Carbon-Decorated Sb for High-Performance Lithium-Ion Batteries. *ChemElectroChem* **2021**, *8*, 1486–1492. [CrossRef]

2. Kim, D.; Kim, H.; Lim, H.; Kim, K.J.; Jung, H.G.; Byun, D.; Kim, C.; Choi, W. A facile control in free-carbon domain with divinylbenzene for the high-rate-performing Sb/SiOC composite anode material in sodium-ion batteries. *Int. J. Energy Res.* **2020**, *44*, 11473–11486. [CrossRef]

3. Le, H.T.T.; Pham, X.-M.; Park, C.-J. Facile citrate gel synthesis of an antimony–carbon nanosponge with enhanced lithium storage. *New J. Chem.* **2019**, *43*, 10716–10725. [CrossRef]

4. Li, Q.; Liang, Z.; Zhang, W.; Lin, D.; Wang, G.; Wang, J.; Guang, C.; Huang, S. Constructing a hierarchical Sb@C nanoarchitectures as free-standing anode for high-performance lithium-ion batteries. *Mater. Lett.* **2021**, *303*, 130563. [CrossRef]

5. Liang, S.; Cheng, Y.-J.; Wang, X.; Xu, Z.; Ma, L.; Xu, H.; Ji, Q.; Zuo, X.; Müller-Buschbaum, P.; Xia, Y. Impact of $CO_2$ activation on the structure, composition, and performance of Sb/C nanohybrid lithium/sodium-ion battery anodes. *Nanoscale Adv.* **2021**, *3*, 1942–1953. [CrossRef]

6. Liang, S.-Z.; Wang, X.-Y.; Xia, Y.-G.; Xia, S.-L.; Metwalli, E.; Qiu, B.; Ji, Q.; Yin, S.-S.; Xie, S.; Fang, K.; et al. Scalable Synthesis of Hierarchical Antimony/Carbon Micro-/Nanohybrid Lithium/Sodium-Ion Battery Anodes Based on Dimethacrylate Monomer. *Acta Metall. Sin.* **2018**, *31*, 910–922. [CrossRef]

7. Ma, Z.; Chen, X.; Wu, H.; Xiao, Y.; Feng, C. Synthesis and electrochemical properties of Pb/Sb@C composite for lithium-ion battery application. *Ionics* **2020**, *26*, 5343–5348. [CrossRef]

8. Mao, Y.; Chen, R.; You, H.; Liu, Y.; Luan, S.; Hou, L.; Gao, F. Advanced performance of S and N co-doped Sb@CNFs with a 3D conductive network as superior lithium-ion battery anodes. *J. Alloy. Compd.* **2022**, *904*, 164000. [CrossRef]

9. Song, J.; Xiao, D.; Jia, H.; Zhu, G.; Engelhard, M.; Xiao, B.; Feng, S.; Li, D.; Reed, D.; Sprenkle, V.L.; et al. A comparative study of pomegranate Sb@C yolk–shell microspheres as Li and Na-ion battery anodes. *Nanoscale* **2019**, *11*, 348–355. [CrossRef]

0. Tian, J.; Yang, H.; Fu, C.; Sun, M.; Wang, L.; Liu, T. In-situ synthesis of microspherical Sb@C composite anode with high tap density for lithium/sodium-ion batteries. *Compos. Commun.* **2020**, *17*, 177–181. [CrossRef]

1. Wang, L.; Zhu, L.; Zhang, W.; Ding, G.; Yang, G.; Xie, L.; Cao, X. Revealing the unique process of alloying reaction in Ni-Co-Sb/C nanosphere anode for high-performance lithium storage. *J. Colloid Interface Sci.* **2020**, *586*, 730–740. [CrossRef]

2. Wang, Z.; Qu, J.; Hao, S.; Zhang, Y.; Kong, F.; Yang, D.; Yu, Z. Sb Nanoparticles Embedded in a Nitrogen-Doped Carbon Matrix with Tuned Voids and Interfacial Bonds for High-Rate Lithium Storage. *ChemElectroChem* **2018**, *5*, 2653–2659. [CrossRef]

3. Wen, J.; Pei, Y.; Liu, L.; Su, D.; Yang, M.; Wang, Q.; Zhang, W.; Dai, J.; Feng, Y.; Wu, T.; et al. Fully encapsulated Sb2Se3/Sb/C nanofibers: Towards high-rate, ultralong-lifespan lithium-ion batteries. *J. Alloy. Compd.* **2021**, *874*, 159961. [CrossRef]

4. Wu, Y.; Pan, Q.; Zheng, F.; Ou, X.; Yang, C.; Xiong, X.; Liu, M.; Hu, D.; Huang, C. Sb@C/expanded graphite as high-performance anode material for lithium ion batteries. *J. Alloy. Compd.* **2018**, *744*, 481–486. [CrossRef]

5. Zeng, T.; Hu, X.; Ji, P.; Peng, Q.; Shang, B.; Gong, S. General synthesis of nano-M embedded $Li_4Ti_5O_{12}$/C composites (M = Sn, Sb and Bi) with high capacity and good cycle stability. *Electrochim. Acta* **2016**, *217*, 299–309. [CrossRef]

6. Zhang, L.; Zhang, G.; Bin Wu, H.; Yu, L.; Lou, X.W. Hierarchical Tubular Structures Constructed by Carbon-Coated $SnO_2$ Nanoplates for Highly Reversible Lithium Storage. *Adv. Mater.* **2013**, *25*, 2589–2593. [CrossRef]

7. Xu, Y.; Zhu, Y.; Liu, Y.; Wang, C. Electrochemical Performance of Porous Carbon/Tin Composite Anodes for Sodium-Ion and Lithium-Ion Batteries. *Adv. Energy Mater.* **2012**, *3*, 128–133. [CrossRef]

8. Li, H.; Zhou, H. Enhancing the performances of Li-ion batteries by carbon-coating: Present and future. *Chem. Commun.* **2011**, *48*, 1201–1217. [CrossRef]

9. Luo, W.; Lorger, S.; Wang, B.; Bommier, C.; Ji, X. Facile synthesis of one-dimensional peapod-like Sb@C submicron-structures. *Chem. Commun.* **2014**, *50*, 5435–5437. [CrossRef]

0. Fan, L.; Zhu, Y.; Zhang, J.; Liang, J.; Wang, L.; Wei, D.; Li, X.; Qian, Y. Uniformly dispersed Sn-MnO@C nanocomposite derived from $MnSn(OH)_6$ precursor as anode material for lithium-ion batteries. *Electrochim. Acta* **2014**, *121*, 21–26. [CrossRef]

1. Wang, Z.; Wang, Z.; Liu, W.; Xiao, W.; Lou, X.W. Amorphous $CoSnO_3$@C nanoboxes with superior lithium storage capability. *Energy Environ. Sci.* **2013**, *6*, 87–91. [CrossRef]

2. Li, L.; Seng, K.H.; Li, D.; Xia, Y.Y.; Liu, H.K.; Guo, Z.P. SnSb@carbon nanocable anchored on graphene sheets for sodium ion batteries. *Nano Res.* **2014**, *7*, 1466–1476. [CrossRef]

3. Sougrati, M.T.; Fullenwarth, J.; Debenedetti, A.; Fraisse, B.; Jumas, J.C.; Monconduit, L. TiSnSb a new efficient negative electrode for Li-ion batteries: Mechanism investigations by operando-XRD and Mössbauer techniques. *J. Mater. Chem.* **2011**, *21*, 10069–10076. [CrossRef]

4. Wang, L.; Wang, C.; Zhang, N.; Li, F.; Cheng, F.; Chen, J. High Anode Performance of in Situ Formed $Cu_2Sb$ Nanoparticles Integrated on Cu Foil via Replacement Reaction for Sodium-Ion Batteries. *ACS Energy Lett.* **2016**, *2*, 256–262. [CrossRef]

5. Xu, J.; Wu, H.; Wang, F.; Xia, Y.; Zheng, G. $Zn_4Sb_3$ Nanotubes as Lithium Ion Battery Anodes with High Capacity and Cycling Stability. *Adv. Energy Mater.* **2013**, *3*, 286–289. [CrossRef]

6. Yang, R.; Huang, J.; Zhao, W.; Lai, W.; Zhang, X.; Zheng, J.; Li, X. Bubble assisted synthesis of Sn–Sb–Cu alloy hollow nanostructures and their improved lithium storage properties. *J. Power Sources* **2010**, *195*, 6811–6816. [CrossRef]

7. Sevilla, M.; Fuertes, A.B. Chemical and Structural Properties of Carbonaceous Products Obtained by Hydrothermal Carbonization of Saccharides. *Chem.–A Eur. J.* **2009**, *15*, 4195–4203. [CrossRef] [PubMed]

8. Li, P.; Yu, L.; Ji, S.; Xu, X.; Liu, Z.; Liu, J.; Liu, J. Facile synthesis of three-dimensional porous interconnected carbon matrix embedded with Sb nanoparticles as superior anode for Na-ion batteries. *Chem. Eng. J.* **2019**, *374*, 502–510. [CrossRef]

49. Liu, X.; Tian, Y.; Cao, X.; Li, X.; Le, Z.; Zhang, D.; Li, X.; Nie, P.; Li, H. Aerosol-Assisted Synthesis of Spherical Sb/C Composite as Advanced Anodes for Lithium Ion and Sodium Ion Batteries. *ACS Appl. Energy Mater.* **2018**, *1*, 6381–6387. [CrossRef]
50. Lannin, J.S.; Calleja, J.M.; Cardona, M. Second-order Raman scattering in the group-Vbsemimetals: Bi, Sb, and As. *Phys. Rev.* **1975**, *12*, 585–593. [CrossRef]
51. Ramireddy, T.; Rahman, M.; Xing, T.; Chen, Y.; Glushenkov, A.M. Stable anode performance of an Sb–carbon nanocomposite i lithium-ion batteries and the effect of ball milling mode in the course of its preparation. *J. Mater. Chem. A* **2014**, *2*, 4282–429 [CrossRef]
52. Ramireddy, T.; Sharma, N.; Xing, T.; Chen, Y.; Leforestier, J.; Glushenkov, A.M. Size and Composition Effects in Sb-Carbo Nanocomposites for Sodium-Ion Batteries. *ACS Appl. Mater. Interfaces* **2016**, *8*, 30152–30164. [CrossRef]
53. Wang, X.; Kunc, K.; Loa, I.; Schwarz, U.; Syassen, K. Effect of pressure on the Raman modes of antimony. *Phys. Rev. B* **2006**, 7 134305. [CrossRef]
54. Rodriguez, J.R.; Hamann, H.J.; Mitchell, G.M.; Ortalan, V.; Pol, V.G.; Ramachandran, P.V. Three-Dimensional Antimon Nanochains for Lithium-Ion Storage. *ACS Appl. Nano Mater.* **2019**, *2*, 5351–5355. [CrossRef]
55. Bodenes, L.; Darwiche, A.; Monconduit, L.; Martinez, H. The Solid Electrolyte Interphase a key parameter of the high performanc of Sb in sodium-ion batteries: Comparative X-ray Photoelectron Spectroscopy study of Sb/Na-ion and Sb/Li-ion batteries. *Power Sources* **2015**, *273*, 14–24. [CrossRef]
56. Marino, C.; Darwiche, A.; Dupre, N.; Wilhelm, H.A.; Lestriez, B.; Martinez, H.; Dedryvère, R.; Zhang, W.; Ghamouss, F Lemordant, D.; et al. Study of the Electrode/Electrolyte Interface on Cycling of a Conversion Type Electrode Material in L Batteries. *J. Phys. Chem. C* **2013**, *117*, 19302–19313. [CrossRef]
57. Pham, X.-M.; Ngo, D.T.; Le, H.T.T.; Didwal, P.N.; Verma, R.; Min, C.-W.; Park, C.-N.; Park, C.-J. A self-encapsulated porous Sb–C nanocomposite anode with excellent Na-ion storage performance. *Nanoscale* **2018**, *10*, 19399–19408. [CrossRef]
58. Dailly, A.; Ghanbaja, J.; Willmann, P.; Billaud, D. Lithium insertion into new graphite–antimony composites. *Electrochim. Act* **2003**, *48*, 977–984. [CrossRef]
59. Sethuraman, V.; Srinivasan, V.; Newman, J. Analysis of Electrochemical Lithiation and Delithiation Kinetics in Silicon. *Electrochem. Soc.* **2012**, *160*, A394–A403. [CrossRef]
60. Sethuraman, V.A.; Srinivasan, V.; Bower, A.F.; Guduru, P.R. In Situ Measurements of Stress-Potential Coupling in Lithiated Silicor *J. Electrochem. Soc.* **2010**, *157*, A1253–A1261. [CrossRef]
61. Kim, H.; Cho, J. Template Synthesis of Hollow Sb Nanoparticles as a High-Performance Lithium Battery Anode Material. *Chem Mater.* **2008**, *20*, 1679–1681. [CrossRef]
62. Zhu, J.; Sun, T.; Chen, J.; Shi, W.; Zhang, X.; Lou, X.; Mhaisalkar, S.; Hng, H.H.; Boey, F.; Ma, J.; et al. Controlled Synthesis o Sb Nanostructures and Their Conversion to $CoSb_3$ Nanoparticle Chains for Li-Ion Battery Electrodes. *Chem. Mater.* **2010**, *2* 5333–5339. [CrossRef]
63. Lv, H.; Qiu, S.; Lu, G.; Fu, Y.; Li, X.; Hu, C.; Liu, J. Nanostructured Antimony/carbon Composite Fibers as Anode Material fo Lithium-ion Battery. *Electrochim. Acta* **2015**, *151*, 214–221. [CrossRef]

*nanomaterials* MDPI

*Article*

# Electrochemical Properties of an Sn-Doped LATP Ceramic Electrolyte and Its Derived Sandwich-Structured Composite Solid Electrolyte

**Aihong Xu [1,2], Ruoming Wang [1,2], Mengqin Yao [1,2], Jianxin Cao [1,2], Mengjun Li [1,2], Chunliang Yang [1,2], Fei Liu [1,2,*] and Jun Ma [1,2,*]**

1 Department of Chemical Engineering, School of Chemistry and Chemical Engineering, Guizhou University, Guiyang 550025, China; 15851959055@163.com (A.X.); w1285459785@126.com (R.W.); mqyao@gzu.edu.cn (M.Y.); jxcao@gzu.edu.cn (J.C.); x1285459785@126.com (M.L.); clyang@gzu.edu.cn (C.Y.)

2 Guizhou Key Laboratory for Green Chemical and Clean Energy Technology, Guiyang 550025, China

* Correspondence: ce.feiliu@gzu.edu.cn (F.L.); jma3@gzu.edu.cn (J.M.)

**Abstract:** An $Li_{1.3}Al_{0.3}Sn_xTi_{1.7-x}(PO_4)_3$ (LATP-xSn) ceramic solid electrolyte was prepared by Sn doping via a solid phase method. The results showed that adding an Sn dopant with a larger ionic radius in a concentration of x = 0.35 enabled one to equivalently substitute Ti sites in the LATP crystal structure to the maximum extent. The uniform Sn doping could produce a stable LATP structure with small grain size and improved relative density. The lattice distortion induced by Sn doping also modified the transport channels of Li ions, which promoted the increase of ionic conductivity from $5.05 \times 10^{-5}$ to $4.71 \times 10^{-4}$ S/cm at room temperature. The SPE/LATP-0.35Sn/SPE composite solid electrolyte with a sandwich structure was prepared by coating, which had a high ionic conductivity of $5.9 \times 10^{-5}$ S/cm at room temperature, a wide electrochemical window of 4.66 V vs. $Li/Li^+$, and a good lithium-ion migration number of 0.38. The Li | | Li symmetric battery test results revealed that the composite solid electrolyte could stably perform for 500 h at 60 °C under the current density of 0.2 $mA/cm^2$, indicating its good interface stability with metallic lithium. Moreover, the analysis of the all-solid-state $LiFePO_4$ | | SPE/LATP-0.35Sn/SPE | | Li battery showed that the composite solid electrolyte had good cycling stability and rate performance. Under the conditions of 60 °C and 0.2 C, stable accumulation up to 200 cycles was achieved at a capacity retention ratio of 90.5% and a coulombic efficiency of about 100% after cycling test.

**Keywords:** NASICON-type LATP; Sn doping; sandwich structure; composite solid electrolyte; lattice distortion

## 1. Introduction

At present, lithium-ion batteries have been widely used in various electronic products, electric/hybrid vehicles, and fixed energy storage systems [1–4]. However, the toxic organic liquid electrolytes commonly employed in traditional lithium ion batteries have many shortcomings, such as undesirable inflammability, easy decomposition at high temperatures, rapid solidification at low temperatures, and fast leakage [5]. Moreover, side reactions of organic liquid electrolytes with positive and negative electrodes are prone to occur [6].

This issue can be successfully solved by introducing all-solid-state batteries in which organic liquid electrolytes can be replaced by inorganic solid electrolytes with high thermal stability [7,8]. In addition to safety, all-solid-state batteries also have many other advantages, such as simplified battery packaging, better electrochemical stability, and wider operating temperature ranges [6,9,10]. Inorganic lithium-ion solid electrolytes include NASICON-type [11,12], garnet-type [13], perovskite-type [14], LISICON [15], LiPON [16], $Li_3N$ [17], sulfides [18], and anti-perovskite [19] systems. Among them, NASICON-type solid electrolytes have attracted much attention due to their excellent electrochemical stability in air and/or water environment, as well as their low manufacturing cost [20,21].

Between various NASICON structures, $LiTi_2(PO_4)_3$ (LTP) has a three-dimensional network assembled by two $TiO_6$ octahedrons and three $PO_4$ tetrahedrons sharing oxygen atoms, which can provide a three-dimensional interconnected conduction pathway for $Li^+$ ion transport [22]. In particular, $Al^{3+}$ doping of LTP ($Li_{1+x}Al_xTi_{1.7-x}(PO_4)_3$ or LATP) enables one to increase the carrier concentration and to reduce the Li-O bond strength, thus improving the ionic conductivity to a certain extent [23]. In recent years, ion doping has still been an effective strategy to upgrade the ionic conductivity of LATP. The LATP frameworks can be modified by intercalating cations with different valence states and ion radii, which causes lattice distortion and enhances ionic conductivity through the adjustment of ion transport channels and the increase of Li ion or Li vacancy concentration. For instance, doping modification of LATP by cations such as $Nb^{5+}$ [24], $Zr^{4+}$ [7], $Y^{3+}$ [25], $Ga^{3+}$ [26], $Sc^{3+}$ [27], $Te^{4+}$ [28], $Si^{4+}$ [29], and $V^{5+}$ [30] has been reported [31,32].

Nevertheless, the crystal structure of LATP after equivalent ion substitution for $Ti^{4+}$ site (0.745 Å) exhibits the better structural stability. Doping modification with larger radius ions can also increase the lattice volume and widen the Li ion channels to a certain extent. In addition, introducing dopants with higher Pauli electronegativity allows one to improve the stability of cations in the LATP structure. Undoubtedly, the selection of dopants that are abundant, inexpensive, and eco-friendly is also a significant factor that cannot be ignored in doping modification.

Taking into account the above favorable factors, we performed doping modification of LATP by substituting $Ti^{4+}$ (0.745 Å) with $Sn^{4+}$ (0.83 Å) ions with larger ionic radius and higher Pauli electronegativity, in order to improve the framework stability and to make channel structures more conducive to ion diffusion. Because of its non-toxicity, low price, and abundant reserves [33], tin (Sn) has been widely concerned in photoelectric, photovoltaic, and energy storage devices [34–37], but its application in the modified LATP structures has not been reported yet.

Solid electrolytes are generally rigid and difficult to be machined [38]. When they contact with solid (either positive or negative) electrodes, there is often severe interfacial impedance [39]. Polymer solid electrolytes usually have good elastic properties and low interfacial impedance with solid electrodes [40,41]. The composite solid electrolyte formed by uniform mixing of polymer precursor solution and ceramic electrolyte powder can combine the respective advantages of both polymers and ceramics [42,43]. However, the reduction of the interfacial impedance between the composite solid electrolyte and the electrode is achieved at the expense of sacrificing conductivity of the ceramic electrolyte. This is because the conductivity of solid polymer electrolytes (SPE) represented by polyethylene oxide (PEO) is generally low ($\leq 10^{-6}$ S/cm) [44–46].

Considering the above two aspects and focusing on the preparation of high-performance composite solid electrolytes, this paper places particular emphasis on the influence of $Sn^{4+}$ doping on the structural properties and electrochemical performance of LATP. Attempts are made to reduce the solid-solid interface impedance through the construction of a sandwich-structured composite solid electrolyte by coating the PEO polymer electrolyte precursor solution on both sides of the modified ceramic electrolyte tablet instead of even mixing. The structure–property relationship of the composite solid electrolyte material was established by means of a series of analytical and electrochemical performance tests.

## 2. Materials and Methods

### 2.1. Materials

The raw materials included $LiNO_3$ (99%, Aladdin, Shanghai, China), $Al_2O_3$ (99%, Aladdin, Shanghai, China), $TiO_2$ (99%, Macklin, Shanghai, China), $SnO_2$ (99.9%, Aladdin, Shanghai, China), $NH_4H_2PO_4$ (99%, Macklin, Shanghai, China), Polyethylene Oxide (PEO) (Mw = 600,000, Macklin, Shanghai, China), and lithium bisimide (LiTFSI) (99.99% purity, Aladdin, Shanghai, China), which were not further purified.

### 2.2. Preparation of an LATP-xSn Ceramic Solid Electrolyte

The Sn-doped $Li_{1.3}Al_{0.3}Sn_xTi_{1.7-x}(PO_4)_3$ (LATP-xSn, x = 0–0.4) solid electrolyte was prepared by a solid-phase reaction method. The stoichiometric quantities of raw materials for preparing an LATP-xSn ceramic solid electrolyte are as follows: $LiNO_3$ (1.3), $Al_2O_3$ (0.3), $SnO_2$ (x, where x = 0, 0.1, 0.2, 0.3, 0.35, 0.4), $TiO_2$ (1.7 − x), and $NH_4H_2PO_4$ (3) are mixed evenly in a ceramic mortar. Then, the powder mixture was calcined in an alumina crucible at 300 °C for 3 h. The temperature was then programmatically increased to 700 °C (5 °C/min) and held for 5 h. After cooling and grinding, the samples were pressed at 16 MPa for 10 min into wafers with a diameter of 13 mm and a thickness of 1.5–2.5 mm. Finally, the green-pressing wafers were sintered at 900 °C for 6 h.

### 2.3. Preparation of a Composite Solid Electrolyte

The PEO and lithium bisimide with a molar ratio of PEO:Li = 8:1 were dissolved in an acetonitrile solvent, and stirred for 12 h in a glovebox under an argon atmosphere to obtain a homogenous precursor solution of polymer electrolyte [47]. The solution was evenly coated on both sides of the LATP-xSn ceramic tablet, and then the sample was transferred to a vacuum oven for drying at 60 °C for 12 h to prepare an SPE/LATP-xSn/SPE sandwich-structured composite solid electrolyte.

### 2.4. Preparation of Electrodes and Assembly of the Battery

To fabricate a $LiFePO_4$ cathode, the $LiFePO_4$ powder, carbon black, and PVDF binder were mixed in N-methyl pyrrolidone at a mass ratio of 8:1:1 and stirred for 12 h to form a slurry of suitable viscosity, which was cast onto the surface of the current collector Al foil. The electrodes were placed in a blast drying oven at 60 °C for 12 h, and then transferred to a vacuum drying oven at 100 °C for 1 h. Afterwards, the dried pole piece was rolled until its surface was smooth and the thickness was 20 μm, and was punched into a circular pole piece with a diameter of 12 mm. The prepared polar pieces were dried in a vacuum drying oven for 12 h and then transferred to a glove box for use. The loading mass of the $LiFePO_4$ in the electrode is 2.4 $mg/cm^2$.

The above sandwich-structured composite solid electrolyte was assembled into a CR2032 button cell for an electrochemical performance test. $LiFePO_4$ and lithium were used as the positive and negative electrodes, respectively, and the sandwich-structured composite solid electrolyte was placed between them (without any auxiliary electrolyte).

### 2.5. Materials Characterization

X-ray diffraction experiments were performed to identify the crystal structure of LATP-xSn (x = 0–0.4) on a Bruker D8 Advance X-ray diffractometer (XRD) equipped with a Cu Kα source (λ = 1.54178 Å) and operated at 40 kV and 40 mA. The XRD data were acquired in the 2θ range of 10–90° and were refined using FullProf software (FullProf Suit, (Version January-2021-JPC-JRC), signed with the "Institut Laue-Langevin" certificate, https://www.ill.eu/sites/fullprof/php/downloads.html (accessed on 11 August 2021)). The microstructure characterization and element distribution analysis were performed by means of a Carl Zeiss Supra 40 scanning electron microscope (SEM) and a Zeiss Gemini 300 energy dispersive X-ray spectrometer (EDS), respectively. The information about the molecular framework of LATP-xSn (x = 0–0.4) was extracted using a Thermo Fisher Nicolet IS50 Fourier transform infrared spectrometer and a LabRam HR Evolution laser microconfocal Raman spectrometer (Horiba Jobin Yvon, Paris, France). The surface elements and their valence states were analyzed on a K-Alpha Plus X-ray photoelectron spectrometer (Thermo Fisher, Waltham, MA, USA). The volumetric density of samples was measured using a Vernier caliper, and the theoretical density was obtained from the refined XRD data. The deviation degree (abbreviated as D) refers to the sum of the relative deviations of the ion radius and electronegativity between the dopant ion and the substituted ion. The calculation formula is as follows [29]:

$$D = \left|\frac{\chi_{Sn} - \chi_i}{\chi_i}\right| + \left|\frac{\gamma_{Sn} - \gamma_i}{\gamma_i}\right| \quad (1)$$

where $\chi_{Sn}$ and $\gamma_{Sn}$ are the ion electronegativity and crystal ion radius of $Sn^{4+}$, respectively, and $\chi_i$ and $\gamma_i$ are the ion electronegativity and crystal ion radius of the substituted ion, respectively.

### 2.6. Electrochemical Performance Test

The ionic conductivity, electrochemical stability window, and lithium-ion migration number of the solid electrolyte were measured using a CHI660E (Chenhua, Shanghai, China) electrochemical workstation.

Ionic conductivities were measured by electrochemical impedance spectroscopy with a signal amplitude of 5 mV over the frequency range of $10^{-2}$–$10^6$ Hz at 25 °C. The ionic conductivity σ (S/cm) is calculated as follows:

$$\sigma = L/RS \quad (2)$$

where $L$ (cm) is the thickness of the electrolyte, $R$ (Ω) is the resistance value of the electrolyte, and $S$ ($cm^2$) is the area of the electrode plate.

By testing the ionic conductivities at different temperatures, the conductive activation energy $Ea$ (eV) of $Li^+$ can be calculated from the Arrhenius formula as follows:

$$\sigma T = Aexp(-\frac{Ea}{K_B T}) \quad (3)$$

where $\sigma$ is the ionic conductivity of the electrolyte, $T$ is the temperature, $A$ is the pre-exponential factor, and $K_B$ is the Boltzmann's constant.

Linear sweep voltammetry (LSV) was performed to investigate the electrochemical stability window of the solid electrolyte with a scanning rate of 0.001 V/s at 60 °C from 2 V to 6 V (vs. $Li^+$/Li). The lithium ion transport numbers ($t_{Li^+}$) of the solid electrolyte were measured with a DC polarization voltage of 10 mV associated with the AC impedance measurement and calculated using Equation (4):

$$t_{Li^+} = I_{SS}(\Delta V - I_0 R_0)/(I_0(\Delta V - I_{ss} R_{ss}) \quad (4)$$

where $I_0$ and $I_{SS}$ are the initial current and steady-state current, both furnished by a direct current (DC) polarization test, $R_0$ and $R_{SS}$ are the charge transfer resistances before and after DC polarization, and $\Delta V$ is the polarization voltage.

The Li | | Li symmetric constant current charge and discharge curves of the composite solid electrolyte, as well as the cycle performance and rate performance of the all-solid-state battery, were measured using a CT3001A battery testing system (Lanhe, Wuhan, China).

## 3. Results

### 3.1. Influences of Sn Doping on the Structural Properties and Electrochemical Performance of an LATP-xSn Solid Electrolyte

Figure 1 shows the XRD patterns of a series of modified LATP-xSn solid electrolytes (x = 0–0.4). The diffraction peaks of the main crystal phase of all modified samples are basically consistent with those of a $LiTi_2(PO_4)_3$ (PDF#35-0754) with an R-3c space group, indicating the successful preparation of a $Li_{1.3}Al_{0.3}Sn_xTi_{1.7-x}(PO_4)_3$ solid electrolyte with a NASICON-type structure. However, a small amount of $AlPO_4$ crystalline phase appears in the LATP-0.1Sn and LATP-0.2Sn samples. With the increase of Sn content, the $AlPO_4$ phase gradually disappears. Moreover, the LATP-0.4Sn sample reveals the emergence of an $SnO_2$ phase. It should be pointed out that the diffraction peak associated with a (113) crystal plane of the modified samples tends to shift to the lower-angle range, and the shift degree becomes more pronounced with the increase of Sn dopant content. This trend can be explained by the effective substitution of smaller ions in the LATP lattice by the larger $Sn^{4+}$ (r = 0.083 nm) ions.

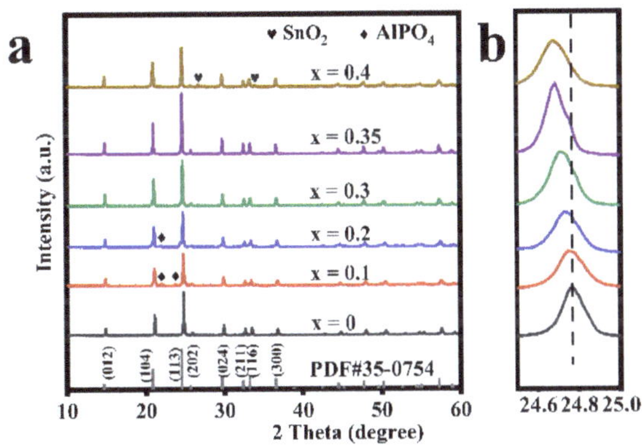


**Figure 1.** (**a**) X-ray diffraction patterns of LATP-xSn (x = 0–0.4) solid electrolytes and (**b**) the magnified view of diffraction patterns at (113) plane for LATP-xSn (x = 0–0.4) solid electrolytes.

This speculation can be further confirmed by the lattice parameters of the refined data (Table 1, Figures S1 and S2). The lattice parameters of modified LATP-xSn (x = 0–0.35) systems increase linearly with the increase of Sn content, which follows the Vegard law, indicating an effective insertion of $Sn^{4+}$ ions [48]. It has been reported that the increase of lattice parameters can facilitate the intercalation and deintercalation of lithium ions during the cycling process [49]. However, the lattice parameters in a and b axes of the modified LATP-0.4Sn exhibit a decreasing trend, which may be related to the appearance of the $SnO_2$ phase in the sample [48]. The theoretical density value of the modified samples also shows a similar variation trend. Rp is the abbreviation of R-pattern, which means graphic variance factor. Rwp is the abbreviation of R-pattern, which means weighted graphic variance factor. Rexp is the abbreviation of R-expected, which means expected variance factor. These are fitting factors, used to judge the result of refinement.

**Table 1.** Lattice parameters of LATP-xSn (x = 0–0.4) solid electrolytes.

| x | a = b (Å) | c (Å) | Density (g/cm$^3$) | Rp | Rwp | Rexp |
|---|---|---|---|---|---|---|
| 0 | 8.509 | 20.864 | 2.92 | 9.8 | 13.5 | 10.3 |
| 0.1 | 8.515 | 20.884 | 2.78 | 10.1 | 14.2 | 9.2 |
| 0.2 | 8.522 | 20.919 | 2.95 | 11.2 | 13.9 | 10.9 |
| 0.3 | 8.526 | 20.939 | 3.31 | 9.3 | 13.1 | 10.2 |
| 0.35 | 8.528 | 20.944 | 3.32 | 10.3 | 14.0 | 11.5 |
| 0.4 | 8.525 | 20.973 | 3.11 | 9.4 | 13.3 | 10.5 |

It is generally believed that doping is more likely to occur at ion sites of similar electronegativity and ion radiuses [50,51]. Thus, the spatial effect and electrostatic interaction can be evaluated from the ion radius and ion electronegativity, respectively [52], and the calculated deviation degree can reflect the most probable Sn doping sites in the lattice. The deviation degree results (Table S1) calculated according to Equation (1) show that the deviation degree value of DTi is the smallest, indicating that Sn is more inclined to occupy octahedral Ti sites in the LATP lattice. This speculation can be verified by the following characterization results.

The Raman-active vibration modes of the LATP-xSn (x = 0–0.4) solid electrolyte are usually divided into internal and external modes. Internal modes occur above ~350 cm$^{-1}$

in the Raman spectra. In Figure 2, the peak at 350 cm$^{-1}$ is ascribed to the Ti-O vibration mode [50], the signal peaks at 435 and 450 cm$^{-1}$ can be attributed to the symmetric bending motion of P-O bonds in $PO_4$ tetrahedrons [51], and the signal peaks at 971, 990, and 1009 cm$^{-1}$ correspond to asymmetric and symmetric stretching vibrations of the P-O bonds [50]. Except for the LATP-0.4Sn sample, all peak intensities of other samples increase with the increase of Sn doping content. Once the Sn content increases to x = 0.35, the peak intensities increase to their maximum values. It is worth mentioning that the band width of the Ti-O peak at 350 cm$^{-1}$ increases after Sn doping, indicating that the lattice distortion around $TiO_6$ is enhanced due to the substitution of Ti sites by Sn ions [53].

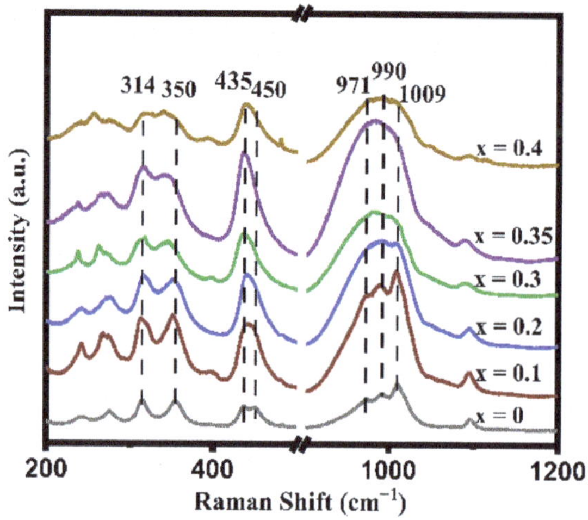


**Figure 2.** Raman spectra of LATP-xSn (x = 0–0.4) solid electrolytes.

In addition, the relative intensities of the peaks at 350 and 314 cm$^{-1}$ in Figure 2 can reflect the alteration of the lithium content in the LATP lattice. When the lithium content is high, the peak intensity at 314 cm$^{-1}$ is higher than that at 350 cm$^{-1}$, and vice versa [54]. Except for the LATP-0.4Sn sample, the peak intensity at 314 cm$^{-1}$ gradually increases with the increase of Sn dopant content, indicating that Sn doping can reduce lithium loss. The largest intensity difference between the two peaks is obtained at the lithium content of x = 0.35.

Figure 3 displays the infrared spectra of the LATP-xSn (x = 0–0.4) solid electrolyte. In the figure, the peaks at 580 and 648 cm$^{-1}$ can be attributed to the Ti-O stretching vibrations of $TiO_6$ octahedra [55], which overlap with the Sn-O vibrations of $SnO_6$ octahedra at 648 cm$^{-1}$ [56]. With the increase of Sn dopant content, the peak intensity at 648 cm$^{-1}$ obviously increases because Sn doping promotes the enhancement of the Ti-O bond strength. This change in structural information can also be verified by XPS results (Figure S3). The characteristic peaks at 487.3 and 495.7 eV in the LATP-0.35Sn sample confirm the successful doping by $Sn^{4+}$ ions. In addition, the shifts of Ti 2p and O 2p characteristic peaks confirm that the O-Sn bonds in Sn-doped LATP are formed by the substitution of Sn for Ti.

The electrochemical impedance spectra of the LATP-xSn (x = 0–0.4) solid electrolyte were measured (Figure 4a). The semicircle displayed in the high frequency region is related to the conductivity of the solid electrolyte, and the straight line in the low frequency region represents the resistance of lithium ion transport between the electrolyte and the blocking electrode. The high frequency region does not display the grain resistance (Rg) due to the limitation of the test frequency. The grain resistance (Rg) and grain boundary resistance (Rgb) can be obtained by fitting the equivalent circuit with Zview software, and CPE is a constant phase element, which is used to replace capacitor in many equivalent circuit models. Rg, Rgb, CPE1, and CPE2 represent grain resistance, grain boundary resistance, grain boundary capacitance, and sample-electrode capacitance respectively. The total resistance (Rt) is the grain resistance (Rg) and the grain boundary resistance (Rgb) [57]. The relevant ionic conductivity data were calculated according to Equation (2) (Figure 4a and Table 2). Due to the low sintering temperature of LATP solid electrolyte, the total ionic conductivity of LATP before being undoped is low ($5.05 \times 10^{-5}$ S/cm), which is significantly lower than the reported value in the previous work [48,55]. The room temperature ionic conductivity of LATP-0.35Sn is $4.71 \times 10^{-4}$ S/cm, which is nearly an order of magnitude higher than that of the undoped one. According to Table 2, it is found that the total ionic conductivity of LATP-0Sn is mainly determined by the ionic conductivity of the grain boundary. With the increase of the Sn content, the grain boundary impedance is greatly reduced, and the total ionic conductivity is significantly improved. The ionic conductivity is the highest ($4.71 \times 10^{-4}$ S/cm), which is higher than that of other typical solid electrolytes (Table S2) [8,58–62]. However, when the Sn content reaches 0.4, both the grain conductivity and grain boundary conductivity decrease.

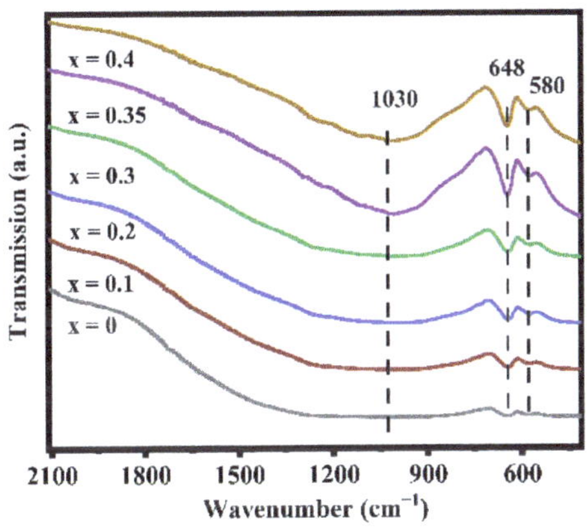


**Figure 3.** FT-IR spectra of LATP-xSn (x = 0–0.4) solid electrolytes.

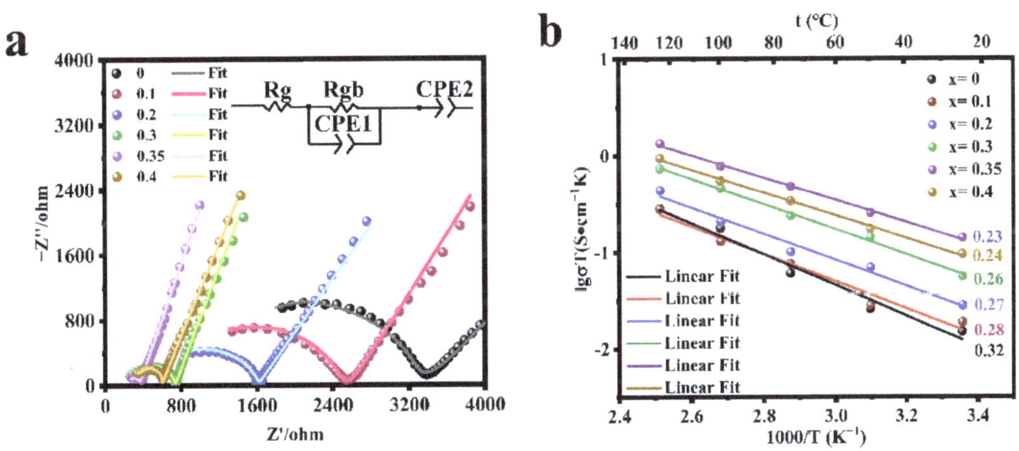


**Figure 4.** (**a**) EIS curves of LATP-xSn (x = 0–0.4) solid electrolytes at 25 °C and (**b**) the Arrhenius plot of the lithium ionic conductivity of LATP-xSn (x = 0–0.4) solid electrolytes.

**Table 2.** Grain impedance (Rg), grain boundary impedance (Rgb), grain conductivities (σg), grain boundary conductivities (σgb), total conductivities (σt), activation energy (Ea), and relative density of LATP-xSn (x = 0–0.4) solid electrolytes.

| x | Rg (Ω) | Rgb (Ω) | σg (mS/cm) | σgb (mS/cm) | σt (mS/cm) | Ea (eV) | Relative Density (%) |
|---|---|---|---|---|---|---|---|
| 0 | 653 | 2798 | 0.267 | 0.0624 | 0.0505 | 0.32 | 87.5 |
| 0.1 | 580 | 1956 | 0.277 | 0.0821 | 0.0633 | 0.28 | 88.2 |
| 0.2 | 375.2 | 1237.4 | 0.402 | 0.122 | 0.0939 | 0.27 | 89.4 |
| 0.3 | 232 | 498.8 | 0.595 | 0.277 | 0.189 | 0.26 | 90.1 |
| 0.35 | 218.5 | 139.6 | 0.772 | 1.21 | 0.471 | 0.23 | 91.8 |
| 0.4 | 339.5 | 253.9 | 0.556 | 0.742 | 0.322 | 0.24 | 92.0 |

The electrochemical impedance spectra of the LATP-xSn (x = 0–0.4) solid electrolyte were tested in the temperature range of 25–125 °C, and the conductive activation energy of lithium ions was calculated according to Equation (3) (Figure 4b and Table 2). All samples show good linear fitting results, among which LATP-0.35Sn exhibits a minimum activation energy of 0.23 eV and a relatively high relative density (91.8%), indicating that $Li^+$ ions can migrate more easily in the LATP-0.35Sn lattice and through the grain boundaries. When the Sn doping content is x = 0.4, the relative density continues to increase to 92.0%, but the ionic conductivity decreases, which is related to the presence of the $SnO_2$ phase in the sample [52].

Through the equivalent substitution of $Ti^{4+}$ sites by $Sn^{4+}$, Sn ions can occupy the LATP lattice uniformly (Figure S4), and the crystal structure does not experience neither excessive lattice distortion nor crystal transformation. More importantly, smaller grains can be obtained (Figure S5) and the relative density is also improved (Table 2), finally enhancing the ionic conductivity.

### 3.2. Electrochemical Performance of the Composite Solid Electrolyte

The solid–solid interface between the LATP-0.35Sn solid electrolyte and lithium has a serious interface impedance problem, and the ionic conductivity of the LATP-0.35Sn all-solid-state battery at 25 °C is only $1.0 \times 10^{-5}$ S/cm (Figure S6 and Table S3). The thickness of the LATP-0.35Sn solid electrolyte is 1.2 mm. After coating the surface with SPE, the SPE film is smooth and dense (Figure S7a), and the thickness of the film is about 24 μm on one side (Figure S7b). The electrolyte resistance (Re) and charge transfer resistance (Rct) were obtained by fitting the equivalent circuit through the software Zview (Figure S7). The interface impedance of the SPE/LATP-0.35Sn/SPE composite solid electrolyte and the electrode was significantly reduced, and its charge transfer resistance Rct (1344 Ω) is much lower than the LATP-0.35Sn charge transfer resistance Rct (8541 Ω), and the difference between the two lies in the PEO-LiTFSI on the surface of SPE/LATP-0.35Sn/SPE. The charge transfer resistance Rct (4408 Ω) of PEO-LiTFSI is much lower than that of the LATP-0.35Sn charge transfer resistance Rct (8541 Ω), indicating that it has better contact with electrodes and is more favorable for charge transfer between electrodes. The electrolyte resistance Re (239 Ω) of SPE/LATP-0.35Sn/SPE is lower than that of LATP-0.35Sn (413 Ω) and that of PEO-LiTFSI (1739 Ω), which is attributed to the fact that PEO-LiTFSI is filled with pores on the surface of the LATP-0.35Sn solid electrolyte and provides more $Li^+$ diffusion pathways for solid–solid interfaces. A salt anion TFSI-LATP-0.35Sn surface with stronger affinity accelerates lithium ion migration [63]. Therefore, the room temperature ionic conductivity of SPE/LATP-0.35Sn/SPE was improved to $5.9 \times 10^{-5}$ S/cm (Figure S6 and Table S3).

The impedance spectra before and after polarization and DC polarization curves of the SPE/LATP-0Sn/SPE and SPE/LATP-0.35Sn/SPE systems at 60 °C were measured for comparison (Figure 5), and the migration number of $Li^+$ ($t_{Li^+}$) were calculated according to Equation (4) (Table 3). As can be seen, the migration number of $Li^+$ ions for SPE/LATP-0Sn/SPE is only 0.27, whereas that of the SPE/LATP-0.35Sn/SPE is increased to 0.38. Combining the LATP-0.35Sn solid electrolyte with high ionic conductivity and the polymer can help to fix $TFSI^-$ and release more $Li^+$ ions for transport, thus increasing the migration number of $Li^+$ [64]. As a result, the increase of $t_{Li^+}$ enables one to reduce the polarization of the all-solid-state battery and promote the uniform deposition of Li [63].

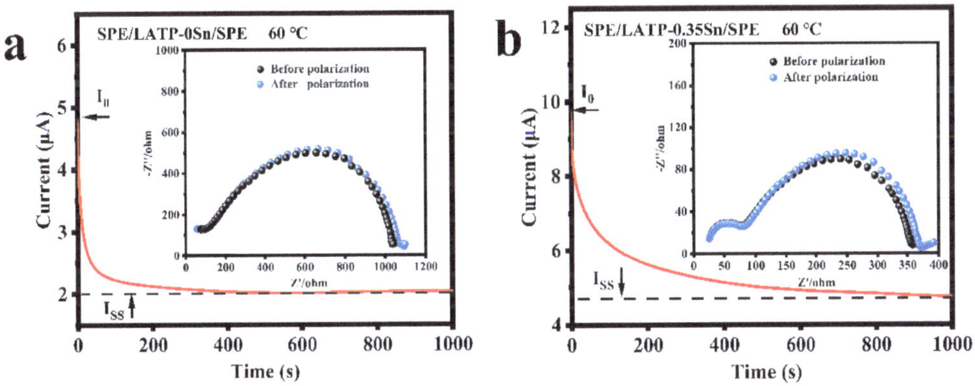


**Figure 5.** Impedance spectra and DC polarization for (**a**) SPE/LATP-0Sn/SPE and (**b**) SPE/LATP-0.35Sn/SPE.

**Table 3.** Lithium ion transport numbers ($t_{Li^+}$) of composite solid electrolytes at 60 °C.

| Electrolytes | $I_0$ (μA) | $I_{ss}$ (μA) | $R_0$ (Ω) | $R_{ss}$ (Ω) | $t_{Li^+}$ |
|---|---|---|---|---|---|
| SPE/LATP-0Sn/SPE | 4.79 | 2.03 | 1039.8 | 1084.1 | 0.27 |
| SPE/LATP-0.35Sn/SPE | 9.72 | 4.73 | 358.2 | 372.3 | 0.38 |

The LSV curves of PEO-LiTFSI, SPE/LATP-0Sn/SPE, and SPE/LATP-0.35Sn/SPE at 60 °C are given in Figure 6. The electrochemical window of PEO-LiTFSI is only 3.76 V vs. $Li/Li^+$. This is mainly attributed to the structural instability of the pure PEO-LiTFSI polymer electrolyte in the electric field, which is prone to the electrochemical oxidation decomposition [65]. The electrochemical window of SPE/LATP-0Sn/SPE is 4.34 V, whereas that of SPE/LATP-0.35Sn/SPE increases to 4.66 V. These mean that the designed sandwich composite structure can not only effectively hindering the transfer of $TFSI^-$ anions and thus reduce the decomposition of PEO, but also block the direct contact between the metallic Li negative electrode and the solid electrolyte, avoiding the occurrence of side reactions [66]. Moreover, the proposed sandwich structure makes full use of the advantage of the high electrochemical performance of an LATP-0.35Sn.

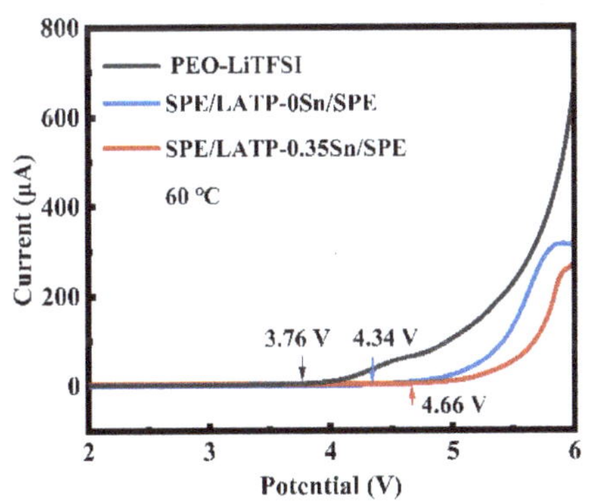


**Figure 6.** LSV curves of PEO-LiTFSI, SPE/LATP-0Sn/SPE, and SPE/LATP-0.35Sn/SPE solid electrolytes.

Figure 7 displays the electrochemical compatibility and stability results of SPE/LATP-0Sn/SPE and SPE/LATP-0.35Sn/SPE composite solid electrolytes and metallic lithium. The voltage of the SPE/LATP-0Sn/SPE begins to increase gradually after 26 h of cycling at a current density of 0.5 mA/cm$^2$, and exceeds the safety voltage of 5 V after 100 h. By contrast, the SPE/LATP-0.35Sn/SPE can perform stably for 500 h at a current density of 0.2 mA/cm$^2$ and show a low polarization voltage (~59 mV), indicating that the migration resistance of lithium ions between the composite solid electrolyte and Li metal is low. The polarization voltage of the lithium symmetric battery decreases slightly during the first cycles at the current density of 0.5 mA/cm$^2$. After cycling for 23 h, the polarization voltage of the lithium symmetric battery enhances stability, which may be due to the interface optimization caused by repeated electroplating/stripping of lithium [67]. At a higher current density of 0.5 mA/cm$^2$, the battery can still circulate for 300 h, and the voltage slightly changes with time. On the whole, the SPE/LATP-0.35Sn/SPE electrolyte exhibits good stability of the metallic Li interface, which seems to be due to the stable contact at the PEO/Li interface. More importantly, the compact structure of the modified LATP−0.35Sn solid electrolyte can enhance the mechanical strength and increase the ionic conductivity of PEO. In addition, the modified LATP-0.35Sn solid electrolyte can make the distribution of lithium ions more uniform and thus effectively prevent the growth of lithium dendrites.

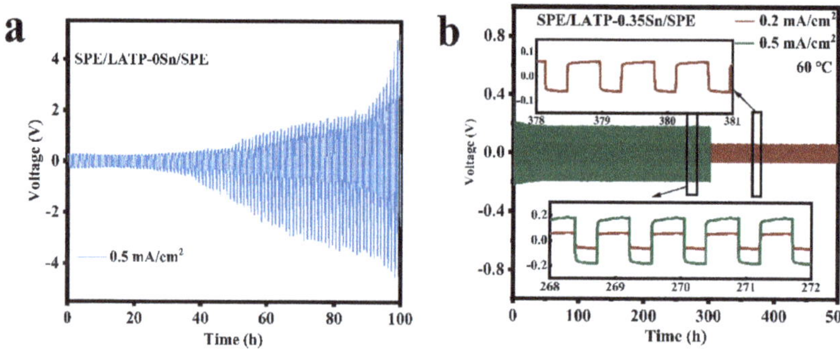


**Figure 7.** Voltage of the Li | | Li symmetrical cells with (**a**) SPE/LATP-0Sn/SPE at the current density of 0.5 mA/cm$^2$ and (**b**) SPE/LATP-0.35Sn/SPE at the current density of 0.2 mA/cm$^2$ and 0.5 mA/cm$^2$.

The sandwich-structured SPE/LATP-0.35Sn/SPE composite solid electrolyte, the cathode of the lithium iron phosphate ($LiFePO_4$), and the anode of the metallic lithium (Li) were afterwards assembled into an all-solid-state battery. The cycling performance and rate performance of the $LiFePO_4$ | | SPE/LATP-0.35Sn/SPE | | Li battery were tested, and the results are shown in Figure 8. At a current density of 0.2 C, the all-solid-state battery can perform stably up to 200 cycles; the battery capacity decreases from 153.5 to 138.9 mAh/g, with a capacity retention rate of 90.5%, and the coulombic efficiency is close to 100% (Figure 8a,b). Furthermore, the $LiFePO_4$ | | SPE/LATP-0.35Sn/SPE | | Li all-solid-state lithium battery can also be cycled stably for 100 cycles at the high rate of 0.5 C. The battery capacity is reduced from 145.1 mAh/g to 132.3 mAh/g with a capacity retention rate of 91.2%, and the coulombic efficiency is close to 100% (Figure 8c). As the number of charge-discharge cycles increases, the voltage and the interface impedance change slightly (Figure 8d,e), indicating that the electrode/electrolyte interface is stable in the long-term cycling process. Figure 8f exhibits the rate performance of the all-solid-state battery in the range from 0.1 to 2 C for every five cycles. The first-cycle discharge capacity of the all-solid-state battery is 155.4 mAh/g at a low rate current of 0.1 C, and decreases to 112.7 mAh/g at a higher rate current of 2 C. After cycling at different rate currents, when the rate current returns to 0.2 C, the discharge capacity recovers a value of 152.4 mAh/g. The excellent rate performance is attributed to the good compact structure of the solid-phase interface layer and its ability of uniformly depositing lithium ions. Therefore, the SPE/LATP-0.35Sn/SPE composite solid electrolyte has excellent cycling performance and high rate charge-discharge characteristics.

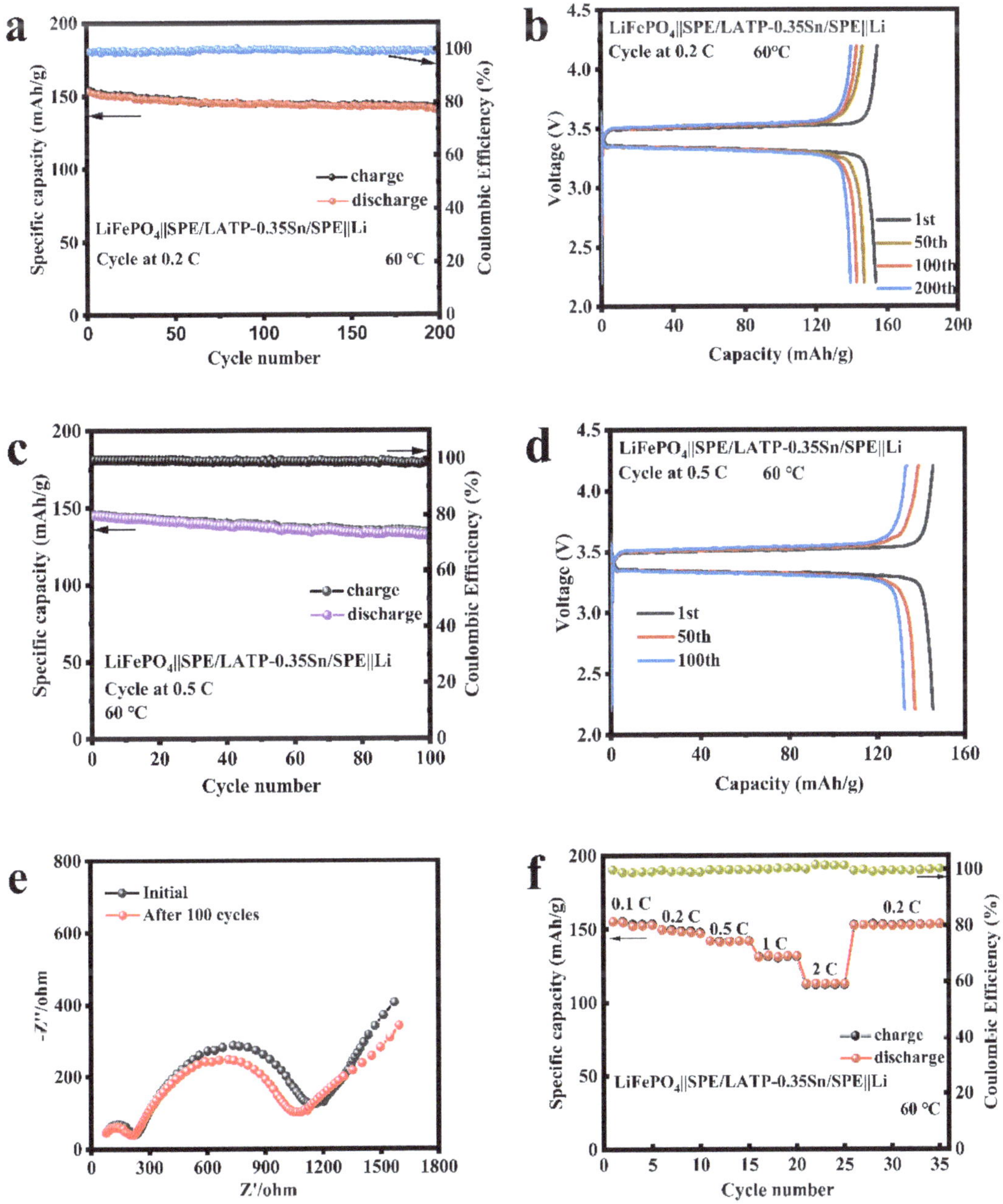


**Figure 8.** Electrochemical performance of LiFePO4 | | SPE/LATP-0.35Sn/SPE | | Li all-solid-state lithium battery: (**a**) cycle performance at 0.2 C; (**b**) charge-discharge curves at 0.2 C; (**c**) cycle performance at 0.5 C; (**d**) charge-discharge curves at 0.5 C; (**e**) Electrochemical impedance plots before and after cycling at a rate of 0.5 C; (**f**) rate performance at various rates ranging from 0.1 C to 2 C.

## 4. Conclusions

A Sn-doped NASICON-type LATP ceramic solid electrolyte was prepared by the solid-phase method. The influences of different Sn dopant contents on the structural properties and electrochemical performance of the LATP solid electrolyte were investigated. After Sn doping, the ionic conductivity of $Li_{1.3}Al_{0.3}Sn_{0.35}Ti_{1.35}(PO_4)_3$ (LATP-0.35Sn) at room temperature could reach $4.71 \times 10^{-4}$ S/cm, which was mainly attributed to the fact that $Sn^{4+}$ with a larger ionic radius could substitute the Ti sites in the LATP crystal structure to the maximum extent. The uniform Sn distribution was conducive to the structural stability and the decrease in grain size of the crystal, thereby improving the relative density. Moreover, the lattice distortion caused by Sn doping also modified the transport channels of Li ions. The prepared sandwich-structured SPE/LATP-0.35Sn/SPE composite solid electrolyte exhibited good electrochemical performance, with an ionic conductivity of $5.9 \times 10^{-5}$ S/cm at room temperature, an electrochemical stability window of 4.66 V vs. $Li/Li^+$, and a lithium-ion migration number of 0.38. The SPE/LATP-0.35Sn/SPE composite solid electrolyte was afterwards used to assembly a Li | | Li symmetric battery, revealing stable cycling performance for 500 h at 60 °C under the current density of 0.2 $mA/cm^2$. Additionally, combining the SPE/LATP-0.35Sn/SPE composite solid electrolyte with a $LiFePO_4$ cathode and a metallic Li anode enabled one to obtain an all-solid-state battery with excellent cycling stability and rate performance. The capacity retention rate of the all-solid-state battery cycled at a low rate of 0.2 C and at high rate of 0.5 C, reaching 90.5% and 91.2%, respectively. The coulombic efficiency at different rates are close to 100%.

**Supplementary Materials:** The following supporting information can be downloaded at: https://www.mdpi.com/article/10.3390/nano12122082/s1, Figure S1: Rietveld refinement of XRD patterns of LATP-xSn (x = 0–0.4) solid electrolytes; Figure S2: Lattice parameters of LATP-xSn (x = 0–0.4) solid electrolytes; Figure S3: XPS spectra of LATP-0Sn and LATP-0.35Sn solid electrolytes (a) survey spectra, (b) Sn 3d spectra, (c) Ti 2p spectra, (d) O 1s spectra; Figure S4: (a) SEM image of LATP-0.35Sn solid electrolyte and (b–f) the corresponding EDS elemental maps; Figure S5: SEM images and grain size distribution curves of (a) LATP-0Sn solid electrolyte and (b) LATP-0.35Sn solid electrolyte; Figure S6: EIS curves of LATP-0.35Sn and SPE/LATP-0.35Sn/SPE solid electrolytes at 25 °C; Figure S7: SEM images of SPE/LATP-0.35Sn/SPE solid electrolytes (a) surface and (b) cross-section; Table S1: The deviation degrees of $Sn^{4+}$ at $P^{5+}$ site, $Al^{3+}$ site and $Ti^{4+}$ site respectively under different coordination environments.; Table S2: Performance comparison of LATP-xSn obtained with other typical solid electrolytes; Table S3: Electrolyte resistance (Re), charge transfer resistance (Rct), total resistance (Rt) and total ionic conductivities ($\sigma$t) of PEO-LiTFSI, LATP-0.35Sn and SPE/LATP-0.35Sn/SPE solid electrolytes at 25 °C.

**Author Contributions:** A.X.: Investigation, methodology, writing—original draft, formal analysis, and data curation; F.L.: Supervision, resources, writing—review & editing, project administration, and funding acquisition; R.W.: Data curation; M.Y.: Methodology; J.M.: Writing—review & editing and funding acquisition; M.L.: Formal analysis; C.Y.: Methodology; J.C.: Conceptualization and funding acquisition. All authors have read and agreed to the published version of the manuscript.

**Funding:** This work was supported by Innovation Group Project of Education Department in Guizhou Province (NO. 2021010), One Hundred Person Project of Guizhou Province (NO. 20165655), Excellent Young Scientific and Technological Talent Program of Guizhou Province (NO. 20195645), and Zunyi City Innovative Talent Team Project (NO. 20195645), Natural Science Special Foundation of Guizhou University (Grant NO. X2021073 Special Post D), Natural Science Basic Research Program of Guizhou Province (Grant NO. ZK [2022]088), and the cultivation project of Guizhou University (Grant NO. [2020]38).

**Data Availability Statement:** The data that support the findings of this study are available upon reasonable request.

**Acknowledgments:** We are thankful to the following authors for their contributions, including for the purchase of materials and conducting of the relative experiments.: Fei Liu, Mengqin Yao, Jun Ma, Ruoming Wang, Mengjun Li, Jianxin Cao, and Xiaodan Wang.

**Conflicts of Interest:** The authors declare no conflict of interest.

## References


1. Mamidi, S.; Gangadharan, A.; Sharma, C.S. Graphite coated pyrolyzed filter paper as a low-cost binder-free and freestanding anode for practical lithium-ion battery application. *Electrochim. Acta* **2019**, *310*, 222–229. [CrossRef]
2. Gangadharan, A.; Mamidi, S.; Sharma, C.S.; Rao, T.N. Urea-modified candle soot for enhanced anodic performance for fast-charging lithium-ion battery application. *Mater. Today Commun.* **2020**, *23*, 100926. [CrossRef]
3. Mamidi, S.; Pandey, A.K.; Pathak, A.D.; Rao, T.N.; Sharma, C.S. Pencil lead powder as a cost-effective and high-performancegraphite-silica composite anode for high performance lithium-ion batteries. *J. Alloys Compd.* **2021**, *87*, 159719. [CrossRef]
4. Zheng, F.; Kotobuki, M.; Song, S.; Lai, M.O.; Lu, L. Review on solid electrolytes for all-solid-state lithium-ion batteries. *J. Power Sources* **2018**, *389*, 198–213. [CrossRef]
5. Zhang, B.; Tan, R.; Yang, L.; Zheng, J.; Zhang, K.; Mo, S.; Lin, Z.; Pan, F. Mechanisms and properties of ion-transport in inorganic solid electrolytes. *Energy Storage Mater.* **2018**, *10*, 139–159. [CrossRef]
6. Luo, C.; Ji, X.; Chen, J.; Gaskell, K.J.; He, X.; Liang, Y.; Jiang, J.; Wang, C. Solid-State Electrolyte Anchored with a Carboxylated Azo Compound for All-Solid-State Lithium Batteries. *Angew. Chem. Int. Ed.* **2018**, *57*, 8567–8571. [CrossRef] [PubMed]
7. Rai, K.; Kundu, S. Fabrication and performances of high lithium-ion conducting solid electrolytes based on NASICON $Li_{1.3}Al_{0.3}Ti_{1.7-x}Zr_x(PO_4)_3$ ($0 \leq x \leq 0.2$). *Ceram. Int.* **2020**, *46*, 23695–23705. [CrossRef]
8. Huang, B.; Xu, B.; Li, Y.; Zhou, W.; You, Y.; Zhon, S.; Wang, C.A.; Goodenough, J.B. Li-Ion Conduction and Stability of Perovskite $Li_{3/8}Sr_{7/16}Hf_{1/4}Ta_{3/4}O_3$. *ACS Appl. Mater. Interfaces* **2016**, *8*, 14552–14557. [CrossRef] [PubMed]
9. Bachman, J.C.; Muy, S.; Grimaud, A.; Chang, H.H.; Pour, N.; Lux, S.F.; Paschos, O.; Maglia, F.; Lupart, S.; Lamp, P.; et al. Inorganic Solid-State Electrolytes for Lithium Batteries: Mechanisms and Properties Governing Ion Conduction. *Chem. Rev.* **2016**, *116*, 140–162. [CrossRef] [PubMed]
10. Kotobuki, M.; Koishi, M. Influence of precursor calcination temperature on sintering and conductivity of $Li_{1.5}Al_{0.5}Ti_{1.5}(PO_4)_3$ ceramics. *J. Asian Ceram. Soc.* **2019**, *7*, 69–74. [CrossRef]
11. Waetzig, K.; Rost, A.; Heubner, C.; Coeler, M.; Nikolowski, K.; Wolter, M.; Schilm, J. Synthesis and sintering of $Li_{1.3}Al_{0.3}Ti_{1.7}(PO_4)_3$ (LATP) electrolyte for ceramics with improved $Li^+$ conductivity. *J. Alloys Compd.* **2020**, *818*, 153237. [CrossRef]
12. Liu, Y.; Sun, Q.; Zhao, Y.; Wang, B.; Kaghazchi, P.; Adair, K.R.; Li, R.; Zhang, C.; Liu, J.; Kuo, L.Y.; et al. Stabilizing the Interface of NASICON Solid Electrolyte against Li Metal with Atomic Layer Deposition. *ACS Appl. Mater. Interfaces* **2018**, *10*, 31240–31248. [CrossRef] [PubMed]
13. Hitz, G.T.; McOwen, D.W.; Zhang, L.; Ma, Z.; Fu, Z.; Wen, Y.; Gong, Y.; Dai, J.; Hamann, T.R.; Hu, L.; et al. High-rate lithium cycling in a scalable trilayer Li-garnet-electrolyte architecture. *Mater. Today* **2019**, *22*, 50–57. [CrossRef]
14. Hu, Z.; Sheng, J.; Chen, J.; Sheng, G.; Li, Y.; Fu, X.Z.; Wang, L.; Sun, R.; Wong, C.P. Enhanced Li ion conductivity in Ge-doped $Li_{0.33}La_{0.56}TiO_3$ perovskite solid electrolytes for all-solid-state Li-ion batteries. *New J. Chem.* **2018**, *42*, 9074–9079. [CrossRef]
15. Deng, Y.; Eames, C.; Fleutot, B.; David, R.; Chotard, J.N.; Suard, E.; Masquelier, C.; Islam, M.S. Enhancing the Lithium Ion Conductivity in Lithium Superionic Conductor (LISICON) Solid Electrolytes through a Mixed Polyanion Effect. *ACS Appl. Mater. Interfaces* **2017**, *9*, 7050–7058. [CrossRef]
16. Sepúlveda, A.; Criscuolo, F.; Put, B.; Vereecken, P.M. Effect of high temperature LiPON electrolyte in all solid state batteries. *Solid State Ion.* **2019**, *337*, 24–32. [CrossRef]
17. Hartley, G.O.; Jin, L.; Bergner, B.J.; Jolly, D.S.; Rees, G.J.; Zekoll, S.; Ning, Z.; Pateman, A.T.R.; Holc, C.; Adamson, P.; et al. Is Nitrogen Present in $Li_3N{\cdot}P_2S_5$ Solid Electrolytes Produced by Ball Milling? *Chem. Mater.* **2019**, *31*, 9993–10001. [CrossRef]
18. Lau, J.; DeBlock, R.H.; Butts, D.M.; Ashby, D.S.; Choi, C.S.; Dunn, B.S. Sulfide Solid Electrolytes for Lithium Battery Applications. *Adv. Energy Mater.* **2018**, *8*, 1800933. [CrossRef]
19. Hanghofer, I.; Redhammer, G.J.; Rohde, S.; Hanzu, I.; Senyshyn, A.; Wilkening, H.M.R.; Rettenwander, D. Untangling the Structure and Dynamics of Lithium-Rich Anti-Perovskites Envisaged as Solid Electrolytes for Batteries. *Chem. Mater.* **2018**, *30*, 8134–8144. [CrossRef]
20. Lu, X.; Wu, Y.; Yang, D.; Wang, R. Increased ionic conductivity of a NASICON lithium ion conductor under the influence of mesoporous materials. *J. Alloys Compd.* **2019**, *794*, 585–593. [CrossRef]
21. Mertens, A.; Yu, S.; Schön, N.; Gunduz, D.C.; Tempel, H.; Schierholz, R.; Hausen, F.; Kungl, H.; Granwehr, J.; Eichel, R.A. Superionic bulk conductivity in $Li_{1.3}Al_{0.3}Ti_{1.7}(PO_4)_3$ solid electrolyte. *Solid State Ion.* **2017**, *309*, 180–186. [CrossRef]
22. Anantharamulu, N.; Rao, K.K.; Rambabu, G.; Kumar, B.V.; Radha, V.; Vithal, M. A wide-ranging review on Nasicon type materials. *J. Mater. Sci.* **2011**, *46*, 2821–2837. [CrossRef]
23. Arbi, K.; Lazarraga, M.G.; Chehimi, D.B.H.; Ayadi-Trabelsi, M.; Rojo, J.M.; Sanz, J. Lithium Mobility in $Li_{1.2}Ti_{1.8}R_{0.2}(PO_4)_3$ Compounds (R = Al, Ga, Sc, In) as Followed by NMR and Impedance Spectroscopy. *Chem. Mater.* **2004**, *16*, 255–262. [CrossRef]
24. Shang, X.; Cheng, S.; Zhang, J.; Wang, Y. Comparison of a high lithium-ion conductivity solid electrolyte prepared by two methods. *IOP Conf. Ser. Mater. Sci. Eng.* **2017**, *242*, 012046. [CrossRef]
25. Kothari, D.H.; Kanchan, D.K.; Sharma, P. Electrical properties of Li-based NASICON compounds doped with yttrium oxide. *Ionics* **2014**, *20*, 1385–1390. [CrossRef]
26. Liang, Y.; Peng, C.; Kamiike, Y.; Kuroda, K.; Okido, M. Gallium doped NASICON type $LiTi_2(PO_4)_3$ thin-film grown on graphite anode as solid electrolyte for all solid state lithium batteries. *J. Alloys Compd.* **2019**, *775*, 1147–1155. [CrossRef]

27. Rao, R.P.; Maohua, C.; Adams, S. Preparation and characterization of NASICON type $Li^+$ ionic conductors. *J. Solid State Electrochem.* **2012**, *16*, 3349–3354. [CrossRef]
28. Wang, Q.; Liu, L.; Zhao, B.; Zhang, L.; Xiao, X.; Yan, H.; Xu, G.; Ma, L.; Liu, Y. Transport and interface characteristics of Te-doped NASICON solid electrolyte $Li_{1.3}Al_{0.3}Ti_{1.7}(PO_4)_3$. *Electrochim. Acta* **2021**, *399*, 139367. [CrossRef]
29. Zhu, J.; Xiang, Y.; Zhao, J.; Wang, H.; Li, Y.; Zheng, B.; He, H.; Zhang, Z.; Huang, J.; Yang, Y. Insights into the Local Structure, Microstructure and Ionic Conductivity of Silicon Doped NASICON-type Solid Electrolyte $Li_{1.3}Al_{0.3}Ti_{1.7}P_3O_{12}$. *Energy Storage Mater.* **2021**, *44*, 190–196. [CrossRef]
30. Vijayan, L.; Govindaraj, G. Structural and electrical properties of high-energy ball-milled NASICON type $Li_{1.3}Ti_{1.7}Al_{0.3}(PO_4)_{2.9}(VO_4)_{0.1}$ ceramics. *J. Phys. Chem. Solids* **2011**, *72*, 613–619. [CrossRef]
31. Li, S.; Huang, Z.; Xiao, Y.; Sun, C. Chlorine-doped $Li_{1.3}Al_{0.3}Ti_{1.7}(PO_4)_3$ as an electrolyte for solid lithium metal batteries. *Mater. Chem. Front.* **2021**, *5*, 5336–5343. [CrossRef]
32. Jeong, H.; Na, D.; Baek, J.; Kim, S.; Mamidi, S.; Lee, C.R.; Seo, H.K.; Seo, I. Synthesis of Superionic Conductive $Li_{1+x+y}Al_xSi_yTi_{2-x}P_{3-y}O_{12}$ Solid Electrolytes. *Nanomaterials* **2022**, *12*, 1158. [CrossRef] [PubMed]
33. Rao, A.V.; Veeraiah, V.; Rao, A.V.P.; Babu, B.K.; Brahmayya, M. Spectroscopic characterization and conductivity of Sn-substituted $LiTi_2(PO_4)_3$. *Res. Chem. Intermed.* **2014**, *41*, 4327–4337.
34. Roose, B.; Baena, J.P.C.; Gödel, K.C.; Graetzel, M.; Hagfeldt, A.; Steiner, U.; Abate, A. Mesoporous $SnO_2$ electron selective contact enables UV-stable perovskite solar cells. *Nano Energy* **2016**, *30*, 517–522. [CrossRef]
35. Wang, K.; Teng, G.; Yang, J.; Tan, R.; Duan, Y.; Zheng, J.; Pan, F. Sn(II,IV) steric and electronic structure effects enableself-selective doping on Fe/Si-sites of $Li_2FeSiO_4$ nanocrystals for high performance lithium ion batteries. *J. Mater. Chem. A* **2015**, *3*, 24437–24445. [CrossRef]
36. Park, G.D.; Lee, J.K.; Kang, Y.C. Design and synthesis of Janus-structured mutually doped $SnO_2$-$Co_3O_4$ hollow nanostructures as superior anode materials for lithium-ion batteries. *J. Mater. Chem. A* **2017**, *5*, 25319–25327. [CrossRef]
37. Mehraz, S.; Kongsong, P.; Taleb, A.; Dokhane, N.; Sikong, L. Large scale and facile synthesis of Sn doped $TiO_2$ aggregates using hydrothermal synthesis. *Sol. Energy Mater. Sol. Cells* **2019**, *189*, 254–262. [CrossRef]
38. Lim, Y.J.; Kim, H.W.; Lee, S.S.; Kim, H.J.; Kim, J.K.; Jung, Y.G.; Kim, Y. Ceramic-Based Composite Solid Electrolyte for Lithium-Ion Batteries. *Chempluschem* **2015**, *80*, 1100–1103. [CrossRef]
39. Zou, Y.; Zheng, H.; Wu, S.; Hei, Z.; Liu, H.; Duan, H. Improving Li/garnet interface by amorphous $SnO_2$ interlayerdeposited via sol–gel method. *Mater. Lett.* **2021**, *297*, 129959. [CrossRef]
40. Chen, L.; Li, Y.; Li, S.P.; Fan, L.Z.; Nan, C.W.; Goodenough, J.B. PEO/garnet composite electrolytes for solid-state lithium batteries: From "ceramic-in-polymer" to "polymer-in-ceramic". *Nano Energy* **2018**, *46*, 176–184. [CrossRef]
41. Jiang, Y.; Yan, X.; Ma, Z.; Mei, P.; Xiao, W.; You, Q.; Zhang, Y. Development of the PEO Based Solid Polymer Electrolytes for All-Solid State Lithium Ion Batteries. *Polymers* **2018**, *10*, 1237. [CrossRef] [PubMed]
42. Guo, Q.; Han, Y.; Wang, H.; Xiong, S.; Li, Y.; Liu, S.; Xie, K. New Class of LAGP-Based Solid Polymer Composite Electrolyte for Efficient and Safe Solid-State Lithium Batteries. *ACS Appl. Mater. Interfaces* **2017**, *9*, 41837–41844. [CrossRef] [PubMed]
43. Li, A.; Liao, X.; Zhang, H.; Shi, L.; Wang, P.; Cheng, Q.; Borovilas, J.; Li, Z.; Huang, W.; Fu, Z.; et al. Nacre-Inspired Composite Electrolytes for Load-Bearing Solid-State Lithium-Metal Batteries. *Adv. Mater.* **2020**, *32*, 1905517. [CrossRef] [PubMed]
44. Sun, C.; Liu, J.; Gong, Y.; Wilkinsone, D.P.; Zhang, J. Recent advances in all-solid state rechargeable lithium batteries. *Nano Energy* **2017**, *33*, 363–386. [CrossRef]
45. Kim, J.G.; Son, B.; Mukherjee, S.; Schuppert, N.; Bates, A.; Kwon, O.; Choi, M.J.; Chung, H.Y.; Park, S. A review of lithium and non-lithium based solid state batteries. *J. Power Sources* **2015**, *282*, 299–322. [CrossRef]
46. Cheng, Z.; Liu, T.; Zhao, B.; Shen, F.; Jin, H.; Han, X. Recent advances in organic-inorganic composite solid electrolytes for all-solid-state lithium batteries. *Energy Storage Mater.* **2021**, *34*, 388–416. [CrossRef]
47. Zhai, H.; Xu, P.; Ning, M.; Cheng, Q.; Mandal, J.; Yang, Y. A Flexible Solid Composite Electrolyte with Vertically Aligned and Connected Ion-Conducting Nanoparticles for Lithium Batteries. *Nano Lett.* **2017**, *17*, 3182–3187. [CrossRef]
48. Yang, J.; Huang, Z.; Zhang, P.; Liu, G.; Xu, X.; Yao, X. Titanium Dioxide Doping toward High-Lithium-Ion-Conducting $Li_{1.5}Al_{0.5}Ge_{1.5}(PO_4)_3$ Glass-Ceramics for All-Solid-State Lithium Batteries. *ACS Appl. Energy Mater.* **2019**, *2*, 7299–7305. [CrossRef]
49. Santagneli, S.H.; Baldacim, H.V.A.; Ribeiro, S.J.L.; Kundu, S.; Rodrigues, A.C.M.; Doerenkamp, C.; Eckert, H. Preparation, Structural Characterization, and Electrical Conductivity of Highly Ion-Conducting Glasses and Glass Ceramics in the System $Li_{1+x}Al_xSn_yGe_{2-(x+y)}(PO_4)_3$. *J. Phys. Chem. C* **2016**, *120*, 14556–14567. [CrossRef]
50. Pérez-Estébanez, M.; Isasi-Marín, J.; Díaz-Guerra, C.; Rivera-Calzada, A.; León, C.; Santamaría, J. Influence of chromium content on the optical and electrical properties of $Li_{1+x}Cr_xTi_{2-x}(PO_4)_3$. *Solid State Ion.* **2013**, *241*, 36–45. [CrossRef]
51. Burba, C.M.; Frech, R. Vibrational spectroscopic study of lithium intercalation into $LiTi_2(PO_4)_3$. *Solid State Ion.* **2006**, *177*, 1489–1494. [CrossRef]
52. Xu, Q.; Tsai, C.L.; Song, D.; Basak, S.; Kungl, H.; Tempel, H.; Hausen, F.; Yu, S.; Eichel, R.A. Insights into the reactive sintering and separated specific grain/grain boundary conductivities of $Li_{1.3}Al_{0.3}Ti_{1.7}(PO_4)_3$. *J. Power Sources* **2021**, *492*, 229631. [CrossRef]
53. Ohta, S.; Kobayashi, T.; Asaoka, T. High lithium ionic conductivity in the garnet-type oxide $Li_{7-X}La_3(Zr_{2-X}, Nb_X)O_{12}$ (X = 0–2). *J. Power Sources* **2011**, *196*, 3342–3345. [CrossRef]
54. DeWees, R.; Wang, H. Synthesis and Properties of NASICON-type LATP and LAGP Solid Electrolytes. *ChemSusChem* **2019**, *12*, 3713–3725. [CrossRef] [PubMed]

5. He, S.; Xu, Y.; Zhang, B.; Sun, X.; Chen, Y.; Jin, Y. Unique rhombus-like precursor for synthesis of $Li_{1.3}Al_{0.3}Ti_{1.7}(PO_4)_3$ solid electrolyte with high ionic conductivity. *Chem. Eng. J.* **2018**, *345*, 483–491. [CrossRef]
6. Cortés-Adasme, E.; Vega, M.; Martin, I.R.; Llanos, J. Synthesis and characterization of $SrSnO_3$ doped with $Er^{3+}$ for up-conversion luminescence temperature sensors. *RSC Adv.* **2017**, *7*, 46796–46802. [CrossRef]
7. Cai, Z.; Huang, Y.; Zhu, W.; Xiao, R. Increase in ionic conductivity of NASICON-type solid electrolyte $Li_{1.5}Al_{0.5}Ti_{1.5}(PO_4)_3$ by $Nb_2O_5$ doping. *Solid State Ion.* **2020**, *354*, 113359. [CrossRef]
8. Dermenci, K.B.; Buluc, A.F.; Turan, S. The effect of limonite addition on the performance of $Li_7La_3Zr_2O_{12}$. *Ceram. Int.* **2019**, *45*, 21401–21408. [CrossRef]
9. Song, S.; Lu, J.; Zheng, F.; Duong, H.M.; Lu, L. A facile strategy to achieve high conduction and excellent chemical stability of lithium solid electrolytes. *RSC Adv.* **2015**, *5*, 6588–6594. [CrossRef]
0. Lee, S.; Jung, S.; Yang, S.; Lee, J.H.; Shin, H.; Kim, J.; Park, S. Revisiting the LiPON/Li thin film as a bifunctional interlayer for NASICON solid electrolyte-based lithium metal batteries. *Appl. Surf. Sci.* **2022**, *586*, 152790. [CrossRef]
1. Ito, S.; Nakakita, M.; Aihara, Y.; Uehara, T.; Machida, N. A synthesis of crystalline Li7P3S11 solid electrolyte from 1,2-dimethoxyethane solvent. *J. Power Sources* **2014**, *271*, 342–345. [CrossRef]
2. Lü, X.; Howard, J.W.; Chen, A.; Zhu, J.; Li, S.; Wu, G.; Dowden, P.; Xu, H.; Zhao, Y.; Ji, Q. Antiperovskite $Li_3OCl$ Superionic Conductor Films for Solid-State Li-Ion Batteries. *Adv. Sci.* **2016**, *3*, 1500359. [CrossRef] [PubMed]
3. Lin, Y.; Liu, K.; Xiong, C.; Wu, M.; Zhao, T. A composite solid electrolyte with an asymmetric ceramic framework for dendrite-free all-solid-state Li metal batteries. *J. Mater. Chem. A* **2021**, *9*, 9665–9674. [CrossRef]
4. Peng, J.; Xiao, Y.; Clarkson, D.A.; Greenbaum, S.G.; Zawodzinski, T.A.; Chen, X.C. A Nuclear Magnetic Resonance Study of Cation and Anion Dynamics in Polymer-Ceramic Composite Solid Electrolytes. *ACS Appl. Polym. Mater.* **2020**, *2*, 1180–1189. [CrossRef]
5. Zhao, E.; Guo, Y.; Xin, Y.; Xu, G.; Guo, X. Enhanced electrochemical properties and interfacial stability of poly(ethylene oxide) solid electrolyte incorporating nanostructured $Li_{1.3}Al_{0.3}Ti_{1.7}(PO_4)_3$ fillers for all solid state lithium-ion batteries. *Int. J. Energy Res.* **2021**, *45*, 6876–6887. [CrossRef]
6. Jin, Y.; Liu, C.; Zong, X.; Li, D.; Fu, M.; Tan, S.; Xiong, Y.; Wei, J. Interface engineering of $Li_{1.3}Al_{0.3}Ti_{1.7}(PO_4)_3$ ceramic electrolyte via multifunctional interfacial layer for all-solid-state lithium batteries. *J. Power Sources* **2020**, *460*, 228125. [CrossRef]
7. He, J.; Manthiram, A. 3D CoSe@C Aerogel as a Host for Dendrite-Free LithiumMetal Anode and Efficient Sulfur Cathode in Li–S Full Cells. *Adv. Energy Mater.* **2020**, *10*, 2002654. [CrossRef]

nanomaterials

MDPI

*Article*

# Enhanced Thermochromic Performance of $VO_2$ Nanoparticles by Quenching Process

**Senwei Wu , Longxiao Zhou, Bin Li, Shouqin Tian * and Xiujian Zhao ***

State Key Laboratory of Silicate Materials for Architectures, Wuhan University of Technology (WUT), No. 122, Luoshi Road, Wuhan 430070, China; senwei_wu@whut.edu.cn (S.W.); zhoulongxiao@whut.edu.cn (L.Z.); libin625@whut.edu.cn (B.L.)

* Correspondence: tiansq@whut.edu.cn (S.T.); opluse@whut.edu.cn (X.Z.)

**Abstract:** Vanadium dioxide ($VO_2$) has been a promising energy-saving material due to its reversible metal-insulator transition (MIT) performance. However, the application of $VO_2$ films has been seriously restricted due to the intrinsic low solar-energy modulation ability ($\Delta T_{sol}$) and low luminous transmittance ($T_{lum}$) of $VO_2$. In order to solve the problems, the surface structure of $VO_2$ particles was regulated by the quenching process and the $VO_2$ dispersed films were fabricated by spin coating. Characterizations showed that the $VO_2$ particles quenched in deionized water or ethanolreserved $VO_2$(M) phase structure and they were accompanied by surface lattice distortion compared to the pristine $VO_2$. Such distortion structure contributed to less aggregation and highly individual dispersion of the quenched particles in nanocomposite films. The corresponding film of $VO_2$ quenched in water exhibited much higher $\Delta T_{sol}$ with an increment of 42.5% from 8.8% of the original $VO_2$ film, because of the significant localized surface plasmon resonance (LSPR) effect. The film fabricated from the $VO_2$ quenched in ethanol presented enhanced thermochromic properties with 15.2% of $\Delta T_{sol}$ and 62.5% of $T_{lum}$. It was found that the excellent $T_{lum}$ resulted from the highly uniform dispersion state of the quenched $VO_2$ nanoparticles. In summary, the study provided a facile way to fabricate well-dispersed $VO_2$ nanocomposite films and to facilitate the industrialization development of $VO_2$ thermochromic films in the smart window field.

**Keywords:** $VO_2$; quenching process; dispersion structure; thermochromic properties

## 1. Introduction

Since Morin F. J. discovered the reversible metal-insulator transition (MIT) behavior of vanadium oxides in 1959 [1], VO, $V_2O_3$, $VO_2$, $V_6O_{13}$, $V_3O_7$, and $V_2O_5$ [1–3] have been reported undergoing MIT at different ambient temperatures and showing significant gaps of the electrical, optical, magnetic properties before and after the MIT. Among them, the phase transition temperature ($T_c$) of $VO_2$, 341 K (68 °C), is the closest to room temperature, which makes $VO_2$ a promising candidate for intelligent film for smart windows. Below $T_c$, the strong electron-correlated $VO_2$ is monoclinic (M phase, $P2_1/c$) and highly transparent to NIR, which helps warm up the indoor room. When the ambient temperature rises above $T_c$, the $VO_2$ crystal transits to be tetragonal (R phase, $P4_2/mnm$) and reflective to the NIR, which is beneficial for blocking out the thermal radiation and cooling down the room temperature. Such characteristic of temperature-sensitive response of $VO_2$ is expected to enable intelligent regulation of indoor temperature, thereby reducing the energy consumption of architectures [2–4].

However, challenges remain in balancing the admirable luminous transmittance ($T_{lum}$) for illuminance and the appreciable solar-energy modulation ability ($\Delta T_{sol}$) for temperature regulations [5–7]. Such problems severely hinder the application of $VO_2$ thermochromic smart windows. Up till now, plenty of strategies have been dedicated to the performance optimization of $VO_2$. Element doping is confirmed to be an effective way to diminish the yellow color and boost the visible transmittance $T_{lum}$ [3,5,8], and F [9], Mg [9,10], Ti [11], and Zr [12] are common dopants. Besides, $VO_2$ nano-/microstructure engineering [4], such as porous structure [13,14], grid structure [15], and biomimetic patterning [16] of $VO_2$ film, has been testified to cater for the optimization of thermochromic properties. In addition, multilayer construction [17–19] based on $VO_2$ film becomes an effective approach to enhance thermochromic properties, including but not limited to oxide buffer layers [20], antireflection layers [21], and noble metal layers [22]. Apart from multilayer structure, $VO_2$ nanocomposite coating by means of distributing $VO_2$ particles into matrix materials is another feasible way to improve $T_{lum}$ and $\Delta T_{sol}$. When the ambient temperature is above $T_c$, the $VO_2$(R) nanoparticles, with grain size much smaller than the wavelength of the incident light, present the localized surface plasma resonance (LSPR) effect around 1200 nm. Such an effect enables to enlarge the absorption and decreased transmittance of the infrared light. On the other hand, $VO_2$(M) nanoparticles below $T_c$ do not possess the characteristic of LSPR and the corresponding films are highly transparent to NIR. Such a NIR transmittance gap between $VO_2$(M) and $VO_2$(R) results in much higher $\Delta T_{sol}$. Additionally, the transparent matrix covering isolated particles gives rise to higher $T_{lum}$ of the $VO_2$ nanocomposite coatings. Typically, $SiO_2$ [23], $SnO_2$ [24], $TiO_2$ [25], and Si-Al oxide [26] are common inorganic matrixes for $VO_2$ nanoparticles. Unfortunately, these matrixes require a strict annealing atmosphere or precisely uniform dispersion of $VO_2$ in the nanocomposite films. Compared to inorganic matrix, organic agents, such as polyvinyl pyrrolidone (PVP) [13,27], polyurethane (PU) [28], polyvinyl butyral (PVB) [29], and acrylic resin [30], have become more favorable hosts due to their low refractive index and facile fabrication process. Gao et al. [31] synthesized a flexible coating by dispersing $SiO_2$ capsuled monocrystal $VO_2$ with good crystallinity, fine grain size ($VO_2$@$SiO_2$) into PU, which showed good optical properties ($T_{lum}$ = 55.3%, $\Delta T_{sol}$ =7.5%). In addition, they further developed a roll-coated $VO_2$ nanocomposite coating with smoothness, uniformity, and high transparency on 1200 × 1000 mm$^2$ large-scale glass. By modulating the viscosity of the host PU, Zou et al. [32] upgraded the solar modulation ability of $VO_2$/PU composite film from 6.6% to 14.5%, whereas the luminous transmittance was maintained 54%. On the other hand, PVP was widely reported to be an alternative matrix. Zhao et al. [13] mixed $VO_2$ nanoparticles, ZnO nanoparticles, and PVP to fabricate a composite film, successfully enhancing $T_{lum}$ from 54.9% to 63.9% and improving $\Delta T_{sol}$ from 9.9% to 11.3%. Recently, Tian et al. [33] demonstrated an in-situ synthesis route to coat $VO_2$ nanocomposite on the glass surface by PVP decomposition and a mass transfer process during annealing, achieving remarkable optical performance with $T_{lum}$ of 72.5% and $\Delta T_{sol}$ of 10.1%. Yet, there were rare reports about the interactions between the $VO_2$ and the matrix.

In this work, high-purity $VO_2$(M) particles obtained by a one-step annealing process were quenched to induce surface structural distortion. Deionized water and ethanol were used as the quenching solvents to modulate the surface structure of $VO_2$ particles. As it turned out, the phase transition temperature of the quenched powders was slightly reduced by around 3 °C. In addition, the quenched $VO_2$ nanoparticles were dispersed individually in the PVP when spin-coated to form the $VO_2$ nanocomposite films. Such dispersion structures contributed to the optimization of the thermochromic properties of the film. It was discovered that the quenched $VO_2$ particles showed high dispersity in the nanocomposite films and presented an evident LSPR effect, which contributed to the enhanced solar-energy modulation ability of the films. Moreover, the film fabricated by the ethanol-quenched $VO_2$ uncovered excellent optical performance, accompanied by $\Delta T_{sol}$ of 15.2% and $T_{lum}$ of 62.5%. The outstanding performance successfully achieved the simultaneous enhancement of both solar energy modulation ability and luminous transmittance of the film, and it was superior to most of the $VO_2$ nanocomposite films in previous reports [21,23,27,31–33] that failed to make a positive balance of $\Delta T_{sol}$ and $T_{lum}$. Herein, this work provided a facile quenching process to benefit the thermochromic performance of $VO_2$ composite film and to drive its application in smart windows.

## 2. Methods

### 2.1. Fabrication of Quenched $VO_2$(M) Nanoparticles

All reagents that were used directly were analytically pure and provided by Sinopharm Chemical Reagent Co., Ltd, Shanghai, China. A certain mass of vanadium pentoxide ($V_2O_5$, 0.8 g) powders and half the amount of ammonium bicarbonate ($NH_4HCO_3$, 0.4 g) particles were loaded together in a quartz crucible and then transferred into the tube furnace. The furnace was pumped below 50 Pa and heated around 550 °C until the powders changed to dark-blue $VO_2$ particles. After that, the particles were processed with quenching treatment as Scheme 1 described. In detail, $VO_2$ particles were placed in a vacuum tube furnace for thermal insulation treatment for half an hour and then they were rapidly moved into the 0 °C quenching solution surrounded by an ice-water system. Extra ice bulks were constantly added to the system so that the quenching solvent was maintained at around 0 °C. Next, the beaker holding the above-mentioned solution mixed with quenched $VO_2$ grains was put in a thermostatic oven to completely evaporate the solvent. Finally, the remaining nanoparticles in the beaker were collected as quenched $VO_2$(M).

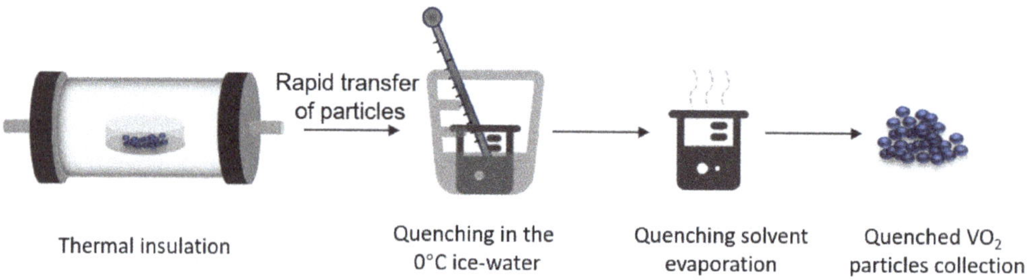


**Scheme 1.** Diagram of the quenching process of the $VO_2$ particles.

### 2.2. Fabrication of $VO_2$-PVP Nanocomposite Film

The obtained $VO_2$(M) nanoparticles were mixed with polyvinylpyrrolidone (PVP, K30) and ethanol in the mill tank. Intermittent ball milling gave rise to particles with smaller sizes to achieve thorough dispersion in the ethanol. Afterward, the mixture liquid was transferred into a centrifuge tube for solid-liquid separation. The upper liquid in the tube remained turbid as $VO_2$ with extreme fine sizes dispersed well in PVP, which dissolved well in ethanol. With the evaporation of liquid in the upper suspension, $VO_2$-PVP nanocomposites were collected and then made into a coating slurry by mixing them with additional ethanol. Continuous stirring and ultrasonic vibration were implemented to make the slurry homogeneous. Then spin-coating was carried out to form a sol/wet film by dropping the solution on a common soda-lime-silica glass substrate. In the end, ethanol was removed by heating to form the $VO_2$-PVP nanocomposite films.

### 2.3. Characterization

X-ray diffraction (XRD, D8DISCOVER, Bruker, Billerica, MA, USA) with Cu K$\alpha$ ($\lambda$ = 0.154056 nm) serving as the source of radiation, and 3 kW of the output power)was adopted to determine the phase structures of the powders over the 2$\theta$ between 10° and 80°. A differential scanning calorimeter (DSC, DSC8500,PerkinElmer, Waltham, MA, USA) was used to examine the phase transition temperature of the powders with the temperature ranging from 0 °C to 100 °C at the rate of 5 °C/min in the heating/cooling loop, as illustrated in Equation (1). $T_c$ refers to the average phase transition temperature of $VO_2$ particles. $T_{c,h}$ and $T_{c,c}$ correspond to the phase transition temperature peak of $VO_2$ in the heating and cooling stages, respectively. A field emission scanning electron microscopy (FE-SEM, Zeiss Ultra Plus, Carl Zeiss CMP GmbH, Oberkochen, Germany) was used to observe the morphology of both the composite powders and films. X-ray photoelectron spectroscopy (XPS, ESCALAB 250Xi, ThermoFisher, Waltham, MA, USA) was utilized to determine the element composition and valence. FT−IR spectrum (Nicolet6700, ThermoFisher, Waltham, MA, USA) was served to identify the functional groups of the samples.

$$T_c = (T_{c,h} + T_{c,c})/2 \quad (1)$$

The thermochromic performance of the films was measured from 300 to 2500 nm by a UV–vis–NIR spectrophotometer (UV-3600) equipped with a temperature-controlling device. The transmittance of the films at 20 °C and 90 °C corresponded to $VO_2$(M) and $VO_2$(R), respectively. The integrated luminous transmittance ($T_{lum}$, 380 nm $\leq \lambda \leq$ 780 nm) and solar transmittance ($T_{sol}$, 300 nm $\leq \lambda \leq$ 2500 nm) can be calculated according to Equations (2) and (3).

$$T_{lum} = \frac{\int \varphi_{lum}(\lambda)T(\lambda)d\lambda}{\int \varphi_{lum}(\lambda)d\lambda} \quad (2)$$

$$T_{sol} = \frac{\int \varphi_{sol}(\lambda)T(\lambda)d\lambda}{\int \varphi_{sol}(\lambda)d\lambda} \quad (3)$$

In the equations, $T(\lambda)$ represents the film transmittance of light at a certain wavelength ($\lambda$), $\varphi_{lum}(\lambda)$ is the standard luminous efficiency function for the photopic vision of human eyes [11], and $\varphi_{sol}(\lambda)$ is the solar irradiance spectrum for air mass 1.5 corresponding to the sun standing 37° above the horizon. $\Delta T_{sol}$ is the difference value of $T_{sol}$ at 20 °C and 90 °C, as given in Equation (4).

$$\Delta T_{sol} = T_{sol}(20\ °C) - T_{sol}(90\ °C) \quad (4)$$

## 3. Results and Discussion

### 3.1. Structure of the $VO_2$ Nanoparticles

$VO_2$ powders were obtained from $V_2O_5$ by a facial annealing reduction reaction. As shown in the TG-DSC pattern (Figure 1a) of the mixture of $V_2O_5$ and $NH_4HCO_3$, $NH_4HCO_3$ kept decomposing to release $NH_3$ at the beginning during the homogeneous heating period, illustrating the continuous mass loss of the raw materials. Till 183.5 °C, $NH_4HCO_3$ decomposed completely and the mass loss in this period was 8.28% in total. The endothermic peak at 226.2 °C indicated that the absorbed $H_2O$ was released, accounting for 3.29% of the reaction agents. With the rising temperature, the reducibility of $NH_3$ contributed to changing the vanadium (V) in $V_2O_5$ to lower valence, accompanied by an exothermic process at 360.0 °C, and finally reached the stable phase until 400 °C, as no mass loss or well as energy exchange could be observed in the pattern. This result suggested that 400 °C was suitable for $V_2O_5$ reduction. Additionally, the small temperature deviation between DSC and DTG could be attributed to the errors caused by the test instrument. Figure 1b showed the XRD patterns of the $VO_2$ samples annealed from $V_2O_5$ and $NH_4HCO_3$ mixture at 400 °C, 450 °C, 550 °C, and 550 °C. Although the main diffraction peaks (27.8°, 35.7°, 37.8°, and 55.8°) matched well with $VO_2$(M) (PDF #44-0252), extra peaks with evident intensity (25.3°, 33.5°, and 49.5°) belonging to $V_6O_{13}$ (PDF #27-1318) could be observed for the samples obtained at 400 °C, which meant the incomplete reduction of $V_2O_5$ to $V_6O_{13}$. Such a result disagreed with the ideal reaction temperature in Figure 1a. The reason could be attributed to the fact that the furnace chamber in the annealing process was too large, and the effective heating interval of the furnace was small, which caused partial energy loss. When raising the annealing temperature to 450 °C, the fabricated samples came across the same situation. Interestingly, the diffraction peaks belonging to $V_6O_{13}$ of the $VO_2$ sample reduced at 450 °C were much weaker than those of $VO_2$ reduced at 400 °C, testifying that the proportion of $V_6O_{13}$ among the $VO_2$ samples annealed at 450 °C was much less than that of the $VO_2$ annealed at 400 °C. Thus, a higher annealing temperature was demanded to overcome the energy barrier for reducing $V_2O_5$ into $VO_2$ thoroughly. In addition, it turned out that the material annealed at 500 °C became pure $VO_2$(M) with sharp diffraction peaks. Such peaks implied that the 500 °C reduced $VO_2$ featured excellent crystallinity. Comparably, the powders annealed at 550 °C remained pure

$VO_2(M)$ structure but exhibited a decreased peak intensity, indicating worse crystallinity of the related $VO_2$ powders. This might be caused by the unexpected grain agglomeration during the thermal insulation at higher temperatures. Therefore, 500 °C was selected as the ideal temperature to produce pristine $VO_2$ (labeled as P-$VO_2$).

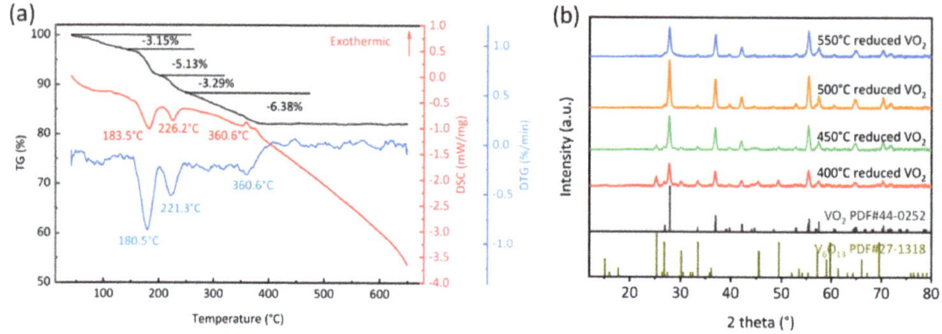

**Figure 1.** (**a**) TG-DSC pattern of the $V_2O_5$ and $NH_4HCO_3$ raw mixture and (**b**) XRD pattern of $VO_2$ powders annealed at different temperatures.

To observe the metal-insulator transition, P-$VO_2$ was characterized through in-situ XRD technique, and the results were shown in Figure 2a,b, and Figure 2b was the enlarged view of the pink area in Figure 2a. The $VO_2$ stayed the monoclinic structure below 45 °C, and the main peak (27.76°) corresponded to the (011) crystal plane of $VO_2(M)$. When the temperature rose above 65 °C, the peak shifted to 27.63°, revealing that the $VO_2$ crystals had transferred to the rutile phase, featured with the (110) crystal plane of $VO_2(R)$ (PDF #73-2362). This phenomenon suggested that the phase transition temperature of P-$VO_2$ ranged between 45 °C and 65 °C. This was in agreement with previous reports [5] for $VO_2$ particles. When the $VO_2$ particles cooled down from 75 °C to 65 °C, $VO_2$ crystals transited back to monoclinic structure from rutile structure. This helped obverse the evident and reversible MIT behavior between $VO_2(M)$ and $VO_2(R)$ in the previous report [34]. Furthermore, the phase transition temperature of the cooling stage was inconsistent with that of the heating stage, confirming the thermal hysteresis loop feature of $VO_2$ in the phase transition stage. It was the MIT behavior that made $VO_2$ a promising material for smart windows.

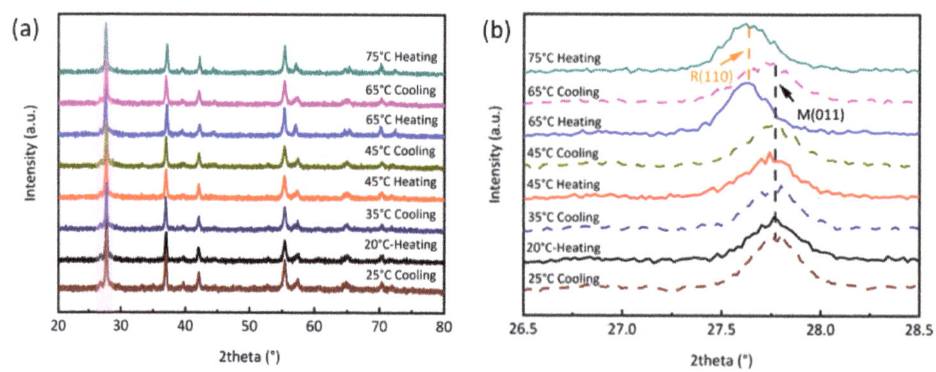

**Figure 2.** (**a**) In-situ XRD pattern of P-$VO_2$ and (**b**) enlarged view of the diffraction angle ranging from 26.5° to 28.5° in the heating and cooling stages.

In order to cause surface distortion for the crystal, P-$VO_2$ featured with the notable MIT behavior was placed in a 0 °C ice-water system for quenching treatment as Scheme 1 illustrated. According to the difference in quenching solvent, the $VO_2$ quenched in deionized water was named H-$VO_2$ and the one quenched in ethanol was noted as E-$VO_2$. The XRD pattern of $VO_2$ under different quenching solvents was shown in Figure 3a. Compared to the pristine $VO_2$ without the quenching process, both the H-$VO_2$ and E-$VO_2$ powders held the phase structure of $VO_2$(M) in Figure 1b. A smaller peak located at 25.3° appeared in the quenched samples, which could be the localized oxidation caused by exposure to air during powder transfer. It is worth noting that quenched particles presented the $VO_2$(M) phase and manifested great crystallinity. In addition, surface lattice distortion of the $VO_2$ was assumed to occur after the quenching process [35], which would be discussed in the next part. According to the Scherrer equation, the average grain size of E-$VO_2$ and H-$VO_2$ were calculated as 30.08 nm and 17.06 nm, respectively, larger than the average size (12.11 nm) of P-$VO_2$. To explain the mean grain size changes of the three samples, an enlarged view of the XRD diffraction angle ranging from 27.0° to 28.5° (Organe area in Figure 3a) was presented to illustrate the half-width changes of the peaks. As shown in Figure 3b, the diffraction curve of P-$VO_2$ exhibited the greatest widening state compared to E-$VO_2$ and H-$VO_2$. Such a state endowed P-$VO_2$ with the largest half-width and thus, the smallest grain size among the three particles. On the other hand, the half-width of the E-$VO_2$ (011) crystal plane diffraction peak was approaching that of H-$VO_2$. However, the diffraction peak of E-$VO_2$ shifted to a higher position at 27.91°, compared to the peak at 27.86° of P-$VO_2$. Such a shift to a higher angle caused the fact that the grain size of E-$VO_2$ was a bit larger than that of H-$VO_2$. This phenomenon might be attributed to the hydroxy group attached to the surface of $VO_2$ when the heated particles encountered an ice-water/ice-ethanol system and thus, leading to surface lattice distortion of the crystals. Moreover, thermal insulation during the quenching process in Scheme 1 was also beneficial to grain growth.

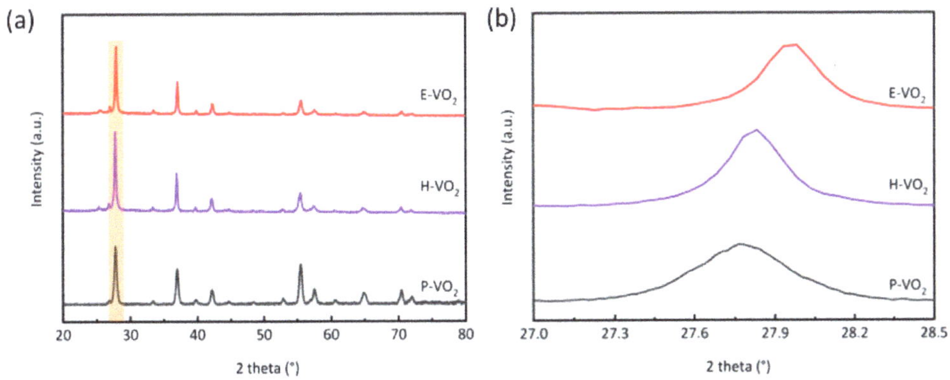


**Figure 3.** (**a**) XRD pattern of P-$VO_2$, H-$VO_2$, and E-$VO_2$ powders and (**b**) enlarged view of the diffraction angle ranging from 27.0° to 28.5°.

With the goal of further determining the effect of the quenching process on $VO_2$, the XPS technique was used to identify the chemical environment of elements, and the results were presented in Figure 4. Regarding the C1s peak at 284.80 eV as the calibration position, the full spectrum (Figure 4a) illustrated the existence of vanadium and oxygen in all the P-$VO_2$, H-$VO_2$, and E-$VO_2$ powders according to the V2p and O1s peaks. The V2p peaks were analyzed by Avantage of Thermo Scientific. In addition, Shirley background subtraction was employed to diminish the influence of heightened peaks due to signals from electrons undergoing inelastic scattering during the XPS characterization, ensuring a convincing quantification analysis of the peaks. As the high-resolution pattern for V2p shown in Figure 4b to d, both the $V2p_{3/2}$ and $V2p_{1/2}$ spectral peaks were significantly asymmetrical and each was split into two peaks, which implied that the vanadium of all the $VO_2$ samples involved two different chemical states, corresponding to $V^{4+}$ (The fitted purple curve) and $V^{5+}$ (The fitted green curve), respectively. $V^{5+}$ owing to $V_2O_5$ consisted of $V2p_{3/2}$ at 517.5 eV and $V2p_{1/2}$ at 525.0 eV. The gap between the $V2p_{3/2}$ orbital and $V2p_{3/2}$ orbital was 7.5 eV, which is in line with previous work [13]. Additionally, it is supposed that the existence of $V^{5+}$ was caused by partial oxidation when the $VO_2$ particles were exposed to air [36]. The $V^{4+}$ attributed to $VO_2$ was made up of $V2p_{3/2}$ at 516.2 eV and $V2p_{1/2}$ at 523.7 eV. In addition, the $V2p_{3/2}$ of P-$VO_2$ was most occupied with $V^{5+}$, and $V^{4+}$ only accounted for a small proportion. Comparably, the $V^{4+}$ in the V2p orbitals of H-$VO_2$ in Figure 4c showed the highest percentage, followed by E-$VO_2$ in Figure 4d, and finally, the P-$VO_2$ declared the least $V^{4+}$ content in Figure 4b. Such results could be attributed to the surface lattice distortion of H-$VO_2$ and E-$VO_2$.

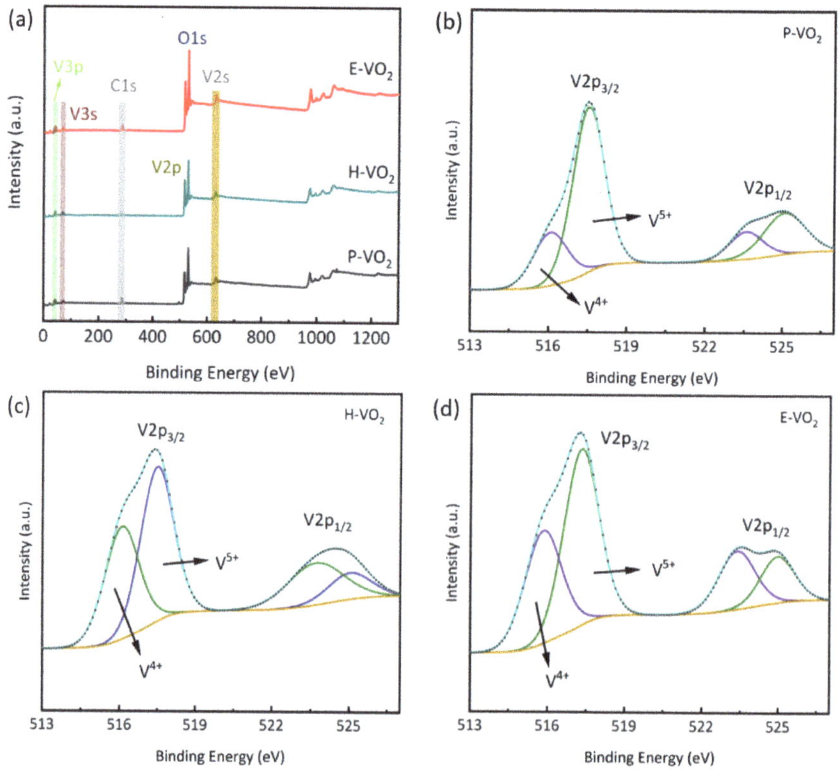


**Figure 4.** (**a**) XPS spectrum comparison of $VO_2$ powders and the high-resolution patterns for V2p of P-$VO_2$ (**b**), H-$VO_2$ (**c**), and E-$VO_2$ (**d**).

SEM was employed to investigate the morphology of the aforementioned $VO_2$ particles. Due to the large size of the commercial precursor $V_2O_5$, the directly annealed P-$VO_2$ particles inherited the large grain size, as shown in Figure 5a. It is evident that the particles aggregated severely to form large-scale clusters, with irregular shapes and different sizes ranging from 0.2 μm to 1.2 μm.Unfortunately, such sizes hindered P-$VO_2$ from dispersing in the PVP matrix for film coating. Compared to the pristine sample, the degree of particle aggregation of quenched $VO_2$ was much improved. In Figure 5b, the big clusters were broken into small H-$VO_2$ parts with varying sizes, showing that the deionized water was able to split the large-size P-$VO_2$ gathering during the quenching process. Especially, when the quenched solvent was replaced with ethanol with a smaller density, E-$VO_2$ interacted more severely with the liquid and turned more entire separation into nanoparticles with approximately 200 nm in size, as presented in Figure 5c. In addition, the voids clearly appeared among the E-$VO_2$ nanoparticles and the linkage knot hinted the E-$VO_2$ separated from large-size P-$VO_2$ particles. The observation that the quenched $VO_2$ nanoparticles were easily separated into small pellets from large-sized clusters resulted from the thermodynamically unstable state of these aggregated clusters. This unstable state was due to the surface crystal distortion caused by the quenching process, confirming the results of XPS spectrums. It is worth mentioning that the particle was much larger than the value calculated in Figure 3a. The fact that the XRD resulted from Cu K$\alpha$ radiation reflected crystalline particles rather than the actual morphology of the powders might explain the difference in the mean particle sizes. As for SEM, the signals of secondary electrons with much smaller De Broglie wavelengths were collected to reflect the topography of the particles. The different wavelengths of Cu K$\alpha$ in XRD and secondary electrons in SEM caused different imaging results. On the other hand, $VO_2$ particles in Figure 5 were assumed to be polycrystalline consisting of substantial crystals.

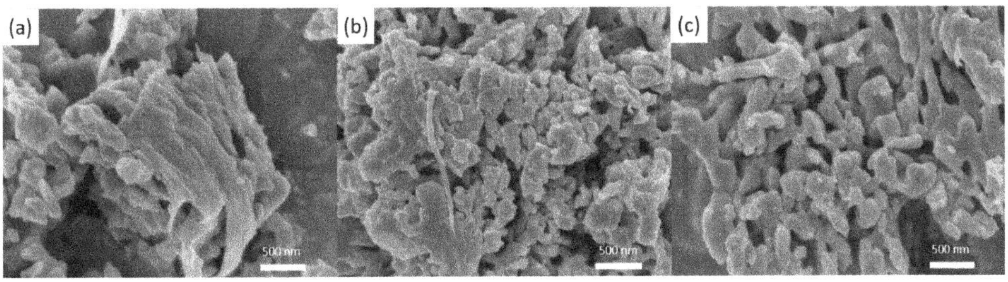


**Figure 5.** SEM images of (**a**) P-$VO_2$, (**b**) H-$VO_2$, and (**c**) E-$VO_2$ powders.

The DSC curves of the above P-$VO_2$ H-$VO_2$ and E-$VO_2$ powders are illustrated in Figure 6a with expected exothermic and endothermic peaks. Such a thermal energy change suggested the phase transition behavior between $VO_2$(M) and $VO_2$(R). The exothermic peak ($T_{c,h}$) at 68.35 °C and endothermic peak ($T_{c,c}$) at 61.09 °C of P-$VO_2$ implied the average $T_c$ was 64.72 °C according to Equation (1). On the other hand, the energy involved in the MIT behavior was 47 J/g, approaching 51 J/g of the bulk $VO_2$, showing good crystallinity of P-$VO_2$ powders as shown in Figure 2. With the quenching process, the H-$VO_2$ powders with phase transition peaks at 65.61 °C and 58.53 °C presented the average $T_c$ as 62.07 °C, while the $T_c$ of E-$VO_2$ nanoparticles was calculated as 62.73 °C from the $T_{c,h}$ peak at 65.73 °C and $T_{c,c}$ peak at 58.24 °C. It is clear that the $T_c$ of the quenched samples was slightly lowered, which could be attributed to the surface lattice distortion of the $VO_2$ crystal structure as the XPS results suggested.

Considering the similar structure of $H_2O$ and $C_2H_6O$, FT−IR was employed to detect the -OH bond of the $VO_2$ particles. In Figure 6b, the -OH bond corresponded to the peaks at 3450 cm$^{-1}$ and hydrogen bonds appeared at 1630 cm$^{-1}$. The integral area of the peak at 1630 cm$^{-1}$ was used to reflect the relative content of the hydroxyl group in the samples. It turned out that the hydroxyl peak area of H-$VO_2$ particles was the largest at 106.4, followed by the peak area of 87.7 of E-$VO_2$, and finally the value of 25.3 of P-$VO_2$. Such results indicated that the absorbed hydroxyl on the surface of both H-$VO_2$ and E-$VO_2$ particles was more than P-$VO_2$. This could be due to the surface lattice distortion of the quenched samples, so there were more absorption sites bound to the hydroxyl group on the surface of these particles. Additionally, the peaks at around 990 cm$^{-1}$ and 890 cm$^{-1}$, respectively, corresponded to the stretching vibration and asymmetric stretching vibration of the V=O bond. The overlapping peaks at 720 cm$^{-1}$ and 660 cm$^{-1}$ were the characteristic peaks of $VO_2$(M) [37]. In addition, the peaks at around 530 cm$^{-1}$ were labeled as the stretching vibration of the V-O-V bond. Apart from the peak offset of the corresponding bonds in $VO_2$, small extra peaks at around 1285 m$^{-1}$ were discovered in H-$VO_2$ and E-$VO_2$, which was attributed to the O-H band after the quenching process. Such a phenomenon confirmed the surface lattice distortion to $VO_2$ crystals of H-$VO_2$ and E-$VO_2$.

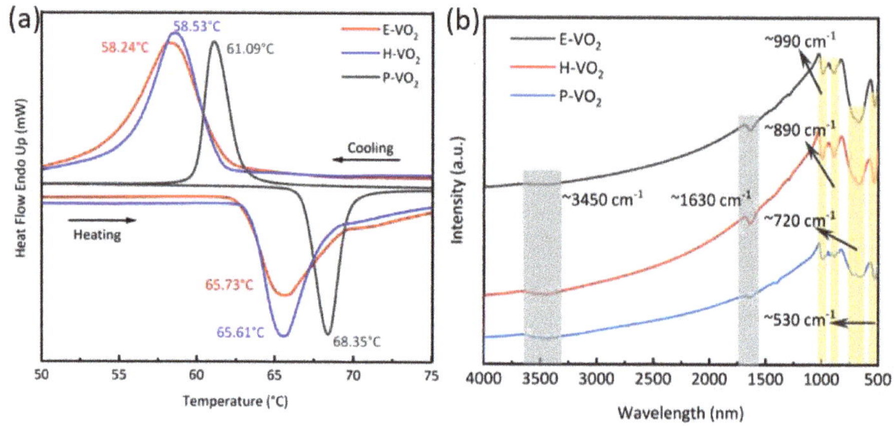


**Figure 6.** (**a**) DSC curve and (**b**) FT−IR spectrum of P-$VO_2$, H-$VO_2$, and E-$VO_2$ powders.

### 3.2. Thermochromic Properties and Morphology of $VO_2$ Nanocomposite Films

To reduce the grain size of the as-synthesized particles, $VO_2$ samples were mixed with PVP and ethanol in the mill tank to conduct thorough ball milling and centrifugation, ending up with a dark liquid mixture. The turbid upper solution was dried to collect the $VO_2$-PVP compound, which was then mixed with additional ethanol to configure the spin-coating slurry. The slurry was then dropped on the soda-lime-silica glass substrate and spun to form the $VO_2$ nanocomposite films. A UV3600 spectrophotometer coupled with a temperature-controlling device was served to characterize the transmittance of the films ranging from 300 nm to 2500 nm, and the thermochromic properties of the $VO_2$ nanocomposite films were integrated by Equations (2) to (4). The solid lines in Figure 7a were tested at 20 °C, revealing the transmittance of M phase $VO_2$ samples, and the dash lines were obtained at 90 °C, appearing the transmittance of R phase $VO_2$ samples. As shown in Figure 7a and Table 1, the $T_{lum}$ of P-$VO_2$ film was 53.2% and $\Delta T_{sol}$ just reached 8.8%. The grain accumulation in Figure 5a was the cause for such poor properties. Comparably, the H-$VO_2$ film sacrificed a small amount of luminous transmittance of 3.1% to achieve as enormous as 42.5% of improvement in solar-energy modulation ability to reach 12.5%. The reason for the enhanced $\Delta T_{sol}$ originated from the LSPR absorption peak of H-$VO_2$ located at 1258 nm, which was stronger than the P-$VO_2$ peak at 1293 nm. Such a phenomenon led to an enlarged gap in the transmittance of $VO_2$ at 20 °C and 90 °C and contributed to higher $\Delta T_{sol}$. The E-$VO_2$ nanocomposite film that came across with the same LSPR effect at 1150 nm revealed an exceeding increment of $\Delta T_{sol}$ of 72.2% (from 8.8% to 15.2%). In addition, the $T_{lum}$ of E-$VO_2$ film was boosted to 62.5%, indicating a better dispersity of the $VO_2$ nanoparticles in the PVP matrix. The optimized luminous transmittance was also credited for less aggregation of the nanoparticles as shown in Figure 5c. In Figure 7b, the thermochromic performance of E-$VO_2$ in this work was compared to previously reported $VO_2$ nanocomposite films, and it was clear that the E-$VO_2$ film exceeded most $VO_2$ films [21,23,27,31–33]. Besides, the excellent $\Delta T_{sol}$ of the E-$VO_2$ film was able to satisfy the demands of effectively regulating room temperature while the great $T_{lum}$ met the requirement of indoor brightness, which was a great achievement for the potential application of $VO_2$ thermochromic smart windows.

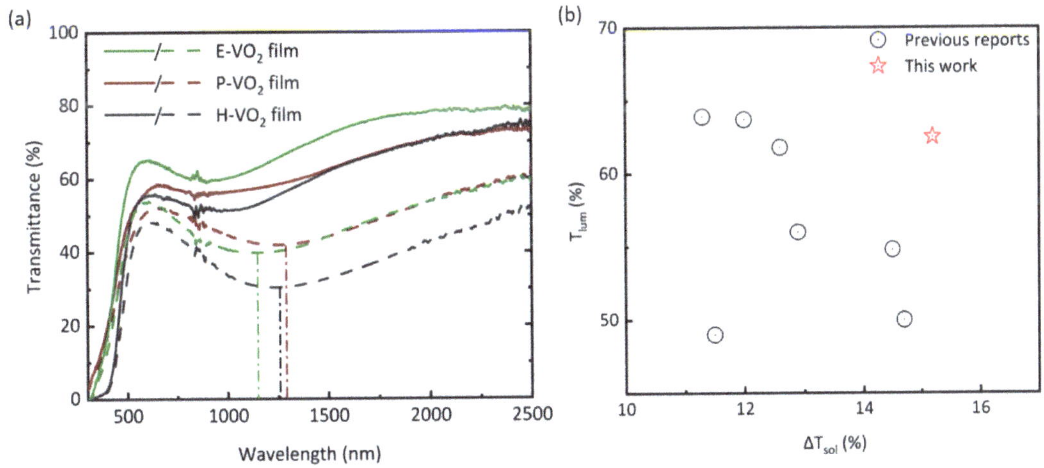


**Figure 7.** (**a**) Transmittance spectrum comparison of $VO_2$ films and (**b**) thermochromic performance comparison between this work and previously reported $VO_2$ nanocomposite films.

**Table 1.** Thermochromic properties of different $VO_2$ films.

| Sample | $T_{sol}$ (%) | | $T_{lum}$ (%) | | $\Delta T_{sol}$ (%) | $\Delta T_{NIR}$ (%) |
|---|---|---|---|---|---|---|
| | 20 °C | 90 °C | 20 °C | 90 °C | | |
| P-$VO_2$ film | 52.5 | 43.7 | 53.2 | 47.3 | 8.8 | 12.8 |
| H-$VO_2$ film | 47.9 | 35.4 | 51.6 | 43.5 | 12.5 | 19.8 |
| E-$VO_2$ film | 58.2 | 43.0 | 62.5 | 50.9 | 15.2 | 21.1 |

The morphology of the films was presented. It is clear that P-$VO_2$ turned into nanoparticles with a size of around 100 nm after a ball milling process as described in Figure 8a. However, particle aggregation in Figure 5a was still common even though they were spin-coated to film. This aggregation structure was the reason for the unsatisfied thermochromic properties of the P-$VO_2$ film. Comparably, grain accumulation also existed in H-$VO_2$ film, which caused low luminous transmittance of the film. Additionally, cracks could be found in the film, and they were the cause of the inadequate performance of $T_{lum}$. It is worth mentioning that most H-$VO_2$ nanoparticles exhibited much smaller sizes than P-$VO_2$ and they were dispersed individually in the PVP in Figure 8b, causing the LSPR effect and the improvement of $\Delta T_{sol}$. In Figure 8c, the E-$VO_2$ nanoparticles showed an average size of tens of nanometers and nearly no sign of particle aggregation. Additionally, these particles with surface lattice distortion were highly isolated from each other and uniformly dispersed in the E-$VO_2$ nanocomposite film. The large-scale surface morphology of E-$VO_2$ film in Figure 8d introduced such dispersity of $VO_2$ nanoparticles in a more intuitive perspective. This dispersion structure became beneficial for the optical properties of E-$VO_2$ nanocomposite film.

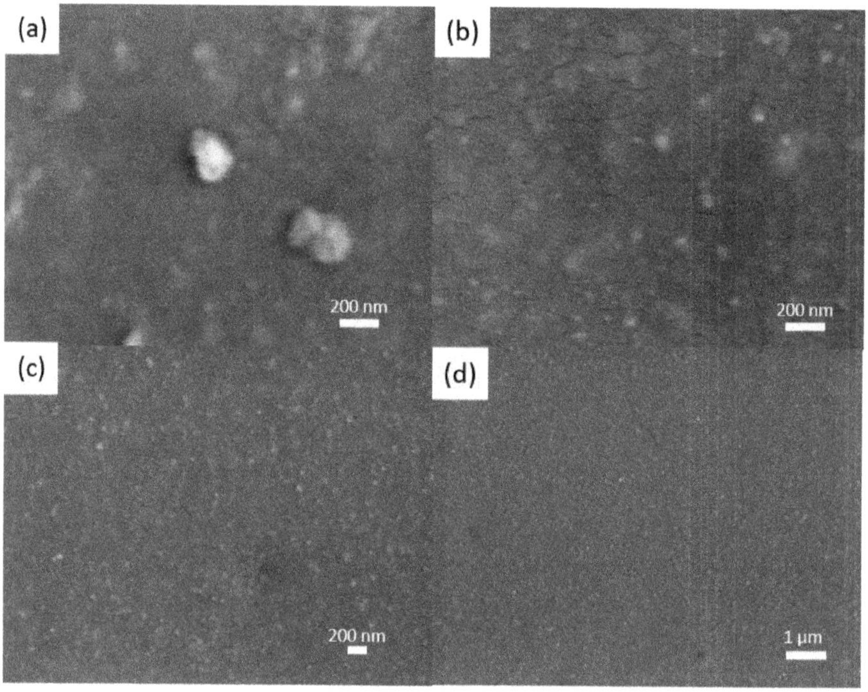


**Figure 8.** SEM images of (**a**) P-$VO_2$, (**b**) H-$VO_2$, and (**c**,**d**) E-$VO_2$ nanocomposite films.

## 4. Conclusions

As a strong electronic associated material, $VO_2$(M) underwent a reversible phase transition between the monoclinic phase and rutile phase, corresponding with an abrupt change of near-infrared light transmittance, thus showing great potential in the smart window application. To improve the thermochromic properties of $VO_2$, this paper applied a facile annealing method to synthesize $VO_2$(M) powders. With the quenching treatment, the $VO_2$ particles presented surface lattice distortion and they were individually dispersed in the PVP host to induce the LSPR effect, contributing to the exciting increment of solar-energy modulation ability. Hereby, the film corresponding to H-$VO_2$ achieved an enormous improvement in $\Delta T_{sol}$ from 8.8% to 12.5%. Moreover, the film fabricated by the ethanol-quenched $VO_2$ with no sign of aggregation showed an exceedingly high $\Delta T_{sol}$ of 15.2%, and the high dispersity of E-$VO_2$ nanoparticles in the film contributed to much enhanced $T_{lum}$ of 62.5%. Therefore, this work provided a new thought to promote the thermochromic performance of $VO_2$ by particle quenching and the strategy was beneficial to the commercialization of $VO_2$ thermochromic smart windows.

**Author Contributions:** S.W. contributed to writing the original draft and conducting the experiment and data analysis. L.Z. contributed to characterizing the structure of the samples and B.L. was responsible for the optical performance characterization of the films. S.T. contributed to analyzing all the data and supervising the experimental routine and data analysis. X.Z. was involved in analyzing all the data and revising the manuscript. All authors have read and agreed to the published version of the manuscript.

**Funding:** This work was supported by the National Natural Science Foundation of China (Grant No. 51772229), the 111 project (No. B18038), the National Key R&D Program of China (No. 2017YFE0192600), Key R&D Project of Hubei Province (No. 2020BAB061), National innovation and entrepreneurship training program for college students (No. 201910497034), Open Foundation of the State Key Laboratory of Silicate Materials for Architectures at WUT (No. SYSJJ2021-05 and SYSJJ2022-01), State Key

Laboratory of Materials Processing and Die & Mould Technology, Huazhong University of Science and Technology (No. P2021-010).

**Institutional Review Board Statement:** Not applicable.

**Informed Consent Statement:** Not applicable.

**Data Availability Statement:** Data are available on request from the authors.

**Acknowledgments:** We also thank the Analytical and Testing Center of WUT for the help with carrying out the materials characterization.

**Conflicts of Interest:** The authors declare no potential conflict of financial interest with respect to the research.

## References


1. Morin, F.J. Oxides which show a metal-to-insulator transition at the neel temperature. *Phys. Rev. Lett.* **1959**, *3*, 34–36. [CrossRef]
2. Hu, P.; Hu, P.; Vu, T.D.; Li, M.; Wang, S.; Ke, Y.; Zeng, X.; Mai, L.; Long, Y. Vanadium oxide: Phase diagrams, structures, synthesis, and applications. *Chem. Rev.* **2023**, *123*, 4353–4415. [CrossRef]
3. Zhang, Z.; Zhang, L.; Zhou, Y.; Cui, Y.; Chen, Z.; Liu, Y.; Li, J.; Long, Y.; Gao, Y. Thermochromic energy efficient windows: Fundamentals, recent advances, and perspectives. *Chem. Rev.* **2023**, *123*, 7025–7080. [CrossRef]
4. Bhupathi, S.; Wang, S.; Ke, Y.; Long, Y. Recent progress in vanadium dioxide: The multi-stimuli responsive material and its applications. *Mater. Sci. Eng. R Rep.* **2023**, *155*, 100747. [CrossRef]
5. Gao, Y.; Luo, H.; Zhang, Z.; Kang, L.; Chen, Z.; Du, J.; Kanehira, M.; Cao, C. Nanoceramic $VO_2$ thermochromic smart glass: A review on progress in solution processing. *Nano Energy* **2012**, *1*, 221–246. [CrossRef]
6. Xu, F.; Cao, X.; Luo, H.; Jin, P. Recent advances in $VO_2$-based thermochromic composites for smart windows. *J. Mater. Chem. C* **2018**, *6*, 1903–1919. [CrossRef]
7. Song, J.; Zhao, Y.; Sun, L.; Luo, Q.; Xu, H.; Wang, C.; Xin, H.; Wu, W.; Ma, F. $VO_2$/ATO nanocomposite thin films with enhanced solar modulation and high luminous transmittance for smart windows. *Ceram. Int.* **2022**, *48*, 15868–15876. [CrossRef]
8. Ding, X.; Li, Y.; Zhang, Y. Sol-gel derived tungsten doped $VO_2$ thin films on Si substrate with tunable phase transition properties. *Molecules* **2023**, *28*, 3778. [CrossRef]
9. Abdellaoui, I.; Merad, G.; Maaza, M.; Abdelkader, H.S. Electronic and optical properties of Mg-, F-doped and Mg\F-codoped M1-$VO_2$ via hybrid density functional calculations. *J. Alloys Compd.* **2016**, *658*, 569–575. [CrossRef]
10. Panagopoulou, M.; Gagaoudakis, E.; Boukos, N.; Aperathitis, E.; Kiriakidis, G.; Tsoukalas, D.; Raptis, Y.S. Thermochromic performance of Mg-doped $VO_2$ thin films on functional substrates for glazing applications. *Sol. Energy Mater. Sol. Cells* **2016**, *157*, 1004–1010. [CrossRef]
11. Zhao, J.; Chen, D.; Hao, C.; Mi, W.; Zhou, L. The optimization and role of Ti surface doping in thermochromic $VO_2$ film. *Opt. Mater.* **2022**, *133*, 112960. [CrossRef]
12. Qin, S.; Fan, Y.; Qiu, X.; Gou, G.; Zhang, K.; Feng, Q.; Gan, G.; Sun, W. Modulation of the phase transition behavior of $VO_2$ nanofilms by the coupling of Zr doping and thickness-dependent band gap. *ACS Appl. Mater. Interfaces* **2022**, *4*, 6067–6075. [CrossRef]
13. Wu, S.; Tian, S.; Liu, B.; Tao, H.; Zhao, X.; Palgrave, R.G.; Sankar, G.; Parkin, I.P. Facile synthesis of mesoporous $VO_2$ nanocrystals by a cotton-template method and their enhanced thermochromic properties. *Sol. Energy Mater. Sol. Cells* **2018**, *176*, 427–434. [CrossRef]
14. Long, S.; Cao, X.; Wang, Y.; Chang, T.; Li, N.; Jin, L.; Ma, L.; Xu, F.; Sun, G.; Jin, P. Karst landform-like $VO_2$ single layer solution: Controllable morphology and excellent optical performance for smart glazing applications. *Sol. Energy Mater. Sol. Cells* **2020**, *209*, 110449. [CrossRef]
15. Ke, Y.; Wen, X.; Zhao, D.; Che, R.; Xiong, Q.; Long, Y. Controllable fabrication of two-dimensional patterned $VO_2$ nanoparticle, nanodome, and nanonet arrays with tunable temperature-dependent localized surface plasmon resonance. *ACS Nano* **2017**, *11*, 7542–7551. [CrossRef] [PubMed]
16. Qian, X.; Wang, N.; Li, Y.; Zhang, J.; Xu, Z.; Long, Y. Bioinspired multifunctional vanadium dioxide: Improved thermochromism and hydrophobicity. *Langmuir* **2014**, *30*, 10766–10771. [CrossRef]
17. Yao, L.; Qu, Z.; Sun, R.; Pang, Z.; Wang, Y.; Jin, B.; He, J. Long-lived multilayer coatings for smart windows: Integration of energy-saving, antifogging, and self-healing functions. *ACS Appl. Energy Mater.* **2019**, *2*, 7467–7473. [CrossRef]
18. Mlyuka, N.R.; Niklasson, G.A.; Granqvist, C.G. Thermochromic multilayer films of $VO_2$ and $TiO_2$ with enhanced transmittance. *Sol. Energy Mater. Sol. Cells* **2009**, *93*, 1685–1687. [CrossRef]
19. Long, S.; Zhou, H.; Bao, S.; Xin, Y.; Cao, X.; Jin, P. Thermochromic multilayer films of $WO_3/VO_2/WO_3$ sandwich structure with enhanced luminous transmittance and durability. *RSC Adv.* **2016**, *6*, 106435–106442. [CrossRef]
20. Panagopoulou, M.; Gagaoudakis, E.; Aperathitis, E.; Michail, I.; Kiriakidis, G.; Tsoukalas, D.; Raptis, Y.S. The effect of buffer layer on the thermochromic properties of undoped radio frequency sputtered $VO_2$ thin films. *Thin Solid Films* **2015**, *594*, 310–315. [CrossRef]

21. Yao, L.; Qu, Z.; Pang, Z.; Li, J.; Tang, S.; He, J.; Feng, L. Three-layered hollow nanospheres based coatings with ultrahig performance of energy-saving, antireflection, and self-cleaning for smart windows. *Small* **2018**, *14*, e1801661. [CrossRef] [PubMec
22. Granqvist, C.G. Transparent conductors as solar energy materials: A panoramic review. *Sol. Energy Mater. Sol. Cells* **2007**, *9* 1529–1598. [CrossRef]
23. Kang, J.; Liu, J.; Shi, F.; Dong, Y.; Song, X.; Wang, Z.; Tian, Z.; Xu, J.; Ma, J.; Zhao, X. Facile fabrication of $VO_2/SiO_2$ aerog nanocomposite films with excellent thermochromic properties for smart windows. *Appl. Surf. Sci.* **2022**, *573*, 151507. [CrossRe
24. Li, D.; Deng, S.; Zhao, Z.; Yang, J.; Wang, B.; Li, J.; Jin, H. $VO_2$(M)@$SnO_2$ core–shell nanoparticles: Improved chemical stabilit and thermochromic property rendered by $SnO_2$ shell. *Appl. Surf. Sci.* **2022**, *598*, 153741. [CrossRef]
25. Samal, A.; Lakshya, A.K.; Dhar Dwivedi, S.M.M.; Dalal, A.; Ghosh, A.; Paul, A.D.; Mahapatra, R.; Gupta, R.K.; Hasan, M.A Dey, A.; et al. Stable and reversible phase change performance of $TiO_2$ coated $VO_2$ nano-columns: Experiments and theoretica analysis. *Ceram. Int.* **2021**, *47*, 14741–14749. [CrossRef]
26. Liu, C.; Cao, X.; Kamyshny, A.; Law, J.Y.; Magdassi, S.; Long, Y. $VO_2$/Si-Al gel nanocomposite thermochromic smart foils: Largel enhanced luminous transmittance and solar modulation. *J. Colloid Interface Sci.* **2014**, *427*, 49–53. [CrossRef]
27. Li, B.; Liu, J.; Tian, S.; Liu, B.; Yang, X.; Yu, Z.; Zhao, X. $VO_2$-ZnO nanocomposite films with enhanced thermochromic propertie for smart windows. *Ceram. Int.* **2020**, *46*, 2758–2763. [CrossRef]
28. Kang, J.; Liu, J.; Shi, F.; Dong, Y.; Jiang, S. The thermochromic characteristics of Zn-doped $VO_2$ that were prepared by th hydrothermal and post-annealing process and their polyurethane nanocomposite films. *Ceram. Int.* **2021**, *47*, 15631–1563 [CrossRef]
29. Cool, N.I.; Larriuz, C.A.; James, R.; Ayala, J.R.; Anita; Al-Hashimi, M.; Banerjee, S. Thermochromic fenestration elements based o the dispersion of functionalized $VO_2$ nanocrystals within a polyvinyl butyral laminate. *ACS Eng. Au* **2022**, *2*, 477–485. [CrossRe
30. Mao, Z.; Huang, W.; Zhou, W.; Tang, L.; Shi, Q. In-situ stirring assisted hydrothermal synthesis of W-doped $VO_2$(M) nanorod with improved doping efficiency and mid-infrared switching property. *J. Alloys Compd.* **2020**, *821*, 153556. [CrossRef]
31. Cao, C.; Yanfeng, G.; Hongjie, L. Pure single-crystal rutile vanadium dioxide powders: Synthesis, mechanism and phas transformation property. *J. Phys. Chem. C* **2008**, *112*, 18810–18814. [CrossRef]
32. Yang, X.; Zou, J. Facile fabrication of $VO_2$ composite film with enhanced significantly solar modulation performance by adjustin viscosity of $VO_2$/PU dispersion. *J. Alloys Compd.* **2023**, *940*, 168868. [CrossRef]
33. Li, B.; Tian, S.; Qian, J.; Wu, S.; Liu, B.; Zhao, X. In situ synthesis of highly dispersed $VO_2$(M) nanoparticles on glass surface fo energy efficient smart windows. *Ceram. Int.* **2023**, *49*, 2310–2318. [CrossRef]
34. Yuan, L.; Hu, Z.; Hou, C.; Meng, X. In-Situ thermochromic mechanism of Spin-Coated $VO_2$ film. *Appl. Surf. Sci.* **2021**, *564*, 15044 [CrossRef]
35. Liu, B.; Cheng, K.; Nie, S.; Zhao, X.; Yu, H.; Yu, J.; Fujishima, A.; Nakata, K. Ice–water quenching induced $Ti^{3+}$ self-doped TiO with surface lattice distortion and the increased photocatalytic activity. *J. Phys. Chem. C* **2017**, *121*, 19836–19848. [CrossRef]
36. Li, B.; Tian, S.; Tao, H.; Zhao, X. Tungsten doped M-phase $VO_2$ mesoporous nanocrystals with enhanced comprehensiv thermochromic properties for smart windows. *Ceram. Int.* **2019**, *45*, 4342–4350. [CrossRef]
37. Mohamed Azharudeen, A.; Karthiga, R.; Rajarajan, M.; Suganthi, A. Enhancement of electrochemical sensor for the determinatio of glucose based on mesoporous $VO_2$/PVA nanocomposites. *Surf. Interfaces* **2019**, *16*, 164–173. [CrossRef]

*nanomaterials*

MDPI

*Article*

# Construction of Electrostatic Self-Assembled 2D/2D $CdIn_2S_4/g\text{-}C_3N_4$ Heterojunctions for Efficient Visible-Light-Responsive Molecular Oxygen Activation

**Hongfei Yin [1], Chunyu Yuan [1], Huijun Lv [1], Xulin He [2], Cheng Liao [2], Xiaoheng Liu [3,*] and Yongzheng Zhang [1,*]**

1 School of Physics and Physical Engineering, Qufu Normal University, Qufu 273165, China; yinhongfei1016@foxmail.com (H.Y.); chunyuyuan.qfnu@gmail.com (C.Y.); hjlv.qfnu@gmail.com (H.L.)
2 Chengdu Science and Technology Development Center, China Academy of Engineering Physics, Chengdu 610000, China; hexl0003@yinhe596.cn (X.H.); cliao@pku.edu.cn (C.L.)
3 Key Laboratory for Soft Chemistry and Functional Materials of Ministry of Education, School of Chemical Engineering, Nanjing University of Science and Technology, Nanjing 210094, China
* Correspondence: xhliu@njust.edu.cn (X.L.); yzzhang@qfnu.edu.cn (Y.Z.)

**Abstract:** Molecular oxygen activated by visible light to generate radicals with high oxidation ability exhibits great potential in environmental remediation The efficacy of molecular oxygen activation mainly depends on the separation and migration efficiency of the photoinduced charge carriers. In this work, 2D/2D $CdIn_2S_4/g\text{-}C_3N_4$ heterojunctions with different weight ratios were successfully fabricated by a simple electrostatic self-assembled route. The optimized sample with a weight ratio of 5:2 between $CdIn_2S_4$ and $g\text{-}C_3N_4$ showed the highest photocatalytic activity for tetracycline hydrochloride (TCH) degradation, which also displayed good photostability. The enhancement of the photocatalytic performance could be ascribed to the 2D/2D heterostructure; this unique 2D/2D structure could promote the separation and migration of the photoinduced charge carriers, which was beneficial for molecular oxygen activation, leading to an enhancement in photocatalytic activity. This work may possibly provide a scalable way for molecular oxygen activation in photocatalysis.

**Keywords:** 2D/2D; electrostatic self-assembled; heterojunction; photocatalytic; molecular oxygen activation

## 1. Introduction

The energy crisis and environmental pollution are serious problems worldwide. Environmental pollution originating from refractory organic pollutants, especially antibiotics, has an extremely negative influence on humans. Photocatalytic technology has been recognized as a potential way to mitigate environmental pollution because titanium dioxide is used as the catalyst for water splitting under UV light irradiation [1]. However, the relatively large bandgap of $TiO_2$ renders it unacceptable in handling the above-mentioned environmental problems with high efficiency. Therefore, it is strongly desirable that photocatalysts with high solar utilization be explored [2–8].

$CdIn_2S_4$, a ternary sulfide of chalcogenide, with an appropriate bandgap and suitable band edge positions, has gained increasing attention in the scope of photocatalysis, due to its potential applications in photocatalytic hydrogen production [9–12], organic conversions [13,14], and organic pollutant degradation [15–17], as well as $CO_2$ reduction [18,19]. However, two issues have greatly restricted the widespread use of pure $CdIn_2S_4$. One is the fast recombination of photogenerated charge carriers, and the other is photo-corrosion. During the photocatalytic process, $S^{2-}$ can be oxidized by photoinduced holes [20], and the generated dissociative $Cd^{2+}$ would have a negative influence on living organisms. Therefore, it is desirable to design an effective $CdIn_2S_4$-based photocatalyst without sacrificing photocatalytic performance and using less Cd source. It was reported that doping heteroatoms [21] or constructing heterojunctions [22–24] were efficient methods of alleviating the two above-mentioned problems, where the construction of heterojunctions could inhibit the speedy recombination of photoexcited charge carriers and alleviate the photo-corrosion more efficiently, owing to the spatial location of photoexcited electrons and holes.

In recent decades, owing to the properties of earth abundance, non-toxicity, a simple preparation process and stable structure, g-$C_3N_4$, which belongs to a type of metal-free photocatalyst, has been widely used for building heterojunctions [25–30]. Nevertheless, bulk g-$C_3N_4$ exhibits low specific surface areas and suffers from the rapid recombination of photoinduced electron-hole pairs; these drawbacks seriously limit photocatalytic efficiency. Exfoliating bulk g-$C_3N_4$ into two-dimensional nanosheets with few layers or single layers is an effective method for improving the photocatalytic activities of g-$C_3N_4$ [31–33]. Wang et al. described an ultrasonic exfoliation route for fabricating g-$C_3N_4$ nanosheets with boosted photocatalytic performance; the abundant reaction active sites and the low recombination rate of charge carriers were attributed to the enhanced photocatalytic performances [34]. Qu et al. combined freeze-dried, ultrasonic and solvothermal process-synthesized g-$C_3N_4$ nanosheets with an atomically thin mesoporous structure that exhibited superior photocatalytic hydrogen evolution performance; the ultrathin nanostructure could promote light absorption as well as shorten the migration time and migration distance of photoexcited charge carriers [35].

It has been reported that the 2D/2D nanostructures have tight interfacial contact and a large contact area, which not only provides more channels for carrier transfer, but also shortens the transfer time and migration distance, leading to improved photocatalytic performance [36–41]. It is expected that coupling 2D $CdIn_2S_4$ with 2D g-$C_3N_4$ is an efficient method to enhance the photocatalytic activities of $CdIn_2S_4$ while using less of the Cd species. In this work, 2D/2D $CdIn_2S_4$/g-$C_3N_4$ nanocomposites with various weight ratios were constructed through a simple and low-cost electrostatic self-assembled method. Various characterization technologies were utilized to fully study the crystallization, morphology, optical and electrochemical properties of the obtained 2D/2D $CdIn_2S_4$/g-$C_3N_4$ heterojunctions. The photocatalytic performance of the obtained 2D/2D $CdIn_2S_4$/g-$C_3N_4$ heterojunctions was estimated through TCH degradation under visible light illumination. The constructed 2D/2D nanostructures could efficiently facilitate the separation and transfer of photoexcited charge carriers between hetero-interfaces, which is favorable for the process of molecular oxygen activation, resulting in improved photocatalytic activity.

## 2. Materials and Methods

### 2.1. Reagents

Urea (AR) and indium chloride ($InCl_3 \cdot 4H_2O$, AR) were bought from Sigma Aldrich (Shanghai, China). Cadmium acetate [Cd $(CH_3COO)_2 \cdot 2H_2O$, AR] and thioacetamide ($C_2H_5N_S$, AR) were provided by Sinopharm Chemical Reagent Co. Ltd. (Shanghai, China). All the chemical reagents were used without further treatment.

### 2.2. Synthesis of the Photocatalysts

#### 2.2.1. Synthesis of g-$C_3N_4$ Nanosheets and Protonated g-$C_3N_4$ Nanosheets

First, bulk g-$C_3N_4$ was fabricated through thermal condensation of urea. Typically, 20 g urea was set in a covered crucible, then heated to 550 °C within 240 min and maintained at this temperature for 240 min to obtain bulk g-$C_3N_4$, denoted as BCN. g-$C_3N_4$ nanosheets were obtained via a secondary calcination of the BCN with the same calcination procedure for BCN, and the generated samples were ground for further use and denoted as CNNSs.

Protonated g-$C_3N_4$ nanosheets were prepared on the basis of previous reports with some modifications [42,43]. Typically, 2 g CNNSs were added to 300 mL 1M HCl aqueous solutions; after ultrasonic treatment for 1 h, a homogeneous suspension was formed, which was further stirred for 4 h to promote the protonation process, the protonated g-$C_3N_4$ nanosheets were obtained via centrifugation, followed by washing with a large amount of distilled water to eliminate the excess HCl until pH = 7. Finally, the protonated $g\text{-}C_3N_4$ nanosheets were dried at 60 °C overnight and denoted as PCNNSs.

#### 2.2.2. Preparation of $CdIn_2S_4$ Nanosheets

The $CdIn_2S_4$ nanosheets were prepared as described previously with some adjustment [44]. In a typical preparation procedure, 399 mg Cd $(CH_3COO)_2{\cdot}2H_2O$ and 880 mg $InCl_3{\cdot}4H_2O$ were added into 300 mL deionized water. After 30 min stirring at room temperature, 600 mg thioacetamide (TAA) was added. Then, the temperature was raised to 100 °C from room temperature after 30 min stirring, and the reaction system was refluxed at this temperature for 12 h under continuous magnetic stirring. After the reaction was finished, the produced samples were collected by centrifugation and then cleaned with deionized water 2 times. Finally, after drying at 60 °C overnight, the final product was denoted as CIS.

#### 2.2.3. Synthesis of 2D/2D $CdIn_2S_4/g\text{-}C_3N_4$ Heterojunctions

$CdIn_2S_4$ nanosheets and PCNNSs were simultaneously dispersed into deionized water with the assistance of ultrasonic treatment for 2 h to form 2 uniform suspensions with a concentration of 0.75 mg/mL. Then, the PCNNSs dispersion was dropwise added into the $CdIn_2S_4$ nanosheet dispersion under continuous magnetic stirring at room temperature; after 6 h, the mixed suspension was separated by centrifugation, then dried at 60 °C in a vacuum oven overnight. The final mass ratios of $CdIn_2S_4$ to PCNNSs were 5:1, 5:2, 5:3, and 5:4, denoted as CISCN-1, CISCN-2, CISCN-3, and CISCN-4, respectively.

### 2.3. *Characterization*

The crystal nature of the obtained $CdIn_2S_4/g\text{-}C_3N_4$ hybrids was examined by powder X-ray diffraction (XRD) performed on a Bruker D8 Advance instrument. The morphology of the fabricated samples was observed by field-emission scanning electron microscope (FESEM, Hitachi S-4800) and transmission electron microscope (TEM, JEOL JEM-2100). The optical properties of the as-prepared samples were measured by ultraviolet-visible (UV-vis) diffuse reflection spectra (DRS) using a Shimadzu UV-3100 spectrophotometer, where $BaSO_4$ was used as reference, with a test range of 200–800 nm. Element composition and chemical state of each element on the surface of the as-prepared photocatalysts were detected on a PHI Quantera II SXM photoelectron spectrometer under Al K$\alpha$ radiation ($\lambda$ = 0.84 nm).

### 2.4. *Electrochemical Analysis*

The photoelectrochemical tests, including transit photocurrent response and electrochemical impedance spectra (EIS), were characterized by a three-electrode electrochemical system on a CHI760E electrochemical workstation. In the test system, Ag/AgCl electrode and Pt wire electrode were utilized as reference electrode and counter electrode, respectively. A 300 W Xe lamp was employed to provide visible light illumination. For transit photocurrent measurement, the electrolyte was $Na_2SO_4$ aqueous solution with a concentration of 0.5 M (pH $\approx$ 6.8). For EIS measurements, the electrolyte was a mixed solution containing 0.5 M KCl and 5 mM $K_3[Fe(CN)_6]/K_4[Fe(CN)_6]$, and the signals were recorded from 100 kHz to 0.01 Hz, respectively. The working electrode was fabricated as follows: 4 mg of the obtained catalysts were dispersed into a mixture solvent of 750 μL water and 250 μL ethanol solution. Then, 10 μL Nafion solution (5 wt%) was added. The mixture was ultrasonicated for 60 min to form a homogeneous slurry. Finally, 100 μL of the dispersions were loaded onto a 1 cm × 3 cm ITO-coated glass substrate with coating area of nearly 1 $cm^2$.

### 2.5. *Catalytic Experiments*

The photocatalytic performance of the fabricated samples was evaluated by TCH degradation under visible light illumination. A 300 W Xe lamp attached with a 400 nm cut-off filter was used to provide visible light. Typically, 30 mg sample was dispersed into 50 mL 50 mg/L TCH aqueous solution. Then, the mixture was stirred for 1 h in darkness to promote adsorption–desorption equilibrium between the sample and the TCH. After the reaction system was exposed to visible light, 3 mL suspension was sucked out every 20 min. After removing the catalyst from the suspension, the remained filtrate was analyzed to determine the concentration of TCH.

### 2.6. Quantitative Analysis of $\bullet O_2^-$

The generation of $\bullet O_2^-$ was measured by nitroblue tetrazolium (NBT) conversion strategy. Due to the reaction between $\bullet O_2^-$ and NBT at a mole ratio of 4:1, the concentration of $\bullet O_2^-$ could be determined by the decrease in NBT. In a typical NBT transformation reaction procedure, 10 mg sample was added into 50 mL 0.05 mM NBT aqueous solution. Then, the mixed solution was stirred continuously in the dark for 60 min to promote the adsorption–desorption equilibrium between the sample and the NBT. After the reaction system was exposed to visible light, 3 mL solution was sucked out every 10 min. After removing the catalyst via a millipore filter (0.22 μm), UV-vis spectrometer (UV-1801) was used to test the concentration of NBT.

## 3. Results and Discussion

The crystalline structure of the as-fabricated $CdIn_2S_4$/g-$C_3N_4$ heterojunctions and single component was studied by powder XRD, as displayed in Figure 1. The apparent diffraction peak at $2\theta = 27.6°$ in PCNNSs could be assigned to the (002) planes of graphitic materials, which represent the interlayer stacking of a conjugated aromatic structure [45]. For $CdIn_2S_4$, the diffraction peaks of $2\theta$ at 14.1°, 23.2°, 27.2°, 28.5°, 33.0°, 40.7°, 43.3°, 47.4°, 55.5°, and 66.1° could be indexed into (111), (220), (311), (222), (400), (422), (511), (440), (533), and (731) crystal planes of $CdIn_2S_4$ (JCPDS NO.27-0060) with cubic phase structure. All the diffraction peaks of $CdIn_2S_4$/g-$C_3N_4$ heterojunctions were similar to those of pure CIS, which indicated the existence of CIS in the CISCN heterojunctions. However, the diffraction peak of PCNNSs could not be clearly observed in the $CdIn_2S_4$/g-$C_3N_4$ nanocomposites, which might have originated from a lower peak intensity than that of CIS in the range of 27.2° and 28.5°.

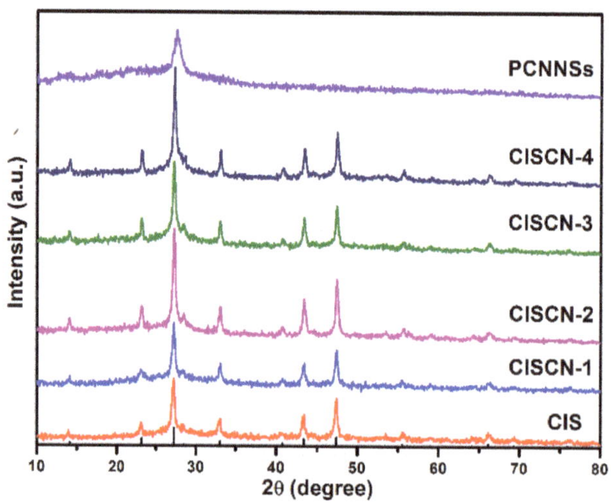


**Figure 1.** XRD patterns of the fabricated catalysts.

The morphologies of the as-prepared PCNNSs, CIS, and CISCN-2 were observed via FESEM and TEM, as shown in Figure 2. The characteristic SEM pattern of the PCNNSs is displayed in Figure 2a; it exhibited a nanostructure of nanosheet, which was further verified by the TEM test, as depicted in Figure 2b,c. Figure 2d shows the SEM image of CIS, which displayed a small nanosheet-like morphology with a size of about 100–200 nm, in accordance with the TEM result (Figure 2e). The high-resolution TEM (HRTEM) pattern of the CIS is illustrated in Figure 2f; the observable lattice space of 0.324 nm marked in Figure 2f could be indexed to the (311) crystal plane of $CdIn_2S_4$. As for the CISCN-2 nanocomposite, from SEM observations (Figure 2g), it displayed a morphology similar to that of PCNNSs, which might be due to the smaller size of the CIS nanosheets. TEM (Figure 2h) and HRTEM (Figure 2i) were used to further investigate its nanostructure, Low-resolution TEM (Figure 2h) revealed that the small CIS nanosheets were stacked on the surface of the PCNNSs nanosheets, exhibiting a sheet-on-sheet morphology. Meanwhile, an obvious interface between CIS and PCNNSs could be observed (Figure 2i), implying the CISCN nanocomposite was successfully prepared by the facile electrostatic self-assembled method.

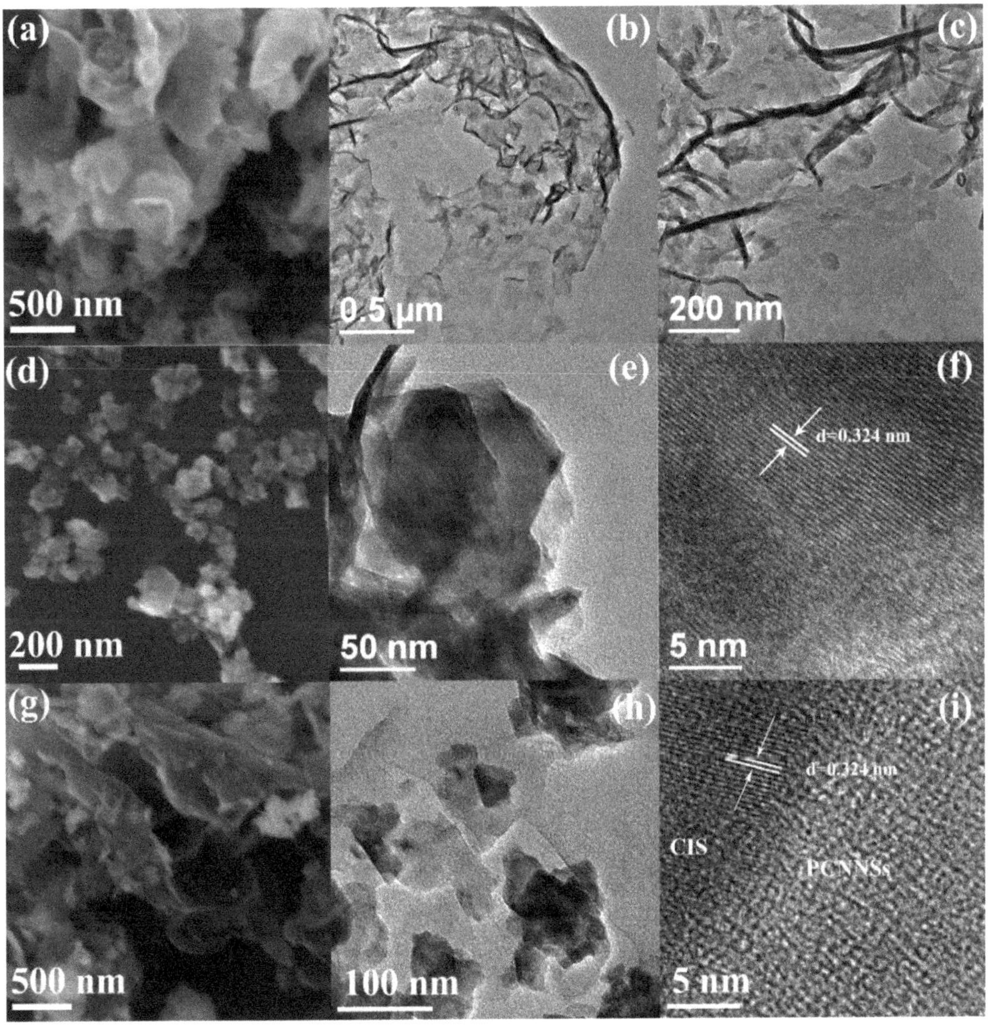


**Figure 2.** SEM of (**a**) PCNNSs, (**d**) $CdIn_2S_4$, (**g**) CISCN-2, TEM of (**b**,**c**) PCNNSs, (**e**) $CdIn_2S_4$, (**h**) CISCN-2 and HRTEM of (**f**) $CdIn_2S_4$, (**i**) CISCN-2.

The composition of surface elements and the chemical state of each element in the obtained photocatalysts were investigated by XPS survey, as shown in Figure 3. It can be seen from Figure 3a that there were characteristic peaks of C, N, Cd, In, and S in the full spectrum of the as-fabricated samples, implying the as-obtained CISCN-2 heterojunction consisted of $CdIn_2S_4$ and $g\text{-}C_3N_4$. The signals of C 1s in PCNNSs were located at 284.8 and 288.2 eV, while the peaks of C 1s in CISCN-2 were located at 284.8 and 288.4 eV (Figure 3b); the former peak could be attributed to the adventitious carbon, while the latter peak could be assigned to N=C-N type carbons [46,47]. As for N 1s (Figure 3c), the characteristic signal in PCNNSs could be divided into three peaks: 398.6 eV (C-N=C), 399.6 eV (N-$(C)_3$), and 401.0 eV (N-H) [18,48], while in CISCN-2, these three peaks had a small shift toward higher binding energy, located at 398.8, 399.8, and 401.2 eV, respectively. For Cd 3d (Figure 3d), two obvious peaks at 405.3 and 412.0 eV could be observed in pure CIS, while in CISCN-2, these two peaks exhibited a small shift toward lower binding energy at 405.2 and 411.9 eV, which corresponded to the Cd $3d_{5/2}$ peak and $3d_{3/2}$ peak, respectively [49]. Meanwhile, this phenomenon also occurred in the case of In 3d (Figure 3e) and S 2p (Figure 3f); compared with the neat $CdIn_2S_4$, the characteristic peaks of In 3d and S 2p in CISCN-2 also exhibited a small shift toward lower binding energy, indicating the change of chemical environment. This might have originated from the bonding interaction between PCNNSs and CIS.

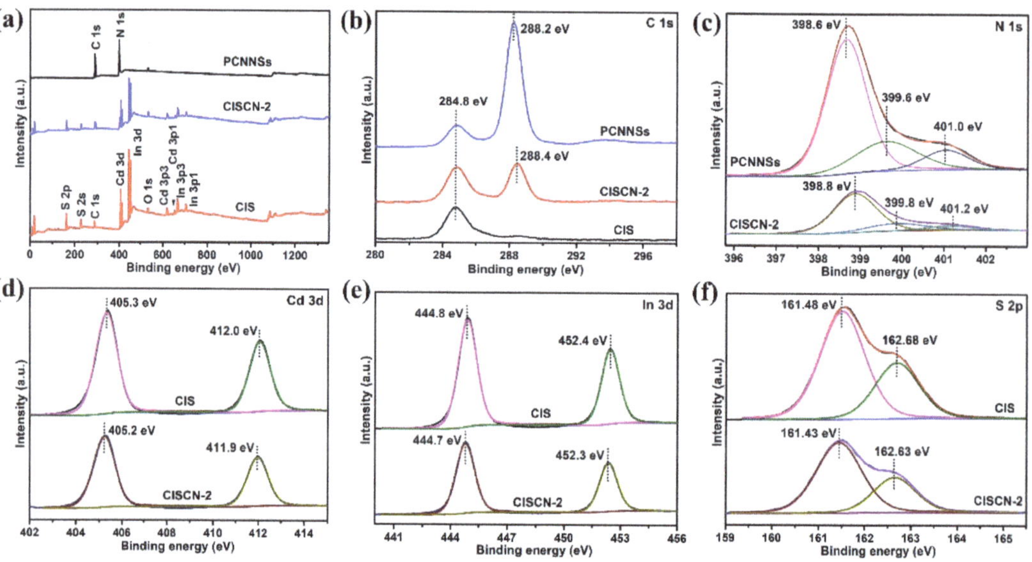


**Figure 3.** XPS spectrum of PCNNSs, CIS, and CISCN-2: (**a**) full-range spectrum, high resolution XPS spectra of (**b**) C1s, (**c**) N 1s, (**d**) Cd 3d, (**e**) In 3d, and (**f**) S 2p.

The optical properties of the attained catalysts were investigated via UV-vis DRS spectra, as displayed in Figure 4a. It can be clearly seen that CIS possessed a higher UV-visible light absorption than pure PCNNSs. The absorption edges of CIS and PCNNSs were about 540 nm and 420 nm, respectively. Compared to the neat CIS, the light absorption over the obtained $CdIn_2S_4/g\text{-}C_3N_4$ heterojunctions exhibited an obvious decrease, indicating that the introduction of PCNNSs was not helpful for light absorption, which may have been due to the microstructural changes. Therefore, the light absorption might not be responsible for the improved photocatalytic performance. However, the introduction of PCNNSs had a positive influence on the formation of a heterogeneous interface between CIS and PCNNSs; the formation of the heterogeneous interface could promote the transfer of photoinduced electrons and holes [37,38]. The bandgap energies of PCNNSs and CIS were calculated on the basis of the Kubelka–Munk equation and estimated to be 2.88 and 2.52 eV, respectively [50]. The relatively higher bandgap energy of PCNNSs compared to bulk g-$C_3N_4$ may possibly originate from the quantum confinement effect [51].

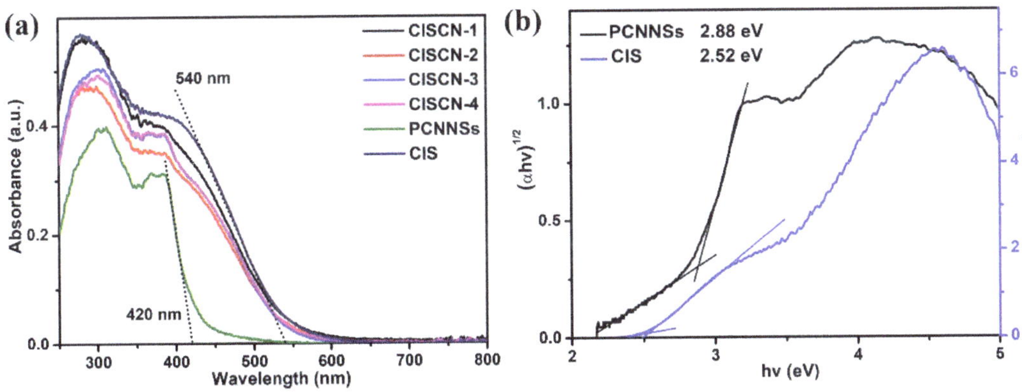


**Figure 4.** (**a**) UV-vis DRS of the fabricated samples and (**b**) calculation of PCNNSs and CIS bandgap energies.

The photocatalytic performance of the attained photocatalysts was assessed via photocatalytic TCH degradation under visible light irradiation, as illustrated in Figure 5. It can be clearly observed from Figure 5a that pure PCNNSs exhibited the poorest photocatalytic performance among the as-prepared photocatalysts; it had a TCH removal ratio of only 39.5%, whereas pure CIS exhibited a removal ratio of 82.1%. After the formation of CISCN heterojunctions, the removal efficiency of TCH over the constructed CISCN-1, CISCN-2, CISCN-3, and CISCN-4 was 79.9%, 83.6%, 80.1%, and 78.0%, respectively. For comparison, the contribution of adsorption over PCNNSs, CIS, CISCN-1, CISCN-2, CISCN-3, and CISCN-4 was 36.8%, 2.1%, 34.7%, 35.7%, 31.5%, and 30.5%, respectively. Obviously, with the increased content of PCNNSs, the photocatalytic performance of the CISCN nanocomposites tended to increase first and then decrease. This might have originated from the excess amount of PCNNSs, which would lead to an insufficient interface formed between CISCN nanocomposites and PCNNSs and may exhibit a negative effect on the interaction between CIS and PCNNSs, inhibiting the separation of the photoexcited electron-hole pairs. The kinetic reaction procedure of TCH degradation was fitted by pseudo-first-order equations, as shown in Figure 5b. Compared to the single component, the apparent kinetic constant of CISCN-2 was 1.14 and 3.05 times as high as that of the $CdIn_2S_4$ and protonated g-$C_3N_4$, respectively, indicating the superior photocatalytic performance of the $CdIn_2S_4$/g-$C_3N_4$ heterojunction. To verify that the decrease in TCH was triggered by photocatalysis during visible light irradiation, a prolonged adsorption experiment with TCH over CISCN-2 under dark conditions was carried out, as shown in Figure 5c. By prolonging the adsorption time, a delayed decrease could be observed, due to the adsorption–desorption equilibrium established between the TCH and the photocatalyst. Figure 5d shows the UV-vis spectra of TCH at different periods of the photocatalytic process over CISCN-2; the absorbance of TCH had an obvious decrease after 2 h of visible light illumination, implying photocatalysis played an important role in the degradation of TCH. Therefore, the decrease in TCH could be attributed to the collaboration of adsorption and photocatalysis, and during the period of visible light irradiation, the decrease in TCH could be attributed to photocatalysis. To highlight the superiority of the fabricated catalyst, a comparison with previous reports is presented in Table 1.

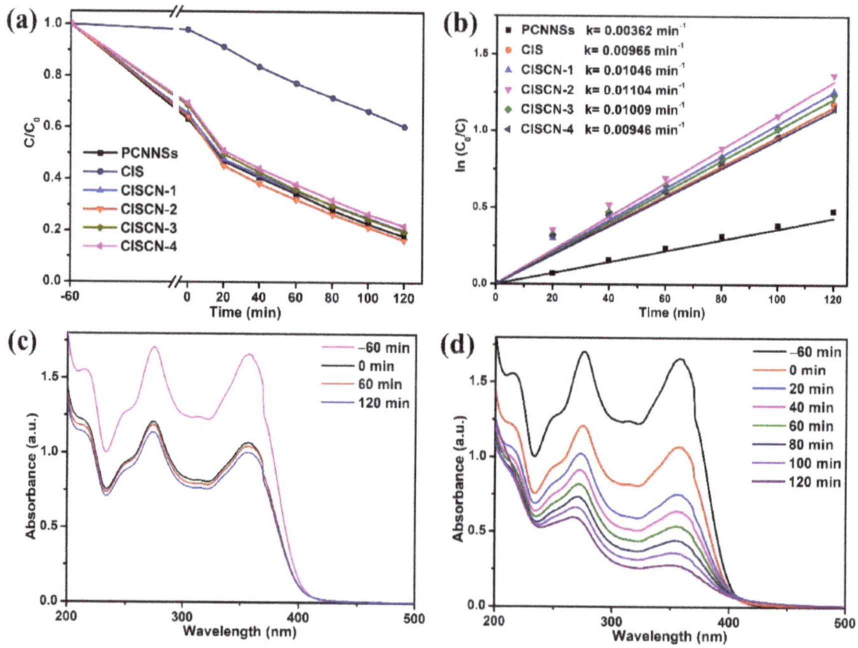


**Figure 5.** (**a**) Photocatalytic performance of various samples toward TCH photodegradation under visible light illumination; (**b**) the kinetic curves fitted by pseudo-first-order equations over different samples; (**c**) UV-vis spectra of TCH adsorbed by CISCN-2 for different time periods; and (**d**) UV-vis spectra of TCH at different periods of photocatalytic process over CISCN-2.

**Table 1.** Degradation efficiency (DE) of TCH over previous reports and CISCN-2 in this work.

| Sample | TC (mg/L) | Dosage (g/L) | t (min) | Light Source | DE (%) | Refs. |
|---|---|---|---|---|---|---|
| Co/V-g-$C_3N_4$ | 10 | 0.5 | 120 | 250 W Xe lamp | 64.3 | [5] |
| 2D/3D g-$C_3N_4$ | 10 | 0.5 | 120 | 250 W Xe lamp | 69.6 | [6] |
| Nitrogen-deficient tubular g-$C_3N_4$ | 10 | 1.0 | 150 | 300 W Xe lamp | 84.3 | [7] |
| BN QDs/g-$C_3N_4$ | 10 | 1.0 | 60 | 300 W Xe lamp | 82 | [8] |
| $WO_3$/g-$C_3N_4$ | 25 | 0.5 | 120 | 300 W Xe lamp | 70 | [28] |
| Bi/$\alpha$-$Bi_2O_3$/g-$C_3N_4$ | 10 | 1.0 | 180 | 300 W Xe lamp | 91.2 | [29] |
| $Nb_2O_5$/g-$C_3N_4$ | 10 | 0.5 | 150 | 250 W Xe lamp | 76.2 | [30] |
| $CdIn_2S_4$/g-$C_3N_4$ | 50 | 0.6 | 120 | 300 W Xe lamp | 83.6 | This work |

The recycling ability of the photocatalyst is a vital factor to evaluate the performance of the photocatalyst; therefore, the progress of the recycling photocatalytic experiment is of great necessity. After each recycling run, the photocatalyst was collected and washed for the next run. It can be observed from Figure 6a that after four recycling runs, the TCH removal ratio over CISCN-2 was still 68.2%; compared to its the first-time usage, there was only an 8.7% reduction, implying the relative stability of the as-prepared CISCN-2. Moreover, XRD of the reused photocatalyst was tested to further confirm the stability of the CISCN-2 composites, as depicted in Figure 6b. Notably, there were no obvious changes in the reused sample as compared to the fresh photocatalyst, suggesting the stability of the crystal structure.

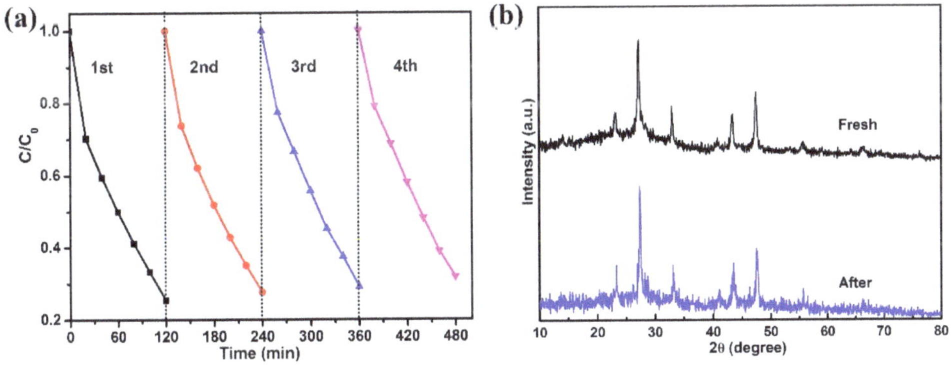


**Figure 6.** (**a**) Four cyclic experiments of TCH photodegradation over CISCN-2. (**b**) XRD patterns of CISCN-2 before and after cyclic experiments.

Generally, three steps are involved in the photocatalytic process: (1) semiconductor photocatalysts were irradiated by light, leading to the formation of photogenerated electrons and holes; (2) photoinduced electrons and holes were separated and transferred to the surface of the photocatalyst; (3) the photoexcited electrons and holes participated in the surface redox reactions [51]. The separation and transfer efficiency of the photoinduced charge carriers are important factors that affect the photocatalytic performance. Transient photocurrent (Figure 7a) was measured to evaluate the separation efficacy of the photoexcited charge carriers, while EIS (Figure 7b) measurement was carried out to estimate the transfer efficiency of the photogenerated electron-hole pairs [52,53]. It can be easily observed from Figure 7a that CISCN-2 displayed the highest photocurrent intensity among all the photocatalysts, implying the highest separation efficacy of the photoexcited electron-hole pairs. Meanwhile, the smallest semicircle of the EIS curve could be observed over CISCN-2, implying the smallest transfer resistance of photoinduced charge carriers and highly interfacial transfer efficacy of the charge carriers. These two results might be reasons for the enhanced photocatalytic performance.

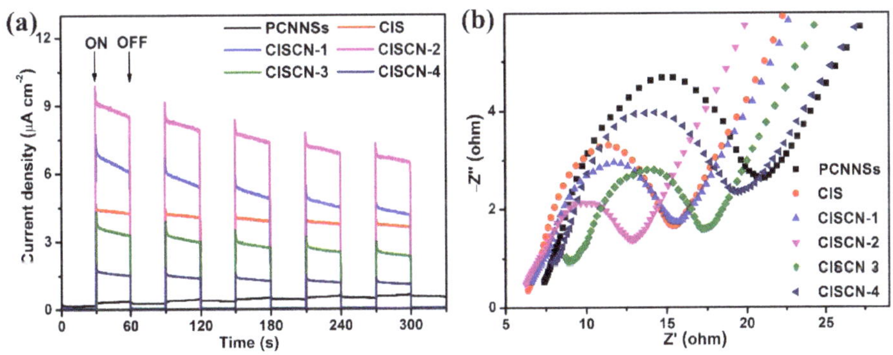


**Figure 7.** (**a**) Transient photocurrent response plots, and (**b**) EIS of the obtained catalysts.

It is worth studying the possible reaction mechanism during the TCH degradation. In general, to distinguish the main active species generated in the photocatalytic reaction process, active species trapping experiments were conducted. Isopropyl alcohol (IPA) and disodium ethylenediaminetetraacetate (EDTA-2Na) were chosen as scavengers for removing •OH and $h^+$, respectively. $N_2$ was continuously bubbled into the reaction system for removing the soluble oxygen, eliminating the formation of $•O_2^-$ [54,55]. As shown in Figure 8, the addition of IPA (1 mM) had an extremely small effect on the photocatalytic performance of the CISCN-2, indicating that the role of •OH formed during the photocatalytic process could be ignored. After the EDTA-2Na (1 mM) was introduced, an apparent decrease in the TCH removal ratio could be observed; it was decreased from 74.6% to 54.6%, implying the photoexcited holes made contributions to the degradation of TCH. Meanwhile, when $N_2$ was continuously bubbled into the reaction system during the whole photocatalytic TCH degradation process, a dramatic decrease could be observed; the removal ratio of TCH was decreased from 74.6% to 25.3%, implying that $•O_2^-$ played the prominent role in TCH removal. Therefore, the generated $•O_2^-$ played a predominant role in the TCH photodegradation procedure, but in the meantime, the contributions of photoexcited holes also could not be overlooked.

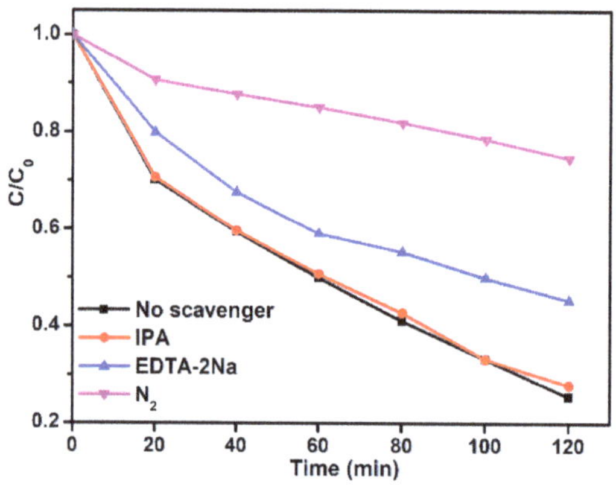


**Figure 8.** Reactive species trapping experiments over CISCN-2.

The generation of $•O_2^-$ was measured by the NBT transformation experiment. Because the reaction between $•O_2^-$ and NBT occurred at a mole ratio of 4:1, the content of the $•O_2^-$ could be estimated by the change in NBT. Figure 9a–c were the typical time courses of absorption variation of NBT over the PCNNSs, CIS, and CISCN-2. For PCNNSs, a delayed change could be observed. However, the intensity of the typical absorbance signal at 260 nm for NBT exhibited a noteworthy decrease over CIS and CISCN-2, indicating the generation of $•O_2^-$ during the photocatalytic reaction. Figure 9d,e displays the change in NBT concentration and the corresponding kinetic fitting curves. Obviously, the reaction between $•O_2^-$ and NBT over the CISCN-2 had the fastest rate. Meanwhile, the $•O_2^-$ generated over CISCN-2 was the largest amount produced among these three samples (Figure 9f), which further verified the enhancement of the photocatalytic activity over CISCN-2 in $•O_2^-$ production.

Moreover, measurements of the band edge positions of the single component were necessary, as they could further verify the possible photocatalytic mechanism. Figure 10a, b exhibits the corresponding VB-XPS spectra of the PCNNSs and CIS. It can be clearly observed that the valence band (VB) position of PCNNSs and CIS was +1.89 and +1.11 eV, respectively. Therefore, based on the evaluation with UV-vis DRS, the conduction band (CB) position of PCNNSs and CIS could be determined by the equation $E_g = |E_{CB} - E_{VB}|$ to be −0.99 and −1.41 eV, respectively. The possible photocatalytic mechanism and the transfer path of photogenerated charge carriers were proposed on the basis of active species trapping experiments and the band edge positions of the PCNNSs and CIS, as displayed in Figure 10c. When the photocatalytic reaction system was exposed to visible light ($\lambda > 400$ nm), both PCNNSs and CIS could be excited to generate electron-hole pairs. Owing to the more negative CB position of CIS than that of PCNNSs, the photoexcited electrons on the CB of CIS would transfer toward the CB of PCNNSs, which could react with soluble oxygen molecules, leading to the formation of $•O_2^-$ with strong oxidation ability that could oxidize organic pollutants. Simultaneously, the photoinduced holes on the VB of PCNNSs with more positive potentials would migrate to the VB of CIS; the photoinduced holes could directly oxidize organic pollutants. This spatial separation of photoinduced charge carriers might be one of the reasons for the enhanced photocatalytic performance.

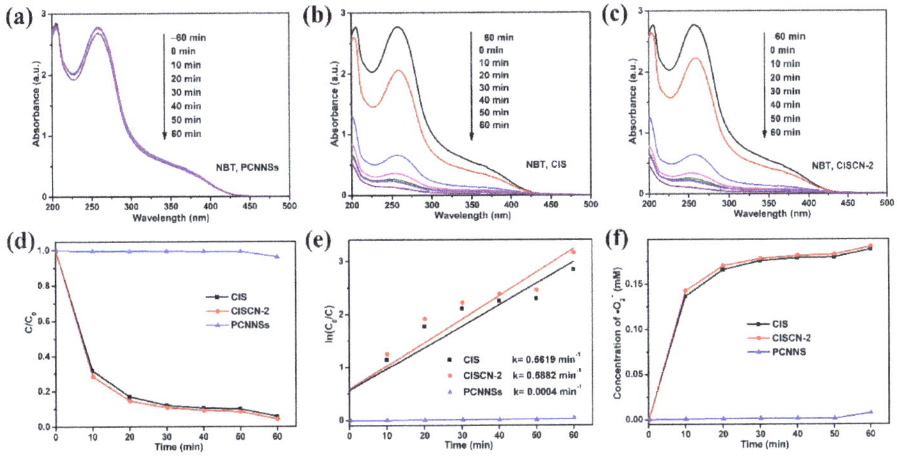


**Figure 9.** Time courses of absorption variation of NBT over (**a**) PCNNSs, (**b**) CIS, and (**c**) CISCN-2. (**d**) Time-dependent photodegradation plots of NBT over different catalysts; (**e**) the corresponding pseudo-first-order kinetic fitting curves, (**f**) time-dependent concentration curves of $\bullet O_2^-$.

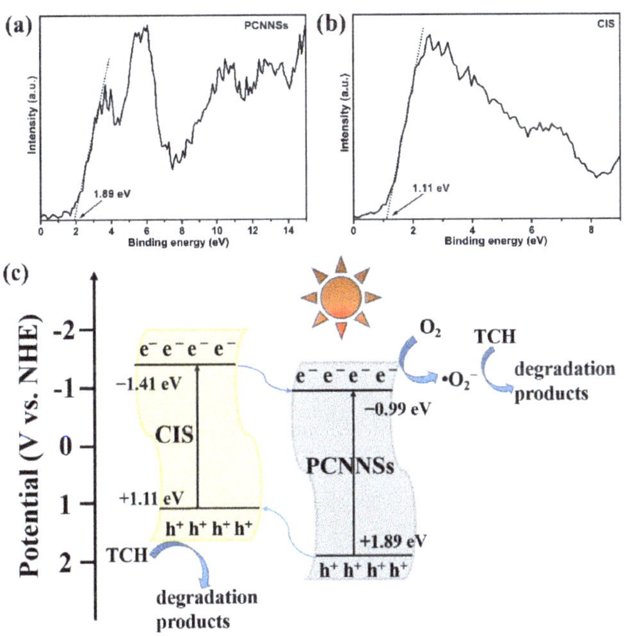


**Figure 10.** (**a**) VB-XPS of PCNNSs, (**b**) VB-XPS of CIS, and (**c**) schematic diagram of photogenerated charge transfer pathway and possible photocatalytic mechanism in $CdIn_2S_4/g\text{-}C_3N_4$ heterojunction toward TCH degradation.

## 4. Conclusions

In summary, 2D/2D $CdIn_2S_4$/g-$C_3N_4$ heterojunctions were productively constructed through a facile electrostatic self-assembled route. The unique 2D/2D nanostructures exhibited tight interfacial contact and a large contact area, which could provide more channels for the transfer of photoinduced charge carriers and shorten the transfer distance of the charge carriers, realizing a higher separation efficacy of photoexcited electron-hole pairs, thus benefiting molecular oxygen activation. The optimized 2D/2D $CdIn_2S_4$/g-$C_3N_4$ heterojunction exhibited the highest photocatalytic performance and photostability toward TCH degradation. The improved photocatalytic activity could be attributed to the high separation and transfer efficacy of the photoexcited charge carriers, achieving a higher molecular oxygen activation efficiency.

**Author Contributions:** Conceptualization, H.Y., X.L. and Y.Z.; Data curation, H.Y., C.Y., H.L., X.H. and C.L.; Funding acquisition, X.H., C.L., Y.Z. and X.L.; Investigation, H.Y., C.Y., H.L., X.H. and C.L.; Project administration, X.L. and Y.Z.; Visualization, H.Y., C.Y. and H.L.; Writing—original draft, H.Y., Writing—review and editing, H.Y., X.L. and Y.Z. All authors have read and agreed to the published version of the manuscript.

**Funding:** This research was funded by Sichuan Science and Technology Program (2020YFG0380), Taishan Scholars Program (tsqn201909107), Shandong Province Natural Science Foundation (ZR2020QF077), National Natural Science Foundation of China (62104131) and National Natural Science Foundation of China (51872141). And the APC was funded by Shandong Province Natural Science Foundation (ZR2020QF077).

**Data Availability Statement:** Data supporting this study are available within the article.

**Acknowledgments:** This work was supported by Sichuan Science and Technology Program (2020YFG0380), Taishan Scholars Program (tsqn201909107), Shandong Province Natural Science Foundation (ZR2020QF077), National Natural Science Foundation of China (62104131) and National Natural Science Foundation of China (51872141).

**Conflicts of Interest:** The authors declare that they have no known competing financial interests or personal relationships that could have appeared to influence the work reported in this paper.

## References


1. Fujishima, A.; Honda, K. Electrochemical photolysis of water at a semiconductor electrode. *Nature* **1972**, *238*, 37–39. [CrossRef] [PubMed]
2. Kadam, A.; Kim, H.; Lee, S. Low-temperature in situ fabrication of porous S-doped g-$C_3N_4$ nanosheets using gaseous-bubble template for enhanced visible-light photocatalysis. *Ceram. Int.* **2020**, *46*, 28481–28489. [CrossRef]
3. Kadam, A.; Moniruzzaman, M.; Lee, S. Dual functional S-doped g-$C_3N_4$ pinhole porous nanosheets for selective fluorescence sensing of $Ag^+$ and visible-light photocatalysis of dyes. *Molecules* **2019**, *24*, 450. [CrossRef]
4. Chen, W.; Huang, G.; Song, H.; Zhang, J. Efficient and stable charge transfer channels for photocatalytic water splitting activity of CdS without sacrificial agents. *J. Mater. Chem. A* **2020**, *8*, 20963–20969. [CrossRef]
5. Dong, H.; Zuo, Y.; Song, N.; Hong, S.; Xiao, M.; Zhu, D.; Sun, J.; Chen, G.; Li, C. Bimetallic synergetic regulating effect on electronic structure in cobalt/vanadium co-doped carbon nitride for boosting photocatalytic performance. *Appl. Catal. B Environ.* **2021**, *287*, 119954. [CrossRef]
6. Dong, H.; Zhang, X.; Li, J.; Zhou, P.; Yu, S.; Song, N.; Liu, C.; Che, G.; Li, C. Construction of morphology-controlled nonmetal 2D/3D homojunction towards enhancing photocatalytic activity and mechanism insight. *Appl. Catal. B Environ.* **2020**, *263*, 118270. [CrossRef]
7. Fang, H.; Guo, H.; Niu, C.; Liang, C.; Huang, D.; Tang, N.; Liu, H.; Yang, Y.; Li, L. Hollow tubular graphitic carbon nitride catalyst with adjustable nitrogen vacancy: Enhanced optical absorption and carrier separation for improving photocatalytic activity. *Chem. Eng. J.* **2020**, *402*, 126185. [CrossRef]
8. Yang, Y.; Zhang, C.; Huang, D.; Zeng, G.; Huang, J.; Lai, C.; Zhou, C.; Wang, W.; Guo, H.; Xue, W.; et al. Boron nitride quantum dots decorated ultrathin porous g-$C_3N_4$: Intensified exciton dissociation and charge transfer for promoting visible-light-driven molecular oxygen activation. *Appl. Catal. B Environ.* **2019**, *245*, 87–99. [CrossRef]
9. Kale, B.; Baeg, J.; Lee, S.; Chang, H.; Moon, S.; Lee, C. $CdIn_2S_4$ nanotubes and marigold nanostructures: A visible-light photocatalyst. *Adv. Funct. Mater.* **2006**, *16*, 1349–1354. [CrossRef]
10. Li, L.; Peng, S.; Wang, N.; Srinivasan, M.; Mhaisalkar, S.G.; Yu, Q.; Ramakrishna, S. A general strategy toward carbon cloth-based hierarchical films constructed by porous nanosheets for superior photocatalytic activity. *Small* **2015**, *11*, 2429–2436. [CrossRef] [PubMed]

11. Xue, C.; An, H.; Yan, X.; Li, J.; Yang, B.; Wei, J.; Yang, G. Spatial charge separation and transfer in ultrathin $CdIn_2S_4$/rGO nanosheet arrays decorated by ZnS quantum dots for efficient visible-light-driven hydrogen evolution. *Nano Energy* **2017**, *39*, 513–523. [CrossRef]
12. Mahadadalkar, M.; Gosavi, S.; Kale, B. Interstitial charge transfer pathways in a $TiO_2/CdIn_2S_4$ heterojunction photocatalyst for direct conversion of sunlight into fuel. *J. Mater. Chem. A* **2018**, *6*, 16064–16073. [CrossRef]
13. Chen, W.; Huang, T.; Hua, Y.; Liu, T.; Liu, X.; Chen, S. Hierarchical $CdIn_2S_4$ microspheres wrapped by mesoporous g-$C_3N_4$ ultrathin nanosheets with enhanced visible light driven photocatalytic reduction activity. *J. Hazard. Mater.* **2016**, *320*, 529–538. [CrossRef] [PubMed]
14. Zhang, Q.; Wang, J.; Ye, X.; Hui, Z.; Ye, L.; Wang, X.; Chen, S. Self-assembly of $CdS/CdIn_2S_4$ heterostructure with enhanced photocascade synthesis of schiff base compounds in an aromatic alcohols and nitrobenzene system with visible light. *ACS Appl. Mater. Interfaces* **2019**, *11*, 46735–46745. [CrossRef]
15. Peng, Z.; Jiang, Y.; Wang, X.; Zhang, R.; Xu, H.; Xiao, Y.; Jing, X.; Zhang, J.; Liu, Y.; Ni, L. Novel $CdIn_2S_4$ nano-octahedra/$TiO_2$ hollow hybrid heterostructure: In-situ synthesis, synergistic effect and enhanced dual-functional photocatalytic activities. *Ceram. Int.* **2019**, *45*, 15942–15953. [CrossRef]
16. Xiao, Y.; Peng, Z.; Zhang, S.; Jiang, Y.; Jing, X.; Yang, X.; Zhang, J.; Ni, L. Z-scheme $CdIn_2S_4$/BiOCl nanosheet face-to-face heterostructure: In-situ synthesis and enhanced interfacial charge transfer for high-efficient photocatalytic performance. *J. Mater. Sci.* **2019**, *54*, 9573–9590. [CrossRef]
17. Wang, J.; Yang, T.; He, R.; Xue, K.; Sun, R.; Wang, W.; Wang, J.; Yang, T.; Wang, Y. Silver-loaded $In_2S_3$-$CdIn_2S_4$@X(X=Ag, $Ag_3PO_4$, AgI) ternary heterostructure nanotubes treated by electron beam irradiation with enhanced photocatalytic activity. *Sci. Total Environ.* **2019**, *695*, 133884. [CrossRef]
18. Jiang, W.; Yin, X.; Xin, F.; Bi, Y.; Liu, Y.; Li, X. Preparation of $CdIn_2S_4$ microspheres and application for photocatalytic reduction of carbon dioxide. *Appl. Surf. Sci.* **2014**, *288*, 138–142. [CrossRef]
19. Wang, S.; Guan, B.; Lu, Y.; Lou, X. Formation of hierarchical $In_2S_3$-$CdIn_2S_4$ heterostructured nanotubes for efficient and stable visible light $CO_2$ reduction. *J. Am. Chem. Soc.* **2017**, *139*, 17305–17308. [CrossRef]
20. Bao, N.; Shen, L.; Takata, T.; Domen, K. Self-templated synthesis of nanoporous CdS nanostructures for highly efficient photocatalytic hydrogen production under visible light. *Chem. Mater.* **2008**, *20*, 110–117. [CrossRef]
21. Chen, Z.; Ren, Z.; Xu, J.; He, Y.; Xiao, G. Hydrothermal synthesis of Cu(X)-$CdIn_2S_4$ microspheres and photocatalytic reduction of aqueous Cr (VI) under visible light irradiation. *Mater. Lett.* **2014**, *117*, 17–20. [CrossRef]
22. Peng, Z.; Jiang, Y.; Xiao, Y.; Xu, H.; Zhang, W.; Ni, L. $CdIn_2S_4$ surface-decorated $Ta_3N_5$ core-shell heterostructure for improved spatial charge transfer: In-situ growth, synergistic effect and efficient dual-functional photocatalytic performance. *Appl. Surf. Sci.* **2019**, *487*, 1084–1095. [CrossRef]
23. Liu, C.; Li, X.; Li, J.; Zhou, Y.; Sun, L.; Wang, H.; Huo, P.; Ma, C.; Yan, Y. Fabricated 2D/2D $CdIn_2S_4$/N-rGO muti-heterostructure photocatalyst for enhanced photocatalytic activity. *Carbon* **2019**, *152*, 565–574. [CrossRef]
24. Liu, C.; Li, X.; Li, J.; Sun, L.; Zhou, Y.; Guan, J.; Wang, H.; Huo, P.; Ma, C.; Yan, Y. Carbon dots modifying sphere-flower $CdIn_2S_4$ on N-rGO sheet muti-dimensional photocatalyst for efficient visible degradation of 2,4-dichlorophenol. *J Taiwan Inst. Chem. E* **2019**, *99*, 142–153. [CrossRef]
25. Li, D.; Shi, F.; Jiang, D.; Chen, M.; Shi, W. $CdIn_2S_4$/g-$C_3N_4$ heterojunction photocatalysts: Enhanced photocatalytic performance and charge transfer mechanism. *RSC Adv.* **2017**, *7*, 231–237. [CrossRef]
26. Babu, B.; Koutavarapu, R.; Shim, J.; Yoo, K. Enhanced visible-light-driven photoelectrochemical and photocatalytic performance of Au-$SnO_2$ quantum dot-anchored g-$C_3N_4$ nanosheets. *Sep. Purif. Technol.* **2020**, *240*, 116652. [CrossRef]
27. Chen, W.; Huang, J.; He, Z.; Ji, X.; Zhang, Y.; Sun, H.; Wang, K.; Su, Z. Accelerated photocatalytic degradation of tetracycline hydrochloride over $CuAl_2O_4$/g-$C_3N_4$ p-n heterojunctions under visible light irradiation. *Sep. Purif. Technol.* **2021**, *277*, 119461. [CrossRef]
28. Xiao, T.; Tang, Z.; Yang, Y.; Tang, L.; Zhou, Y.; Zou, Z. In situ construction of hierarchical $WO_3$/g-$C_3N_4$ composite hollow microspheres as a Z-scheme photocatalyst for the degradation of antibiotics. *Appl. Catal. B Environ.* **2018**, *220*, 417–428. [CrossRef]
29. Chen, D.; Wu, S.; Fang, J.; Lu, S.; Zhou, G.; Feng, W.; Yang, F.; Chen, Y.; Fang, Z. A nanosheet-like $\alpha$-$Bi_2O_3$/g-$C_3N_4$ heterostructure modified by plasmonic metallic Bi and oxygen vacancies with high photodegradation activity of organic pollutants. *Sep. Purif. Technol.* **2018**, *193*, 232–241. [CrossRef]
30. Hong, Y.; Li, C.; Zhang, G.; Meng, Y.; Yin, B.; Zhao, Y.; Shi, W. Efficient and stable $Nb_2O_5$ modified g-$C_3N_4$ photocatalyst for removal of antibiotic pollutant. *Chem. Eng. J.* **2016**, *299*, 74–84. [CrossRef]
31. Xia, P.; Zhu, B.; Yu, J.; Cao, S.; Jaroniec, M. Ultra-thin nanosheet assemblies of graphitic carbon nitride for enhanced photocatalytic $CO_2$ reduction. *J. Mater. Chem. A* **2017**, *5*, 3230–3238. [CrossRef]
32. Xu, J.; Zhang, L.; Shi, R.; Zhu, Y. Chemical exfoliation of graphitic carbon nitride for efficient heterogeneous photocatalysis. *J. Mater. Chem. A* **2013**, *1*, 14766–14772. [CrossRef]
33. Lin, Q.; Li, L.; Liang, S.; Liu, M.; Bi, J.; Wu, L. Efficient synthesis of monolayer carbon nitride 2D nanosheet with tunable concentration and enhanced visible-light photocatalytic activities. *Appl. Catal. B Environ.* **2015**, *163*, 135–142. [CrossRef]
34. Yang, S.; Gong, Y.; Zhang, J.; Zhan, L.; Ma, L.; Fang, Z.; Vajtai, R.; Wang, X.; Ajayan, P. Exfoliated graphitic carbon nitride nanosheets as efficient catalysts for hydrogen evolution under visible light. *Adv. Mater.* **2013**, *25*, 2452–2456. [CrossRef]

35. Han, Q.; Wang, B.; Gao, J.; Cheng, Z.; Zhao, Y.; Zhang, Z.; Qu, L. Atomically thin mesoporous nanomesh of graphitic $C_3N_4$ for high-efficiency photocatalytic hydrogen evolution. *ACS Nano* **2016**, *10*, 2745–2751. [CrossRef] [PubMed]
36. Zhang, M.; Lai, C.; Li, B.; Huang, D.; Zeng, G.; Xu, P.; Qin, L.; Liu, S.; Liu, X.; Yi, H.; et al. Rational design 2D/2D BiOBr/CDs/g-$C_3N_4$ Z-scheme heterojunction photocatalyst with carbon dots as solid-state electron mediators for enhanced visible and NIR photocatalytic activity: Kinetics, intermediates, and mechanism insight. *J. Catal.* **2019**, *369*, 469–481. [CrossRef]
37. Su, T.; Shao, Q.; Qin, Z.; Guo, Z.; Wu, Z. Role of interfaces in two-dimensional photocatalyst for water splitting. *ACS Catal.* **2018**, *8*, 2253–2276. [CrossRef]
38. Yang, H.; Cao, R.; Sun, P.; Yin, J.; Zhang, S.; Xu, X. Constructing electrostatic self-assembled 2D/2D ultra-thin $ZnIn_2S_4$/protonated g-$C_3N_4$ heterojunctions for excellent photocatalytic performance under visible light. *Appl. Catal. B Environ.* **2019**, *256*, 117862 [CrossRef]
39. Xie, L.; Ni, J.; Tang, B.; He, G.; Chen, H. A self-assembled 2D/2D-type protonated carbon nitride-modified graphene oxide nanocomposite with improved photocatalytic activity. *Appl. Surf. Sci.* **2018**, *434*, 456–463. [CrossRef]
40. Tonda, S.; Kumar, S.; Bhardwaj, M.; Yadav, P.; Ogale, S. g-$C_3N_4$/NiAl-LDH 2D/2D hybrid heterojunction for high-performance photocatalytic reduction of $CO_2$ into renewable fuels. *ACS Appl. Mater. Interface* **2018**, *10*, 2667–2678. [CrossRef]
41. Ao, Y.; Wang, K.; Wang, P.; Wang, C.; Hou, J. Synthesis of novel 2D-2D pn hetero-junction BiOBr/$La_2Ti_2O_7$ composite photocatalyst with enhanced photocatalytic performance under both UV and visible light irradiation. *Appl. Catal. B Environ.* **2016**, *194*, 157–168. [CrossRef]
42. Ong, W.; Putri, L.; Tan, Y.; Tan, L.; Li, N.; Ng, Y.; Wen, X.; Chai, S. Unravelling charge carrier dynamics in protonated g-$C_3N_4$ interfaced with carbon nanodots as co-catalysts toward enhanced photocatalytic $CO_2$ reduction: A combined experimental and first-principles DFT study. *Nano Res.* **2017**, *10*, 1673–1696. [CrossRef]
43. Ong, W.; Tan, L.; Chai, S.; Yong, S.; Mohamed, A. Surface charge modification via protonation of graphitic carbon nitride (g-$C_3N_4$) for electrostatic self-assembly construction of 2D/2D reduced graphene oxide (rGO)/g-$C_3N_4$ nanostructures toward enhanced photocatalytic reduction of carbon dioxide to methane. *Nano Energy* **2015**, *13*, 757–770. [CrossRef]
44. Yang, M.; Xu, Y.; Lu, W.; Zeng, K.; Zhu, H.; Xu, Q.; Ho, G. Self-surface charge exfoliation and electrostatically coordinated 2D hetero-layered hybrids. *Nat. Commun.* **2017**, *8*, 14224. [CrossRef]
45. Han, Q.; Wang, B.; Zhao, Y.; Hu, C.; Qu, L. A graphitic-$C_3N_4$ seaweed architecture for enhanced hydrogen evolution. *Angew. Chem. Int. Ed.* **2015**, *54*, 11433–11437. [CrossRef]
46. Kumar, A.; Kumar, P.; Borkar, R.; Bansiwal, A.; Labhsetwar, N.; Jain, S.L. Metal-organic hybrid: Photoreduction of $CO_2$ using graphitic carbon nitride supported heteroleptic iridium complex under visible light irradiation. *Carbon* **2017**, *123*, 371–379. [CrossRef]
47. Yu, H.; Shi, R.; Zhao, Y.; Bian, T.; Zhao, Y.; Zhou, C.; Waterhouse, G.; Wu, L.; Tung, C.; Zhang, T. Alkali-assisted synthesis of nitrogen deficient graphitic carbon nitride with tunable band structures for efficient visible-light-driven hydrogen evolution. *Adv. Mater.* **2017**, *29*, 1605148. [CrossRef] [PubMed]
48. Li, Y.; Zhang, H.; Liu, P.; Wang, D.; Li, Y.; Zhao, H. Cross-linked g-$C_3N_4$/rGO nanocomposites with tunable band structure and enhanced visible light photocatalytic activity. *Small* **2013**, *9*, 3336–3344. [CrossRef]
49. Fan, L.; Guo, R. Fabrication of novel $CdIn_2S_4$ hollow spheres via a facile hydrothermal process. *J. Phys. Chem. C* **2008**, *112*, 10700–10706. [CrossRef]
50. Xiao, P.; Jiang, D.; Ju, L.; Jing, J.; Chen, M. Construction of RGO/$CdIn_2S_4$/g-$C_3N_4$ ternary hybrid with enhanced photocatalytic activity for the degradation of tetracycline hydrochloride. *Appl. Surf. Sci.* **2018**, *433*, 388–397. [CrossRef]
51. Di, J.; Xiong, J.; Li, H.; Liu, Z. Ultrathin 2D photocatalysts: Electronic-structure tailoring, hybridization, and applications. *Adv. Mater.* **2018**, *30*, 1704548. [CrossRef] [PubMed]
52. Han, Q.; Bai, X.; Man, Z.; He, H.; Li, L.; Hu, J.; Alsaedi, A.; Hayat, T.; Yu, Z.; Zhang, W.; et al. Convincing synthesis of atomically thin, single-crystalline $InVO_4$ sheets toward promoting highly selective and efficient solar conversion of $CO_2$ into CO. *J. Am. Chem. Soc.* **2019**, *141*, 4209–4213. [CrossRef] [PubMed]
53. Yin, H.; Fan, T.; Cao, Y.; Li, P.; Yao, X.; Liu, X. Construction of three-dimensional $MgIn_2S_4$ nanoflowers/ two-dimensional oxygen-doped g-$C_3N_4$ nanosheets direct Z-scheme heterojunctions for efficient Cr(VI) reduction: Insight into the role of superoxide radicals. *J. Hazard. Mater.* **2021**, *420*, 126567. [CrossRef] [PubMed]
54. Yi, X.; Ma, S.; Du, X.; Zhao, C.; Fu, H.; Wang, P.; Wang, C. The facile fabrication of 2D/3D Z-scheme g-$C_3N_4$/UiO-66 heterojunction with enhanced photocatalytic Cr (VI) reduction performance under white light. *Chem. Eng. J.* **2019**, *375*, 121944. [CrossRef]
55. Yin, H.; Cao, Y.; Fan, T.; Qiu, B.; Zhang, M.; Yao, J.; Li, P.; Liu, X.; Chen, S. Construction of carbon bridged $TiO_2$/CdS tandem Z-scheme heterojunctions toward efficient photocatalytic antibiotic degradation and Cr (VI) reduction. *J. Alloys Compd.* **2020**, *824*, 153915. [CrossRef]

nanomaterials

MDPI

Article

# Magnetic Properties in Mn-Doped $\delta$-MoN: A Systematic Density Functional Theory Study

Keda Wang [1,2], Jing Yu [1,2,*], Caixia Chi [1,2] and Guiling Zhang [3,*]

1 College of Food and Pharmaceutical Engineering, Suihua University, Suihua 152061, China; shxywangkd@163.com (K.W.); chicaixia617@163.com (C.C.)
2 Key Laboratory of Environmental Catalysis and Energy Storage Materials of Heilongjiang Province, Suihua 152061, China
3 College of Materials Science and Chemical Engineering, Harbin University of Science and Technology, Harbin 150080, China
* Correspondence: yujinghrb@163.com (J.Y.); zglhrb@163.com (G.Z.)

**Abstract:** Due to the potential applications of transition metal nitrides in modern electronic and spintronic devices, we have systematically studied the magnetic properties of $\delta$-MoN induced by the Mn dopant, with the goal of identifying the origin of magnetism and figuring out the magnetic coupling mechanism between the Mn dopants. Based on the density functional theory, one Mn atom doped at different Mo sites (2*a* and 6*c* in the International Tables) in the unit cell of $\delta$-MoN was firstly studied. It was found that the Mn dopant located at the 2*a* or 6*c* site leads to significant spin splitting of the density of states, suggesting that the Mn doping induces magnetism in $\delta$-MoN. The calculations were then extended to a $2 \times 1 \times 2$ supercell, which contains two impurity Mn atoms. Detailed analysis reveals that the different couplings of the Mn–Mn pair cannot be simply attributed to the different Mn–Mn distances but are closely related to the electronic processes that take place in the segment (–N– or –N–Mo–N–) that connects two Mn dopants. The mechanisms responsible for the FM/AFM coupling of the Mn–Mn pairs are the superexchange and the *p*–*d* exchange mediated by the N atoms, and the *d*–*d* coupling between the host Mo atom and the Mn dopant.

**Keywords:** Mn-doped; $\delta$-MoN; magnetic properties; theoretical study

## 1. Introduction

Transition metal nitrides, by virtue of their unique electronic structure and exact nature of bonding (mixed metallic, ionic, and covalent bonding), possess excellent physical properties, such as high hardness [1,2], high melting point [3], and high electronic conductivity [4,5], which have made them attract tremendous attentions in recent years. In order to enhance the performance of this class of materials, much effort has been devoted to investigating the possibility of improving the energetic [6–8], mechanical [9,10], and electrical [11] properties by doping foreign atoms. Recent research has confirmed that introduction of foreign transition metal atoms (for instance, Fe, V, Mn, Cr, etc.) into the host material could effectively enhance the magnetism of transition metal nitrides, such as ScN [12,13], CrN [14,15], and ZnN [16,17]. It has been found that the V, Cr, Mn, Fe, Co, and Ni doping can induce magnetism in ScN [12], and the Mn-doped ScN is a dilute magnetic semiconductor with the Curie temperature above 400 K [18,19]. The CrN undergoes a relative increase in the magnetic order with the substitution of Mn atom [14], and V dopant in CrN could introduce holes into the host material and produce a series of inhomogeneous magnetic or electronic states [15]. The ZnN doped with Cr impurity is found to exhibit half-metallic ferromagnetism, which is a good candidate for spintronic applications [16]. It is known that the enhancement or the inducement of magnetism caused by the transition metal doping is closely related to the modification of electronic states; so, if we want to clearly understand the origin of magnetism, it is necessary to investigate the modified electronic structure of the doped transition metal nitrides.

As a typical type of transition metal nitrides, molybdenum nitrides are known to have a set of interesting properties, such as low compressibility [20], high melting point [21], and excellent catalytic activity [22,23], which are very attractive for a wide range of technological applications. Furthermore, their multiple transport properties (including the metallic and magnetic properties [24,25] and, especially, the high $T_C$ superconductivity [26,27]), also support the potential applications in electronic devices. Molybdenum nitrides can crystallize in different phases, including a stoichiometric structure, hexagonal $\delta$-MoN, and two nonstoichiometric structures, cubic $\gamma$-$Mo_2N$ and tetragonal $\beta$-$Mo_2N$ [27–30]. In particular, hexagonal $\delta$-MoN is considered the hardest superconducting metal nitride [1], whose bulk modulus is measured to be 345 GPa [20] and $T_C$ is up to 12–15 K [31,32]. A number of experimental works have been performed which mainly focus on the synthesis of pure $\delta$-MoN with desired superconducting properties [27,31,33]. Only a few theoretical studies have been devoted to $\delta$-MoN with the purpose of investigating its structural, electronic, and mechanical properties [34,35]. Up to now, research concerning the magnetism of $\delta$-MoN has been really scarce. Our previous work has theoretically studied the Cr-doping effect on the magnetic and spin transport properties of $\delta$-MoN [36]. However, to the best of our knowledge, no theoretical works have been related to the magnetism of $\delta$-MoN induced by the Mn dopant.

In this paper, we systematically studied the electronic and magnetic properties of the Mn-doped $\delta$-MoN, with the goal of identifying the origin of magnetism and figuring out the magnetic coupling mechanism between the Mn dopants. This paper is organized as follows: Firstly, one Mn atom, with $d^5$ structure, was used as dopant to substitute one Mo atom located at two different nonequivalent positions (2*a* and 6*c* in the International Tables) in the unit cell, respectively, to investigate the Mn-doping effects on the structural, electronic, and magnetic properties of $\delta$-MoN. Then, the calculations were extended to a large supercell with two substitutional Mn atoms to explore the favorable coupling between the Mn dopants. Based on the detailed analysis of magnetic ordering, electronic structure, and spin charge density distribution, the origin of magnetism and the coupling mechanism between the Mn dopants were revealed.

## 2. Computational Details

The calculations were performed using the projector augmented wave (PAW) pseudopotentials as implemented in Vienna ab initio simulation package (VASP) [37]. The exchange correlation functional was treated by Perdew–Burke–Ernzerhof form generalized gradient approximation (GGA-PBE) [38]. Considering the electron correlations in the transition metal *d* shell, we introduce the GGA + $U$ method by a simplified approach of Dudarev et al. [39], where the effective Hubbard parameter $U_{eff} = U - J$. The Hubbard parameter $U$ measures the increase in energy caused by placing an additional electron into a particular site, and $J$ is a screened Stoner-like exchange parameter. Here, we choose $U_{eff}$ = 4 eV for Mn-3*d* electrons [40,41] and $U_{eff}$ = 2 eV for Mo-4*d* electrons [42], which are taken from the literature. The tetrahedron method with Blöchl corrections was used to determine the partial occupancies for setting each wave function [43]. The Kohn–Sham orbitals were expanded in a plane-wave basis with a cutoff energy of 500 eV. The Brillouin-zone integration was performed on well-converged $7 \times 7 \times 7$ and $3 \times 5 \times 3$ Monkhorst–Pack *k*-point meshes for $1 \times 1 \times 1$ unit cell and $2 \times 1 \times 2$ supercell, respectively. Both the atomic positions and the unit cell parameters were fully relaxed until the forces on each atom were less than 0.01 eV·Å$^{-1}$. All the calculations were performed based on the optimized geometries.

## 3. Results and Discussions

We begin our discussion with the structural, electronic, and magnetic properties of $\delta$-MoN, with a single substitutional Mn atom. Hexagonal $\delta$-MoN crystallizes in a distorted NiAs-type structure with a space group of $P6_3mc$ (186), which has been determined by experiment investigations [20,32,44], as well as theoretical calculations [34]. In $\delta$-MoN, Mo atoms have two kinds of lattice points: one is labeled 2*a* and the other is labeled 6*c* in the International Tables. Each unit cell contains eight Mo atoms and eight N atoms. The unit cell of $\delta$-MoN is presented in Figure 1a, in which the Mo atoms (green balls) are numbered to facilitate our discussion. $Mo_1$ and $Mo_2$ are in the 2*a* (0, 0, 0) sites, while $Mo_3$–$Mo_8$ are in the 6*c* (0.5082, 0.0165, −0.0064) sites [32]. To find the preferable location of Mn impurity in $\delta$-MoN, we substitute one Mo atom with one Mn atom at two different nonequivalent Mo sites (2*a* and 6*c*), respectively. Here, the side $Mo_1$ atom (representing for the 2*a* site) and the central $Mo_6$ atom (representing for the 6*c* site) were substituted, respectively, by one Mn atom to carry out VASP calculations for investigating the Mn-doping effects on the electronic and magnetic properties of $\delta$-MoN. Such a Mn-doped unit cell contains one Mn atom, seven Mo atoms, and eight N atoms, corresponding to the Mn doping concentration of 12.5%. Following that, the two doped systems with Mn doping at 2*a* and 6*c* sites are written as Mn-MoN(2*a*) and Mn-MoN(6*c*), respectively.

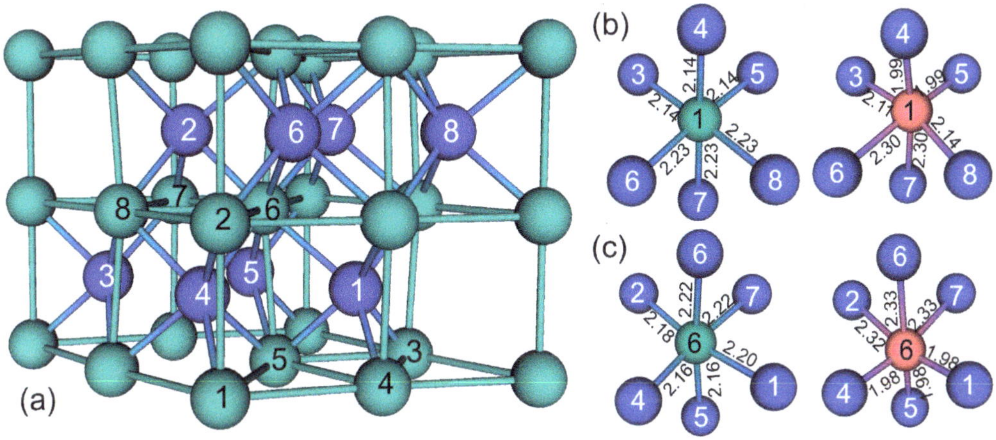


**Figure 1.** (**a**) Schematic structure of the unit cell of $\delta$-MoN, and optimized bond lengths of (**b**) $Mn_1$–N and (**c**) $Mn_6$–N in Mn-MoN(2*a*) and Mn-MoN(6*c*), respectively. For comparison, the relaxed bond lengths of $Mo_1$–N and $Mo_6$–N in the pure $\delta$-MoN are also presented in (**b**,**c**), respectively. The pink, green, and blue balls denote the Mn, Mo, and N atoms, respectively.

Firstly, we performed a geometry optimization for the pure $\delta$-MoN. It is found that the calculated lattice constants of $\delta$-MoN ($a$ = 5.757 Å, $c$ = 5.668 Å) are in good agreement with the experimental values [32] with deviations within 0.9%. On the basis of the equilibrium structure of $\delta$-MoN, the cells of Mn-MoN(2*a*) and Mn-MoN(6*c*) were constructed and relaxed to the minimum energy configurations. The optimized lattice parameters and bond lengths are presented in Table 1 and Figure 1b,c, respectively. The presence of Mn has a marked effect upon the geometry of $\delta$-MoN. After relaxation, the substitutional $Mn_1/Mn_6$ atom is found to pull some N atoms closer and push some N atoms farther, suggesting that the lattice structure at the Mn site is distorted and forms a new local structure. The bond lengths of $Mn_1$–N and $Mn_6$–N are in the range of 1.99–2.30 Å and 1.98–2.33 Å for Mn-MoN(2*a*) and Mn-MoN(6*c*), respectively, which are different from those of $Mo_1$–N (2.14–2.23 Å) and $Mo_6$–N (2.16–2.22 Å) in the pure $\delta$-MoN. The shortest bond lengths occur at $Mn_1$–$N_{4,5}$ (1.99 Å) for Mn-MoN(2*a*) and $Mn_6$–$N_{1,4,5}$ (1.98 Å) for Mn-MoN(6*c*), which are very close to the sum of the covalent radius of $Mn^{3+}$ and $N^{3-}$ (1.92 Å), indicating a covalent bonding feature between Mn and its nearest N atoms. The average bond lengths of $Mn_1$–N and $Mn_6$–N are 2.14 Å and 2.15 Å, respectively, which are about 0.05 Å and 0.04 Å shorter than those of $Mo_1$–N and $Mo_6$–N. The shortening of the average $Mn_1$–N/$Mn_6$–N bond length consequently leads to the shrinkage of the local structure, and thus results in the decrease of the lattice parameters of Mn-MoN(2*a*)/Mn-MoN(6*c*), compared to that of $\delta$-MoN (cf. Table 1). This case is probably due to the fact that the radius of $Mn^{3+}$ is smaller than that of $Mo^{3+}$.

**Table 1.** Optimized lattice parameters of $\delta$-MoN before and after Mn substitution, and formation energies of Mn-MoN(2*a*) and Mn-MoN(6*c*).

| Configurations | *a* (Å) | *c* (Å) | *c*/*a* | Volume (Å$^3$) | $E_f$ (eV) |
|---|---|---|---|---|---|
| $\delta$-MoN | 5.757 | 5.668 | 0.985 | 162.7 | |
| Mn-MoN(2*a*) | 5.726 | 5.626 | 0.983 | 159.5 | −1.71 |
| Mn-MoN(6*c*) | 5.714 | 5.644 | 0.988 | 159.6 | −1.68 |

The charge transfer between Mn dopant and $\delta$-MoN can be examined by the charge density difference, which can be determined by subtracting the charge densities of pristine $\delta$-MoN and isolated Mn atom from the total charge density of the Mn-doped system. Figure 2 shows the charge density difference isosurfaces of Mn-MoN(2*a*) and Mn-MoN(6*c*), which describe redistribution of the valence charge density of atoms caused by chemistry bonding. The yellow and light blue regions refer to electron accumulation and depletion, respectively. Evident charge depletion around the Mn atom can be observed from the isosurfaces. Integrating the density of states up to the Fermi energy, the estimated charges transferred from Mn to $\delta$-MoN are 1.22 and 1.12 electrons for Mn-MoN(2*a*) and Mn-MoN(6*c*), respectively. The charge accumulation between Mn and N suggests a strong covalent character of the Mn–N bonds, which is in line with the calculated bond lengths, as discussed above. This may be generated by the *p*–*d* hybridization between Mn and N, which can be further confirmed by the partial density of states.

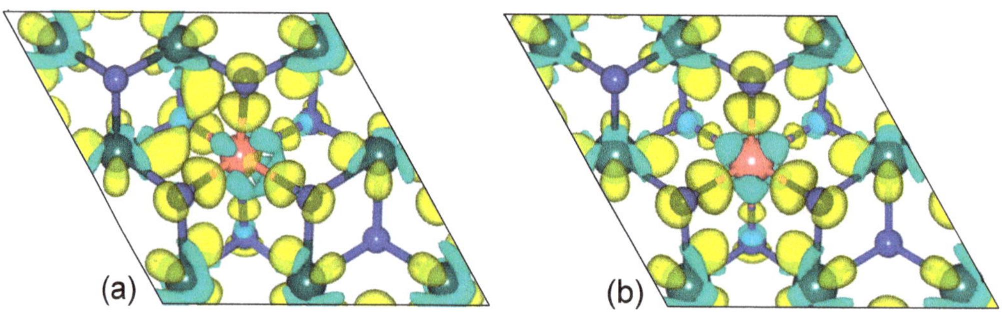


**Figure 2.** Charge density difference isosurfaces for (**a**) Mn-MoN(2*a*) and (**b**) Mn-MoN(6*c*). The pink, green, and blue balls denote the Mn, Mo, and N atoms, respectively.

The preferable site for the Mn dopant in $\delta$-MoN can be determined by estimating the formation energies of different doping configurations. The formation energies of Mn-MoN(2*a*) and Mn-MoN(6*c*), $E_f$, are calculated by the following formula [45]:

$$E_f = E_{tot}(\text{doped}) - E_{tot}(\text{pure}) + m\mu_{Mo} - n\mu_{Mn} \quad (1)$$

where $E_{tot}$(doped) and $E_{tot}$(pure) are the total energies of the Mn-doped and undoped $\delta$-MoN, respectively. $\mu_{Mo}(\mu_{Mn})$ is the chemical potential of the Mo (Mn) atom. The $n$, $m$ are the numbers of the doped Mn atoms and the substituted Mo atoms, respectively. The calculated formation energies of the Mn-doped systems are listed in Table 1. Note that the formation energies of Mn-MoN(2*a*) and Mn-MoN(6*c*) decrease relative to the pure $\delta$-MoN. A negative energy means that the formation of Mn-MoN(2*a*)/Mn-MoN(6*c*) is spontaneous, that is, the Mn dopant will easily occupy the 2*a*/6*c* Mo lattice site in $\delta$-MoN. The relative stability of two Mn-doped configurations can be evaluated by comparing their formation energies. The smaller the formation energy, the more stable the structure. As seen from Table 1, the formation energy of Mn-MoN(2*a*) is slightly lower than that of Mn-MoN(6*c*). It seems that the Mn dopant prefers to substitute the 2*a* Mo lattice site in δ-MoN, and Mn-MoN(2*a*) is a more energetically favorable structure.

To investigate the effect of Mn dopant on the modification of electronic structure, the spin-polarized density of states (DOS) were calculated for both of the undoped and doped systems, as given in Figure 3. The characteristic of the density of states around the Femi level demonstrates the metallic nature of these systems. It is seen that the majority and minority spin carriers of the pure δ-MoN exhibit mirror symmetry, demonstrating its nonmagnetic characteristic. However, evident spin-polarization of the density of states is observed in the Mn-doped systems. This asymmetrical distribution of the wave functions of the spin-up and spin-down channels on the total DOS of Mn-MoN(2*a*) and Mn-MoN(6*c*) suggests that the Mn dopant induces magnetism in δ-MoN. The Mn substitutions at 2*a* and 6*c* sites result in 2.74 and 2.56 $\mu_B$ total magnetic moments for Mn-MoN(2*a*) and Mn-MoN(6*c*), respectively. The partial DOS analysis reveals that the magnetism of the Mn-doped systems is mainly due to the presence of Mn impurity in the δ-MoN host, as well as a partial contribution from the neighboring Mo and N atoms around the dopant. A local magnetic moment of about 4 $\mu_B$ (4.03 $\mu_B$ for $Mn_1$ and 3.99 $\mu_B$ for $Mn_6$) per Mn atom obtained from the calculations indicates that the Mn ion presents a $Mn^{3+}$ valence state ($3d^4$). It is notable from the partial DOS that significant hybridization occurs between the orbitals of Mn-3*d*, Mo-4*d*, and N-2*p*, which is a good proof of the much more delocalized Mn impurity bands in comparison to an isolated Mn atom. The Mn-3*d* states spread throughout the valence band and the conduction band, with almost completely filled majority *d* bands and nearly unoccupied minority *d* bands. The strong interactions between Mn and Mo induce evident spin polarization of the Mo atoms with large magnetization. The Mo atoms adjacent to the Mn dopant get as large as −0.51 and −0.47 $\mu_B$-induced magnetic moments for Mn-MoN(2*a*) and Mn-MoN(6*c*), respectively. The neighboring N atoms to the Mn dopant are also slightly spin-polarized with small magnetic moments of −0.03 to 0.02 $\mu_B$.

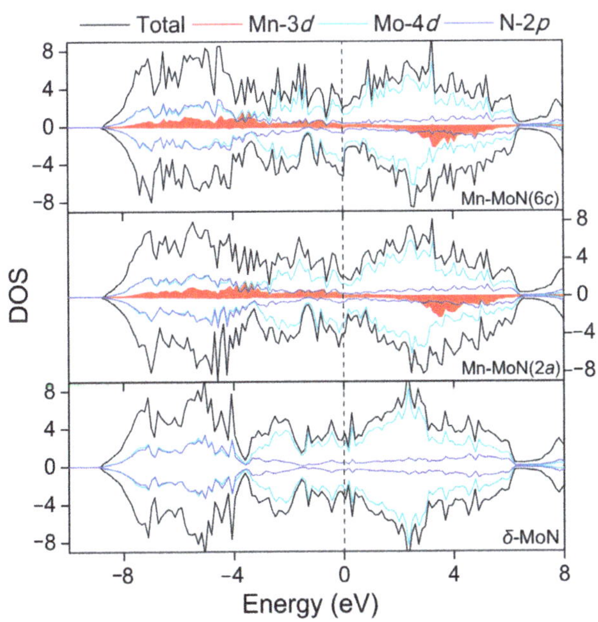


**Figure 3.** Total and partial DOS of δ-MoN, Mn-MoN(2*a*), and Mn-MoN(6*c*). The Fermi level is set as zero.

It is necessary to investigate the favorable coupling between the Mn dopants. Thus, we extend the analysis to a pair of Mn atoms in δ-MoN. Here, a 2 × 1 × 2 δ-MoN supercell, as shown in Figure 4, was used to investigate the magnetic coupling between two impurity Mn atoms. This supercell contains 32 formula units of MoN, in which two of the Mo atoms are replaced by two magnetic Mn atoms, to form a supercell containing 2 Mn atoms, 30 Mo atoms, and 32 N atoms, amounting to 6.25% Mn doping concentration. Such a large supercell adopted in the calculations allows us to simulate various distributions of two Mn dopants. In brief, we fixed one dopant Mn atom and varied the possible positions of the second Mn atom. As shown in Figure 4, the first Mn atom is fixed at a Mo lattice site (marked 0), and the second Mn atom changes its doping position from 1 to 11 to form the Mn–Mn pair in the supercell. The substituted Mo atoms numbered in 1–11 follow the sequence of the distance to $Mo_0$. This generates eleven different doping configurations, in which the Mn–Mn separation within the supercell varies from 2.73 to 8.08 Å. Hereafter, these arrangements are referred to as the (0, *n*) configurations.

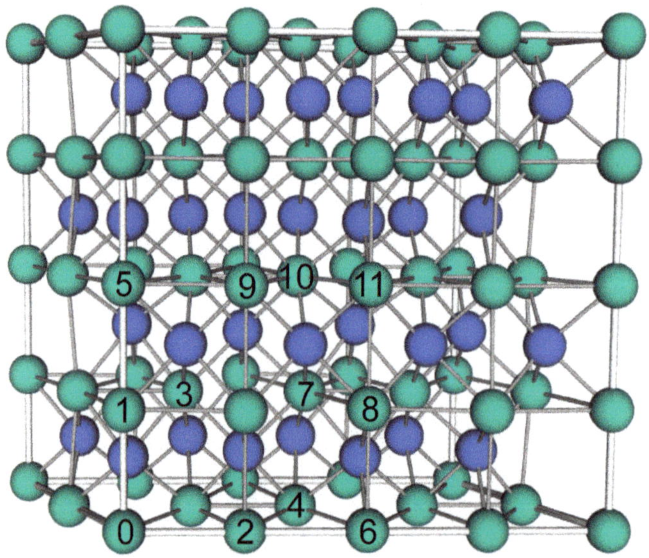


**Figure 4.** A 2 × 1 × 2 supercell structure of δ-MoN. With respect to one Mn atom (labeled as 0), we select eleven nonequivalent positions for the other Mn atom (as numbered from 1 to 11) based on the symmetry of the supercell. The green and blue balls denote the Mo and N atoms, respectively.

After geometry optimization, we calculated the formation energies for all the eleven nonequivalent configurations based on Equation (1), and the results are presented in the eighth column of Table 2. The formation energy for the single-doped system, in which is one single Mn atom doped at 0 Mo lattice site, is also calculated for comparison. It is observed from Table 2 that the formation energies of these double-doped systems vary from −3.06 to −3.48 eV, which is much lower than that of the single-doped system (−1.73 eV). This indicates that the formation energy of Mn decreases with its doping concentration, and it is highly favorable to form Mn substitution in δ-MoN.

**Table 2.** The relaxed Mn–Mn distance ($d$), the energy difference ($\Delta E$) between FM and AFM states, the energy relative to the ground state ($\Delta E_{ground}$), the total magnetic moment of the supercell ($M_{total}$), the magnetic moments of two Mn dopants ($M_1$ and $M_2$), and the formation energies ($E_f$) for various double Mn-doped configurations.

| Configurations | $d$ (Å) | $\Delta E$ (meV) | $\Delta E_{ground}$ (meV) | $M_{total}$ ($\mu_B$) | $M_1/M_2$ ($\mu_B$) | Coupling | $E_f$ (eV) |
|---|---|---|---|---|---|---|---|
| (0, 1) | 2.73 | 44 | 283 | 4.49 | 3.88/4.01 | FM | −3.20 |
| (0, 2) | 3.18 | −46 | 422 | −0.34 | 4.03/−4.05 | AFM | −3.06 |
| (0, 3) | 3.88 | −52 | 265 | −0.41 | 4.01/−4.05 | AFM | −3.22 |
| (0, 4) | 5.01 | −39 | 236 | 0.07 | 4.08/−4.06 | AFM | −3.25 |
| (0, 5) | 5.65 | 11 | 210 | 3.93 | 3.98/3.98 | FM | −3.27 |
| (0, 6) | 5.77 | 382 | 68 | 3.67 | 3.99/3.99 | FM | −3.41 |
| (0, 7) | 6.07 | −19 | 381 | 0.05 | 4.05/−4.06 | AFM | −3.10 |
| (0, 8) | 6.44 | 77 | 0 | 3.38 | 4.00/4.00 | FM | −3.48 |
| (0, 9) | 6.47 | −15 | 376 | −0.35 | 4.03/−4.06 | AFM | −3.11 |
| (0, 10) | 7.25 | −36 | 220 | −0.96 | 4.00/−4.03 | AFM | −3.26 |
| (0, 11) | 8.08 | 54 | 146 | 3.50 | 3.99/3.99 | FM | −3.34 |

The interactions responsible for the magnetic ground state of the double Mn-doped systems can be theoretically estimated by calculating the interatomic-exchange constant $J$, by making use of the total energies in the ordered ferromagnetic (FM) and antiferromagnetic (AFM) structures, $E_{FM}$ and $E_{AFM}$. Based on the nearest-neighbor Heisenberg model, the value of $J$ can be approximately deduced by $\Delta E = 4JS^2$, where $\Delta E = E_{AFM} - E_{FM}$ and $S$ is the net spin per Mn dopant [46]. For a positive exchange parameter $J$, the ground state is FM, while the ground state is AFM at a negative exchange parameter $J$. Table 2 lists the main results of our work concerning the Mn doping for all the eleven (0, $n$) configurations, and the magnetic moment on each Mn atom is presented in the sixth column. It is seen that the magnetic moments of Mn atoms are almost independent of Mn distributions and the values are close to 4 $\mu_B$, implying that each Mn atom doping in the supercell possesses about +3e charge with a spin $S = 2$. Accordingly, the interatomic-exchange constants $J$ were calculated and the variation of $J$ value with the Mn–Mn distance is plotted in Figure 5. Interestingly, a complex oscillatory behavior of the $J$ value as a function of Mn–Mn distance is observed. Among the eleven configurations, five positions ((0, 1), (0, 5), (0, 6), (0, 8), (0, 11)) result in FM and six positions ((0, 2), (0, 3), (0, 4), (0, 7), (0, 9), (0, 10)) result in AFM coupling between two dopant Mn atoms. The lowest energy configuration is found to be an FM state with a $J$ value of 4.8 meV and Mn–Mn distance of 6.44 Å (configuration (0, 8)). Other configurations are higher than the ground state in energy by 68 to 422 meV (cf. Table 2). The oscillation of the $J$ values in Figure 5 indicates that the different magnetic couplings of the Mn–Mn pair cannot be simply attributed to the different Mn–Mn distances. As for the case of FM orientation, the maximum $J$ coupling is reached at the (0, 6) configuration with the $Mn_0$–$Mn_6$ separation of 5.77 Å. Although this Mn–Mn distance is longer than the nearest $Mn_0$–$Mn_1$ distance by 3.04 Å, the coupling strength of $Mn_0$–$Mn_6$ ($J$ = 23.9 meV) is about 8.5 times stronger than that of $Mn_0$–$Mn_1$ ($J$ = 2.8 meV). Compared with the FM orientation, the oscillation magnitude of $J$ value for the AFM orientation is relatively weaker, and the largest $J$ value (−3.3 meV) is achieved at the third-nearest $Mn_0$–$Mn_3$ configuration in which the Mn atoms are separated by 3.88 Å. The above evidence suggests that long-range magnetic coupling exists between the Mn–Mn pair in the supercell. The delocalized character of the Mn-3$d$ states may offer some clues for this long-range magnetic interaction between the Mn dopants. As can be seen from the partial DOS of the double-doped systems in Figure 6, the Mn-3$d$ states are relatively delocalized and strongly hybridize with the electronic states of the host atoms in the supercell, which translates into a long-range effective interaction between the dopant states.

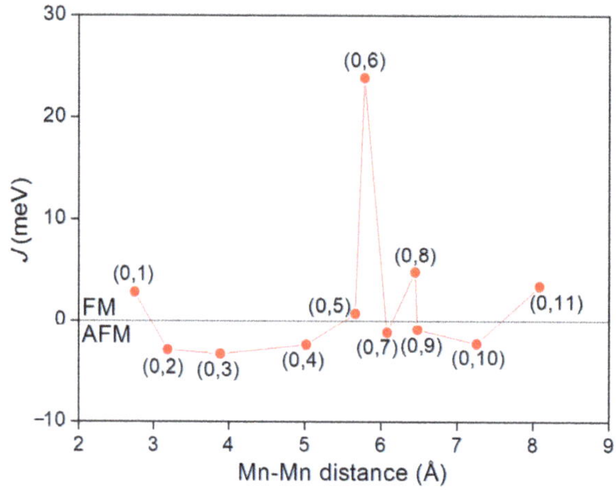


**Figure 5.** Variation of $J$ value as a function of Mn–Mn distance. The positive value of $J$ represents the FM state, while the negative value represents the AFM state.

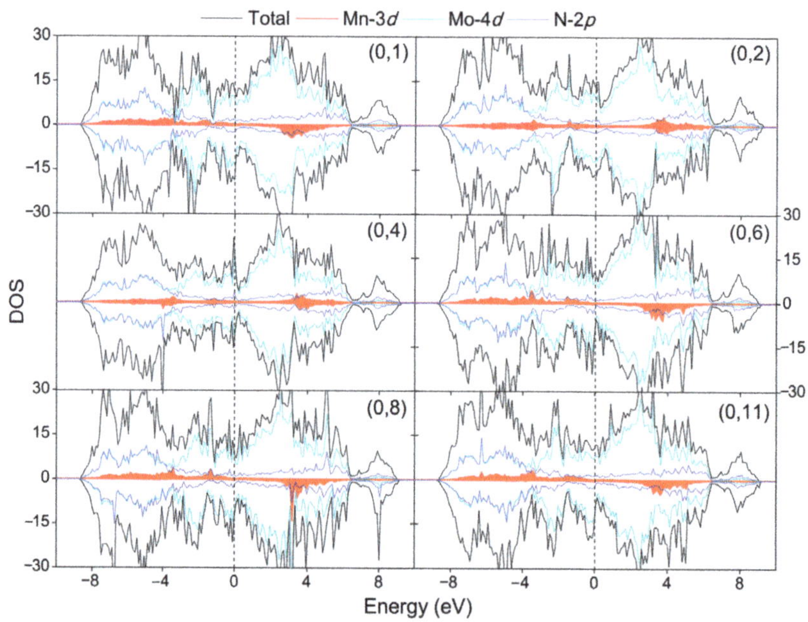


**Figure 6.** Total and partial DOS of the selected double-doped systems. The Fermi level is set as zero.

Ferromagnetic materials are characterized by their coercivity ($H_c$), the magnetic field necessary to reduce the magnetization to zero, and their Curie temperature ($T_C$), the temperature at which random motions cause the magnetization to vanish [47]. Strong spin exchange coupling is a prerequisite for high coercivity [48]. This is because strong exchange interaction makes it hard to flip the spins under an applied magnetic field [48,49]. The strength of Mn−Mn spin exchange can be rationalized by the parameter $J$. A large positive $J$ value (such as 23.9 meV of (0, 6) configuration) indicates a strong FM spin exchange coupling of the Mn–Mn pair, which may lead to a high coercivity. The Curie temperature

$T_C$ can be roughly estimated by means of the mean-field approximation and calculated according to the following equation [50]:

$$T_C = \frac{2\Delta E}{3k_B} \quad (2)$$

where $\Delta E$ is the total energy difference between AFM and FM states and $k_B$ is the Boltzmann constant. Using Equation (2), the $T_C$ of (0, 1), (0, 6), (0, 8) and (0, 11) configurations are estimated to be higher than the room temperature; especially, the $T_C$ of (0, 6) configuration is up to 2951 K. Such high Curie temperatures make the material promising for practical applications.

To gain more insight into the magnetic nature of the double Mn-doped systems, we calculated the spin charge density distributions, as shown in Figure 7. For the (0, 1) configuration, the two Mn dopants are ferromagnetically coupled to each other with the nearest Mn–Mn distance of 2.73 Å considered in our calculations. We find that this Mn–Mn distance is comparable to that of the free $Mn_2$ cluster (2.62 Å) [51]. Thus, a direct interaction of Mn–Mn *d*-orbital core spins is nonnegligible in the (0, 1) configuration, which gives rise to a bonding state between $Mn_0$ and $Mn_1$ (cf. Figure 7). Furthermore, the three N atoms, which simultaneously connect $Mn_0$ and $Mn_1$, contribute their 2*p* orbitals to form *p*–*d* hybridization with two Mn dopants. Evident overlap of N-2*p* and Mn-3*d* states could be observed from the partial DOS in Figure 6, demonstrating an indirect Mn–Mn exchange interaction mediated by the N atoms. This *p*–*d* exchange interaction between N and dopant Mn can also be confirmed by the magnetic moment of about $-0.05\ \mu_B$ for the mediated N atoms. According to the Zener model, the direct interaction between *d* orbitals of adjacent Mn atoms tends to result in an AFM configuration of the *d* spins. Only when the indirect interaction dominates over the direct coupling between adjacent *d* orbitals is ferromagnetism present [52]. Therefore, the coupling of the $Mn_0$–$Mn_1$ pair is determined by a competition between the direct *d*–*d* interaction and the indirect *p*–*d* exchange. Due to the fact that $Mn_0$ and $Mn_1$ are ferromagnetically coupled, the indirect interaction of *d* states on different Mn atoms mediated by the *p* states of N atom is dominant in the (0, 1) configuration.

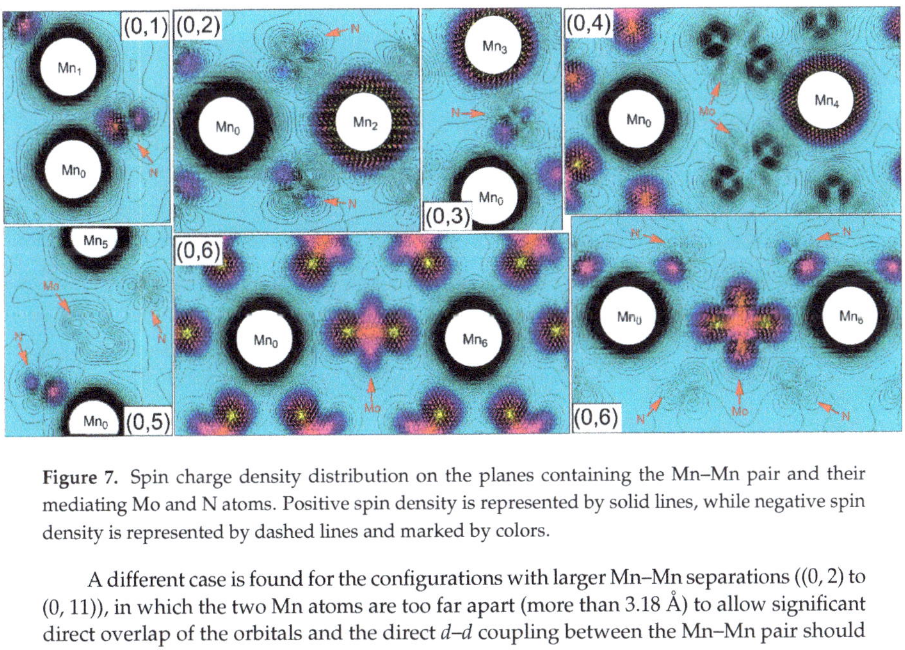


**Figure 7.** Spin charge density distribution on the planes containing the Mn–Mn pair and their mediating Mo and N atoms. Positive spin density is represented by solid lines, while negative spin density is represented by dashed lines and marked by colors.

A different case is found for the configurations with larger Mn–Mn separations ((0, 2) to (0, 11)), in which the two Mn atoms are too far apart (more than 3.18 Å) to allow significant direct overlap of the orbitals and the direct *d*–*d* coupling between the Mn–Mn pair should be ignored. In such case, the key role underlying the mediated Mn–Mn interaction (FM or AFM) in these configurations should be sought in the electronic processes that take place in the segments connecting two Mn dopants. Here, the mediating segments for the Mn–Mn pairs in the (0, 2) to (0, 11) configurations can be divided into the following two cases:

$$\text{Mn-N-Mn}, \quad (1)$$

and

$$\text{Mn-N-Mo-N-Mn} \quad (2)$$

That is, in segment (1) the mediation takes place only by N atom, while in segment (2) the mediated segment is N–Mo–N.

We first focus on the case of the Mn–Mn pair mediated only by the N atom (i.e., two Mn atoms sharing a common N neighbor, forming the segment of Mn–N–Mn), which refers to the configurations of (0, 2) and (0, 3). An indirect Mn–Mn exchange interaction mediated by the N atom is dominant in the two configurations. As seen from Figure 7, the spin polarizations of the mediating N atoms in the two configurations are antiparallel to those of the Mn atoms, and *p* character of the spin-polarized N orbitals is evident. The mediating N atom provides two different *p*-orbitals to hybridize with $Mn_0$ and $Mn_2$ (or $Mn_0$ and $Mn_3$), respectively. That is, the spin-up N *p*-orbital makes a bond with a spin-down Mn orbital, and the spin-down N *p*-orbital bonds to a spin-up Mn orbital. Superexchange is naturally developed between the Mn dopants along the segment of Mn–N–Mn, thus resulting in the AFM alignment of $Mn_0$ and $Mn_2$ (or $Mn_0$ and $Mn_3$). Superexchange is a mechanism which describes an interaction between moments on atoms too far apart to be connected by direct exchange but coupled over a relatively long distance through the mediating atoms [53]. The extent to which a mediating atom orbital contributes to the overall interaction is governed by the magnitude of its orbital overlap with the coupled metal orbitals, which is influenced by the factors such as the symmetry and the distance between the mediating atom and metal. The weaker magnetic coupling of $Mn_0$–$Mn_2$ ($J$ = –2.9 meV) as compared to that of $Mn_0$–$Mn_3$ ($J$ = –3.3 meV) is mainly attributed to the nearly 90° exchange path ($\angle Mn_0\text{–}N\text{–}Mn_2 \approx 95^\circ$) and the longer Mn–N distances in the mediating segment (the bond lengths of $Mn_0$–N and N–$Mn_2$ are about 0.12 Å and 0.07 Å longer than those of $Mn_0$–N and N–$Mn_3$, respectively).

The situation is more complicated in the case of the mediating segment (2), i.e., Mn–N–Mo–N–Mn, which corresponds to the configurations of (0, 4) to (0, 11). In this case, the mediated Mn–Mn interaction is mainly due to the electronic process that takes place in the segment of N–Mo–N connecting two Mn atoms. Detailed analysis reveals that the presence of Mo atom in the segment plays a critical role in mediating the Mn–Mn magnetic interaction. In fact, owing to the anisotropy of electronic structure and the directional nature of chemical bonding, the magnetic coupling of the Mn–Mn pair differs from one configuration to another. Here, we take (0, 4), (0, 5), and (0, 6) configurations as examples to illustrate the magnetic coupling mechanism of the Mn–Mn pair mediated by the segment of N–Mo–N.

We first pay our attention to the (0, 6) configuration, in which the two Mn atoms are strongly coupled with each other with a relatively large $J$ value of 23.9 meV, indicating a strong FM interaction of $Mn_0$–$Mn_6$ pair. From Figure 4, we notice that the two Mn dopants in the (0, 6) configuration are both located at the *ab* plane of the supercell, and each substitutional Mn atom is surrounded by six Mo atoms, which forms a slightly deformed hexagon structure. It is evident from Figure 7 that the six surrounding Mo neighbors are all antiferromagnetically coupled to the Mn dopant, and notable magnetic moments locate at the Mo sites, giving rise to an AFM order in the neighborhood of the Mn atom. This AFM environment around the Mn dopant is the origin of the reduction of total magnetic moment, as reflected from Table 2. In particular, the induced magnetic moment of the mediating Mo atom (connecting $Mn_0$ and $Mn_6$) is up to $-0.54\ \mu_B$. The large magnetization of the mediating Mo atom is partially due to the indirect superexchange interaction mediated by the N atom that connects Mn and Mo, i.e., two different N *p*-orbitals antiferromagnetically hybridize with Mn and Mo, respectively, leading to the induced spin of Mo that is antiparallel to that of Mn. This indirect Mn–N–Mo superexchange interaction resembles the case of Mn–N–Mn, as discussed above. Most importantly, the significant spin polarization of the Mo atom is arising from the direct interaction between the Mo-4*d* and Mn-3*d* orbitals, which can be clearly seen from the spin charge density distributions in Figure 7. The three transition metal *d*-orbitals (one Mo-4*d* orbital and two Mn-3*d* orbitals) along the chain of Mn–Mo–Mn couple well with each other to reduce the total energy. The large $J$ value of the (0, 6) configuration is good proof that the direct *d*–*d* interaction between the dopant Mn and the mediating Mo atom is beneficial to enhance the superexchange interaction of the Mn–Mn pair.

Although the $Mn_0$–$Mn_4$ pair in the (0, 4) configuration both locate at the *ab* plane as well, the local symmetry around the doping site and the spin polarization of the neighboring Mo atoms to the Mn dopants are different from those of the (0, 6) configuration. Because of the loss of symmetry at the doping site upon geometry optimization, the two mediating Mo atoms move away from their original positions and connect $Mn_0$ and $Mn_4$ with nonequivalent Mn–Mo distances (i.e., the $Mn_0$–Mo distance is more than 3.18 Å, while the $Mn_4$–Mo distance is less than 2.72 Å). It is seen from Figure 7 that the magnetic moments of two mediating Mo atoms are antiparallel to the nearer Mn dopant ($Mn_4$), and, at the same time, are parallel to the farther Mn dopant ($Mn_0$). For the two mediating Mo atoms, the direct *d*–*d* interaction to $Mn_4$ is significant, but it is really weak to $Mn_0$. The interaction between $Mn_0$ and the mediating Mo atom is mainly mediated by the N atoms. Such nonequilibrium interactions along the mediating segment consequently lead to the AFM coupling of the $Mn_0$–$Mn_4$ pair. By examining the spin charge density distributions of (0, 4) and (0, 6) configurations, we find that the exhibited magnetism of the Mo atoms parallel/antiparallel to that of the Mn dopants is not only correlated with the longer/shorter Mn–Mo distances, but also the coordination orientation of Mo *d*-orbital that participates in bonding.

As for the (0, 5) configuration, the two Mn dopants are FM coupled to each other along the *c* direction of the supercell. It is visible from Figure 7 that this long-ranged FM coupling of the $Mn_0$–$Mn_5$ pair is established by the indirect *p*–*d* exchange between the N *p*-orbital and the transition metal *d*-orbitals along the chain of Mn(↑)–N(↓)–Mo(↑)–N(↓)–Mn(↑). One Mn dopant ($Mn_0$) develops a strong bonding with its nearest N atom and induces AFM polarization of the N atom with a magnetic moment of about $-0.05\ \mu_B$, which further leads to a $0.13\ \mu_B$ spin of the mediating Mo atom, reflecting the effect of through-bond spin polarization. The other Mn dopant ($Mn_5$) in turn couples to the mediating atoms (N and Mo) in the same way for an energy gain, resulting in an indirect FM coupling among the two Mn dopants. However, due to the weak hybridization between Mo *d*-orbital and N *p*-orbital along the *c* axis, the mediated interaction through the segment of N–Mo–N in (0, 5) configuration is rather feeble, which leads to a relatively small $J$ value of 0.7 meV.

As stated in the above, the mediated Mn–Mn interaction should be sought in the electronic processes that take place in the segment that connects two Mn dopants. From the observations of (0, 4), (0, 5), and (0, 6) configurations, we find that the spin of the mediating N atom (which makes a bond to the Mn dopant) is always antiparallel to that of Mn regardless of the Mn doping position. The mediating Mo atom can couple to the Mn dopant either ferromagnetically or antiferromagnetically, which mainly depends on the different coupling mechanisms (*p*–*d* exchange, superexchange, and *d*–*d* coupling). The *p*–*d* exchange mediated by the N atom could induce the Mo–Mn FM coupling, while the superexchange through the N atom can result in the Mo–Mn AFM coupling. The substantial *d*–*d* coupling between Mo and Mn could effectively enhance the magnetic coupling of the Mn–Mn pair, which plays a critical role in mediating the Mn–Mn indirect interaction. The above analysis provides some insight into the origin of magnetic coupling between the Mn dopants. For the configurations in which the two Mn dopants are connected by the segment of N–Mo–N (configurations of (0, 4) to (0, 11)), the indirect FM/AFM Mn–Mn interaction can be described as follows: One Mn dopant induces AFM polarization of its neighboring

N atom (one of the N atoms belongs to Mn–N–Mo–N–Mn), and this spin-polarized N atom may polarize the mediating Mo atom in a parallel way or an antiparallel way (through *p–d* exchange or superexchange). Then, by the similar mechanism, the mediating Mo atom interacts with the other N atom connected to it, and, through which, ferromagnetically or antiferromagnetically couples to the other Mn dopant. That is, one of the Mn dopants dictates the spin polarization of its neighboring N atom, then this N atom affects the spin direction of the mediating Mo atom, and, through which, interacts with the other N atom, and eventually determines the FM or AFM coupling of the Mn–Mn pair. The different coupling mechanisms, i.e., the *p–d* exchange and the superexchange mediated by the N atom, and the *d–d* coupling between Mo and Mn, are associated with the geometry structure along the segment (such as bond length and bond angle) and the coordination orientation of the bonding orbital. Additionally, it should be emphasized that, except for the mediating atoms, other host atoms located next to the dopants also play important roles: first they become spin-polarized and then they can polarize their nearest neighbors, which facilitates the development of delocalized spin-polarized orbitals over a more extended range.

## 4. Conclusions

Based on the density functional theory, the structural, electronic, and magnetic properties of $\delta$-MoN with one Mn atom substituted at different Mo sites (2*a* and 6*c* in the International Tables) were firstly studied. The substitution of Mn atom results in the local distortion of the structure and the shrink of the unit cell. Formation energy calculations confirm that it is energetically favorable to form Mn substitution in $\delta$-MoN. It is noteworthy that the Mn dopants located at 2*a* and 6*c* sites lead to significant spin splitting of the density of states and induce 2.74 and 2.56 $\mu_B$ total magnetic moments, respectively, suggesting that the Mn dopant induces magnetism in $\delta$-MoN. The calculations were then extended to a $2 \times 1 \times 2$ supercell, which contains two impurity Mn atoms, to investigate the favorable coupling between the Mn dopants. Among the eleven different magnetic configurations, five positions result in FM and six positions result in AFM coupling between the Mn–Mn pair. Strong evidence indicate that long-range magnetic coupling exists between the Mn dopants in the supercell. The different couplings of the Mn–Mn pair cannot be simply attributed to the different Mn–Mn distances, but closely related to the electronic processes that take place in the mediating segment (–N– or –N–Mo–N–) that connects two Mn dopants. The mechanisms responsible for the FM/AFM coupling of the Mn–Mn pairs are the superexchange and the *p–d* exchange mediated by the N atoms, and the *d–d* coupling between the Mo and Mn atoms.

**Author Contributions:** Conceptualization, K.W. and J.Y.; methodology, K.W.; software, K.W., J.Y. and C.C.; validation, K.W. and J.Y.; formal analysis, K.W.; investigation, K.W.; resources, J.Y.; data curation, K.W. and J.Y.; writing—original draft preparation, K.W.; writing—review and editing, K.W., G.Z. and J.Y.; visualization, K.W.; supervision, J.Y. and G.Z.; project administration, J.Y.; funding acquisition, J.Y. and C.C. All authors have read and agreed to the published version of the manuscript.

**Funding:** This research was funded by the Natural Science Foundation of Heilongjiang Province (LH2021B034), the PhD Research Startup Foundation of Suihua University, and the Suihua University Scientific Research Innovation Team Project Funding (SIT05001).

**Institutional Review Board Statement:** Not applicable.

**Informed Consent Statement:** Not applicable.

**Data Availability Statement:** Not applicable.

**Conflicts of Interest:** The authors declare no conflict of interest.

## References


1. Wang, S.; Antonio, D.; Yu, X.; Zhang, J.; Cornelius, A.L.; He, D.; Zhao, Y. The hardest superconducting metal nitride. *Sci. Rep.* **2015**, *5*, 1–8. [CrossRef] [PubMed]
2. Zhang, M.; Cheng, K.; Yan, H.; Wei, Q.; Zheng, B. Electronic bonding analyses and mechanical strengths of incompressible tetragonal transition metal dinitrides TMN2 (TM = Ti, Zr, and Hf). *Sci. Rep.* **2016**, *6*, 1–10. [CrossRef] [PubMed]
3. Andrievski, R.A. High-melting-point compounds: New approaches and new results. *Phys. Uspekhi* **2017**, *60*, 276–289. [CrossRef]
4. Jauberteau, I.; Bessaudou, A.; Mayet, R.; Cornette, J.; Jauberteau, J.L.; Carles, P.; Merle-Méjean, T. Molybdenum nitride films: Crystal structures, synthesis, mechanical, electrical and some other properties. *Coatings* **2015**, *5*, 656–687. [CrossRef]
5. Simmendinger, J.; Pracht, U.S.; Daschke, L.; Proslier, T.; Klug, J.A.; Dressel, M.; Scheffler, M. Superconducting energy scales and anomalous dissipative conductivity in thin films of molybdenum nitride. *Phys. Rev. B* **2016**, *94*, 064506. [CrossRef]
6. Saha, B.; Garbrecht, M.; Perez-Taborda, J.A.; Fawy, M.H.; Koh, Y.R.; Shakouri, A.; Martin-Conzalez, M.; Hultman, L.; Sands, T.D. Compensation of native donor doping in ScN: Carrier concentration control and p-type ScN. *Appl. Phys. Lett.* **2017**, *110*, 252104. [CrossRef]
7. Tang, H.; Luo, J.; Tian, X.L.; Dong, Y.; Li, J.; Liu, M.; Liu, L.; Song, H.; Liao, S. Template-free preparation of 3D porous Co-doped VN nanosheet-assembled microflowers with enhanced oxygen reduction activity. *ACS Appl. Mater. Interfaces* **2018**, *10*, 11604–11612. [CrossRef]
8. Luo, J.; Tian, X.; Zeng, J.; Li, Y.; Song, H.; Liao, S. Limitations and improvement strategies for early-transition-metal nitrides as competitive catalysts toward the oxygen reduction reaction. *ACS Catal.* **2016**, *6*, 6165–6174. [CrossRef]
9. Mu, Y.; Liu, M.; Zhao, Y. Carbon doping to improve the high temperature tribological properties of VN coating. *Tribol. Int.* **2016**, *97*, 327–336. [CrossRef]
10. Ye, Y.; Wang, Y.; Chen, H.; Li, J.; Yao, Y.; Wang, C. Doping carbon to improve the tribological performance of CrN coatings in seawater. *Tribol. Int.* **2015**, *90*, 362–371. [CrossRef]
11. Ningthoujam, R.S.; Gajbhiye, N.S. Synthesis, electron transport properties of transition metal nitrides and applications. *Prog. Mater. Sci.* **2015**, *70*, 50–154. [CrossRef]
12. Sukkabot, W. Structural and magnetic properties of transition-metal doped scandium nitride (ScN): Spin density functional theory. *Phys. B* **2019**, *570*, 236–240. [CrossRef]
13. Benissad, F.; Houari, A. Electronic and magnetic properties of transition-metal-doped ScN for spintronics applications. *Phys. Status Solidi B* **2021**, *258*, 2000241. [CrossRef]
14. Lone, I.U.N.; Mohamed, S.S.M.; Banu, I.S.; Basha, S.S. Structural, elastic and magnetic properties of Mn and Sb doped chromium nitride—An ab initio study. *Mater. Chem. Phys.* **2017**, *192*, 291–298. [CrossRef]
15. Quintela, C.X.; Rivadulla, F.; Rivas, J. Electronic and magnetic phase diagram of $Cr_{1-x}V_xN$. *Phys. Rev. B* **2010**, *82*, 245201. [CrossRef]
16. Sirajuddeen, M.M.S.; Banu, I.S. An ab initio study on electronic and magnetic properties of Cr, V doped Cd and Zn nitrides for spintronic applications. *J. Magn. Magn. Mater.* **2016**, *406*, 48–59. [CrossRef]
17. Sirajuddeen, M.M.S.; Banu, I.S. Electronic and magnetic properties of Fe (Mn)-doped Cd and Zn nitrides for spintronic applications: A first-principles study. *J. Mater. Sci.* **2015**, *50*, 1446–1456. [CrossRef]
18. Herwadkar, A.; Lambrecht, W.R. Mn-doped ScN: A dilute ferromagnetic semiconductor with local exchange coupling. *Phys. Rev. B* **2005**, *72*, 235207. [CrossRef]
19. Herwadkar, A.; Lambrecht, W.R.; van Schilfgaarde, M. Linear response theoretical study of the exchange interactions in Mn-doped ScN: Effects of disorder, band gap, and doping. *Phys. Rev. B* **2008**, *77*, 134433. [CrossRef]
20. Soignard, E.; McMillan, P.F.; Chaplin, T.D.; Farag, S.M.; Bull, C.L.; Somayazulu, M.S.; Leinenweber, K. High-pressure synthesis and study of low-compressibility molybdenum nitride (MoN and $MoN_{1-x}$) phases. *Phys. Rev. B* **2003**, *68*, 132101. [CrossRef]
21. Lévy, F.; Hones, P.; Schmid, P.E.; Sanjinés, R.; Diserens, M.; Wiemer, C. Electronic states and mechanical properties in transition metal nitrides. *Surf. Coat. Technol.* **1999**, *120*, 284–290. [CrossRef]
22. Tagliazucca, V.; Leoni, M.; Weidenthaler, C. Crystal structure and microstructural changes of molybdenum nitrides traced during catalytic reaction by in situ X-ray diffraction studies. *Phys. Chem. Chem. Phys.* **2014**, *16*, 6182–6188. [CrossRef] [PubMed]
23. Zhong, Y.; Xia, X.; Shi, F.; Zhan, J.; Tu, J.; Fan, H.J. Transition metal carbides and nitrides in energy storage and conversion. *Adv. Sci.* **2016**, *3*, 1500286. [CrossRef] [PubMed]
24. Papaconstantopoulos, D.A.; Pickett, W.E.; Klein, B.M.; Boyer, L.L. Electronic properties of transition-metal nitrides: The group-V and group-VI nitrides VN, NbN, TaN, CrN, MoN, and WN. *Phys. Rev. B* **1985**, *31*, 752. [CrossRef] [PubMed]
25. Gajbhiye, N.S.; Ningthoujam, R.S. Structural, electrical and magnetic studies of nanocrystalline δ-MoN and γ-$Mo_2N$. *Phys. Status Solidi C* **2004**, *1*, 3449–3454. [CrossRef]
26. Ihara, H.; Hirabayashi, M.; Senzaki, K.; Kimura, Y.; Kezuka, H. Superconductivity of B1-MoN films annealed under high pressure. *Phys. Rev. B* **1985**, *32*, 1816. [CrossRef]
27. Inumaru, K.; Nishikawa, T.; Nakamura, K.; Yamanaka, S. High-pressure synthesis of superconducting molybdenum nitride δ-MoN by in situ nitridation. *Chem. Mater.* **2008**, *20*, 4756–4761. [CrossRef]
28. Maoujoud, M.; Binst, L.; Delcambe, P.; Offergeld-Jardinier, M.; Bouillon, F. Deposition parameter effects on the composition and the crystalline state of reactively sputtered molybdenum nitride. *Surf. Coat. Technol.* **1992**, *52*, 179–185. [CrossRef]

29. Inumaru, K.; Baba, K.; Yamanaka, S. Superconducting molybdenum nitride epitaxial thin films deposited on MgO and α-$Al_2C$ substrates by molecular beam epitaxy. *Appl. Surf. Sci.* **2006**, *253*, 2863–2869. [CrossRef]
30. Inumaru, K.; Baba, K.; Yamanaka, S. Synthesis and characterization of superconducting β-$Mo_2N$ crystalline phase on a S substrate: An application of pulsed laser deposition to nitride chemistry. *Chem. Mater.* **2005**, *17*, 5935–5940. [CrossRef]
31. Bezinge, A.; Yvon, K.; Muller, J.; Lengaeur, W.; Ettmayer, P. High-pressure high-temperature experiments on δ-MoN. *Solid Sta Commun.* **1987**, *63*, 141–145. [CrossRef]
32. Bull, C.L.; McMillan, P.F.; Soignard, E.; Leinenweber, K. Determination of the crystal structure of δ-MoN by neutron diffraction. *Solid State Chem.* **2004**, *177*, 1488–1492. [CrossRef]
33. Zhang, Y.; Haberkorn, N.; Ronning, F.; Wang, H.; Mara, N.A.; Zhuo, M.; Chen, L.; Lee, J.H.; Blackmore, K.J.; Bauer, E.; et a Epitaxial superconducting δ-MoN films grown by a chemical solution method. *J. Am. Chem. Soc.* **2011**, *133*, 20735–2073 [CrossRef] [PubMed]
34. Sahu, B.R.; Kleinman, L. Theoretical study of structural and electronic properties of δ-MoN. *Phys. Rev. B* **2004**, *70*, 07310 [CrossRef]
35. Zhao, E.; Wang, J.; Wu, Z. Displacive phase transition, structural stability, and mechanical properties of the ultra-incompressibl and hard MoN by first principles. *Phys. Status Solidi B* **2010**, *247*, 1207–1213. [CrossRef]
36. Yu, J.; Wang, K.; Qiao, X.; Tian, J.; Zhang, G.; Guo, Q. Effects of Cr doping in δ-MoN: Structural, magnetic and spin transpo properties. *Theor. Chem. Acc.* **2020**, *139*, 1–9. [CrossRef]
37. Madsen, G.K.; Blaha, P.; Schwarz, K.; Sjöstedt, E.; Nordström, L. Efficient linearization of the augmented plane-wave metho *Phys. Rev. B* **2001**, *64*, 195134. [CrossRef]
38. Perdew, J.P.; Burke, K.; Ernzerhof, M. Generalized gradient approximation made simple. *Phys. Rev. Lett.* **1996**, *77*, 3865. [CrossRe
39. Dudarev, S.L.; Botton, G.A.; Savrasov, S.Y.; Humphreys, C.J.; Sutton, A.P. Electron-energy-loss spectra and the structural stabilit of nickel oxide: An LSDA+ U study. *Phys. Rev. B* **1998**, *57*, 1505. [CrossRef]
40. Wang, L.; Maxisch, T.; Ceder, G. Oxidation energies of transition metal oxides within the GGA+ U framework. *Phys. Rev. B* **200** *73*, 195107. [CrossRef]
41. Krcha, M.D.; Janik, M.J. Examination of oxygen vacancy formation in Mn-doped $CeO_2$ using DFT+ U and the hybrid functiona HSE06. *Langmuir* **2013**, *29*, 10120–10131. [CrossRef] [PubMed]
42. Wang, Y.; Puggioni, D.; Rondinelli, J.M. Assessing exchange-correlation functional performance in the chalcogenide lacuna spinels $GaM_4Q_8$ ($M$ = Mo, V, Nb, Ta; $Q$ = S, Se). *Phys. Rev. B* **2019**, *100*, 115149. [CrossRef]
43. Blöchl, P.E.; Jepsen, O.; Andersen, O.K. Improved tetrahedron method for Brillouin-zone integrations. *Phys. Rev. B* **1994**, 4 16223. [CrossRef] [PubMed]
44. Cendlewska, B.; Morawski, A.; Misiuk, A. Superconducting $MoN_x$ prepared by isostatic direct nitriding at high pressure an high temperature. *J. Phys. F Met. Phys.* **1987**, *17*, L71. [CrossRef]
45. Cantele, G.; Degoli, E.; Luppi, E.; Magri, R.; Ninno, D.; Iadonisi, G.; Ossicini, S. First-principles study of n-and p-doped silico nanoclusters. *Phys. Rev. B* **2005**, *72*, 113303. [CrossRef]
46. Yang, K.; Dai, Y.; Huang, B.; Whangbo, M.H. Density functional studies of the magnetic properties in nitrogen doped $TiO_2$. *Chem Phys. Lett.* **2009**, *481*, 99–102. [CrossRef]
47. Oyama, S.T. Introduction to the chemistry of transition metal carbides and nitrides. In *The Chemistry of Transition Metal Carbide and Nitrides*; Oyama, S.T., Ed.; Springer: Dordrecht, The Netherlands, 1996; pp. 1–27.
48. Zhang, Y.; Miller, G.J.; Fokwa, B.P. Computational design of rare-earth-free magnets with the $Ti_3Co_5B_2$-type structure. *Chem Mater.* **2017**, *29*, 2535–2541. [CrossRef]
49. Ruiz-Díaz, P.; Stepanyuk, O.V.; Stepanyuk, V.S. Effects of interatomic coupling on magnetic anisotropy and order of spins o metallic surfaces. *J. Phys. Chem. C* **2015**, *119*, 26237–26241. [CrossRef]
50. Seña, N.; Dussan, A.; Mesa, F.; Castaño, E.; González-Hernández, R. Electronic structure and magnetism of Mn-doped GaSb fo spintronic applications: A DFT study. *J. Appl. Phys.* **2016**, *120*, 051704. [CrossRef]
51. Guo, Y.; Chen, M.; Guo, Z.; Yan, X. First-principles calculations for magnetic properties of Mn-doped GaN nanotubes. *Phys. Let A* **2008**, *372*, 2688–2691. [CrossRef]
52. Pearton, S.J.; Norton, D.P.; Ivill, M.P.; Hebard, A.F.; Zavada, J.M.; Chen, W.M.; Buyanova, I.A. Ferromagnetism in transition-meta doped ZnO. *J. Electron. Mater.* **2007**, *36*, 462–471. [CrossRef]
53. Hurd, C.M. Varieties of magnetic order in solids. *Contemp. Phys.* **1982**, *23*, 469–493. [CrossRef]

*nanomaterials*

*Review*

# Recent Progress in the Design, Characterisation and Application of $LaAlO_3$- and $LaGaO_3$-Based Solid Oxide Fuel Cell Electrolytes

**Elena Filonova [1,*] and Dmitry Medvedev [2,3,*]**

1 Department of Physical and Inorganic Chemistry, Institute of Natural Sciences and Mathematics, Ural Federal University, 620002 Ekaterinburg, Russia
2 Laboratory of Electrochemical Devices Based on Solid Oxide Proton Electrolytes, Institute of High Temperature Electrochemistry, 620660 Ekaterinburg, Russia
3 Hydrogen Energy Laboratory, Ural Federal University, 620002 Ekaterinburg, Russia
* Correspondence: elena.filonova@urfu.ru (E.F.); dmitrymedv@mail.ru (D.M.)

**Abstract:** Solid oxide fuel cells (SOFCs) are efficient electrochemical devices that allow for the direct conversion of fuels (their chemical energy) into electricity. Although conventional SOFCs based on YSZ electrolytes are widely used from laboratory to commercial scales, the development of alternative ion-conducting electrolytes is of great importance for improving SOFC performance at reduced operation temperatures. The review summarizes the basic information on two representative families of oxygen-conducting electrolytes: doped lanthanum aluminates ($LaAlO_3$) and lanthanum gallates ($LaGaO_3$). Their preparation features, chemical stability, thermal behaviour and transport properties are thoroughly analyzed in terms of their connection with the target functional parameters of related SOFCs. The data presented here will serve as a starting point for further studies of La-based perovskites, including in the fields of solid state ionics, electrochemistry and applied energy.

**Keywords:** SOFCs; solid oxide fuel cells; oxygen-ion electrolytes; perovskite; $LaAlO_3$; $LaGaO_3$; LSGM

## 1. Introduction

The long-term goal of a large body of relevant scientific research is to find a solution to the problem of providing industrial and domestic human needs with renewable and environmentally friendly energy [1,2]. The main fields of sustainable energy concern both the search for renewable energy sources [3–5] and methods for the production of ecological types of energy [6–9], which differ from traditional types based on hydrocarbon fuel [10–12]. The tasks relating to sustainable energy also include the development of technologies for the use of non-renewable energy sources: efficient waste-processing [13–15], the construction of nuclear mini-reactors [16], and the creation of energy devices based on the direct conversion of various types of energy into electrical and thermal energy [17–19]. A well-known device for directly converting the chemical energy of fuels into electrical energy is a fuel cell [19–21]. If the electrolyte in the fuel cell is a ceramic material that is permeable to oxygen ions, it is referred to as a solid oxide fuel cell (SOFC) [21–25].

The advantages of SOFCs are the absence of noble metals in their composition and the flexibility of fuel types [24,26,27], while the disadvantages include high operating temperatures, which lead to chemical interactions between the parts of the SOFCs [28,29] and fast degradation [30–32]. The high temperatures required to operate SOFCs with conventional electrolytes on the basis of yttria-stabilized zirconia (YSZ) lead to the formation of metastable phases, sealing, and thermal and chemical incompatibility with electrode materials [33–35].

One of the ways to solve the described problem is to decrease the operating temperature of SOFCs and develop fuel cells operating at medium- [36–38] and low-temperature ranges [39,40]. This has resulted in investigations into new classes of electrolytes [41–44] and the development of SOFCs enhanced with nanostructured materials [45,46]. The utilization of nanotechnologies, energy production and energy storage devices is extremely prospective due to their durability, sustainability, long lifetime, and low cost [47]. Among the alternative electrolytes used in low- and intermediate-temperature SOFCs, complex oxides with an $ABO_3$-type perovskite structure have attracted specific attention due to their high efficiency in energy conversion [48–50]. Sr, Mg-doped lanthanum gallate ($LaGaO_3$), possessing a high oxide ionic conductivity, which was established originally by Ishihara et al. in 1994 [51], was first used in SOFCs by Feng and Goodenough in 1996 [52]. Later, much more economical materials based on doped lanthanum aluminate $LaAlO_3$ were reported by Fung and Chen in 2011 [53].

It is worth noting that previous generalizing works on lanthanum aluminate were aimed at the synthesis and characterization of $LaAlO_3$ phosphors (published by Kaur et al. in 2013 [54]) and at some properties and applications of $LaAlO_3$ not concerned with SOFCs (observed by Rizwan et al., in 2019) [55]. There is only one overview dedicated to Sr, Mg-doped $LaGaO_3$ oxides as electrolytes for intermediate-temperature solid oxide fuel cells: this was published by Morales et al. in 2016 [56]. The present overview is dedicated to recent progress in the design, characterization and application of electrolyte materials for SOFCs based on the $LaGaO_3$ and $LaAlO_3$ complex oxides with a perovskite structure. Both these phases constitute a family of oxygen-conducting electrolytes, while other La-based perovskites ($LaScO_3$, $LaInO_3$, $LaYO_3$, $LaYbO_3$) exhibit protonic conductivity as well [49]. For this reason, scandates, indates, yttrates, and ytterbates are not considered within the present review.

A schematic image of an $ABO_3$ perovskite structure is shown in Figure 1a,b. Typically, the size of A-site cations is larger than that of B-site cations, but is roughly close to that of the oxygen ions. The A-site cations are surrounded by 12 oxygen-ions in a cubo-octahedral coordination; the B-site cations are surrounded by 6 oxygen-ions in an octahedral coordination. In an ideal perovskite structure, $BO_6$ octahedrons are linked at the corners, thus exhibiting the cubic *Pm3m* space group.

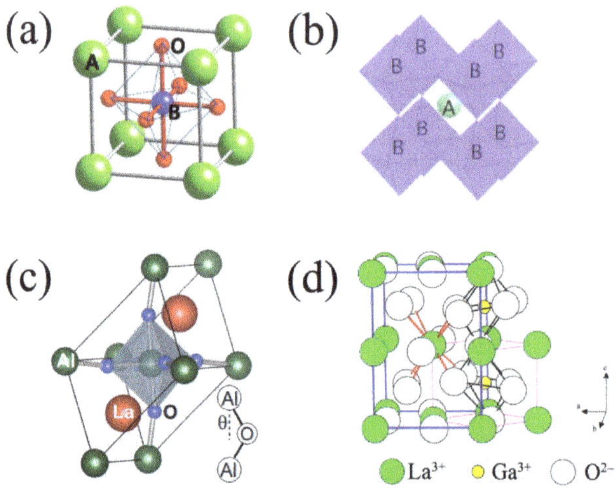


**Figure 1.** $ABO_3$ perovskite structure: (**a**) B-cation centered and (**b**) A-cation centered representations; the perovskite structure of the $ABO_3$ complex oxide with the B (**a**) and A (**b**) central ions [57]; (**c**) a rhombohedral crystal structure (for example, $LaAlO_3$). Reproduced from [58] with permission from the American Physical Society, 2016; (**d**) an orthorhombic crystal structure (for example, $LaGaO_3$) Reproduced from [59] with permission by Elsevier Ltd. (Amsterdam, The Netherlands), 2004.

If the complex oxide structure differs from the ideal perovskite structure by having rhombohedral or orthorhombic distortions due to the $BO_6$ octahedron arrangement, the stability of this oxide can be evaluated with the Goldsmith tolerance factor $t$ equation [60] as follows:

$$t = \frac{r_A + r_O}{\sqrt{2}(r_B + r_O)}, \quad (1)$$

where $r_A$, $r_B$, $r_O$ are the ionic radii of the A-, B-cations, and oxygen ions, respectively. If $t$ is equal to 1, an ideal cubic-type perovskite structure is formed. If $t$ deviates from 1, various distortions occur in the ideal perovskite structure. The first reason for such distortions is the rotation of the $BO_6$ octahedron without axis deformation, which causes tilting around the large A-cations. Take, for example, the rhombohedral structure of $LaAlO_3$ at room temperature presented in Figure 1c. The second reason consists of the appearance of the irregularity in the $BO_6$ octahedrons due to the non-centrality of the B-site cations. Consider, for example, the orthorhombic structure of $LaGaO_3$ at room temperature presented in Figure 1d.

## 2. Electrolyte Materials Based on $LaAlO_3$

### *2.1. Synthesis, Structure and Morphology*

For the synthesis of doped $LaAlO_3$ oxides, several well-developed techniques are usually used: solid-state reaction technology [61–64], the mechanochemical route [65], co-precipitation [66,67] and organic-nitrate precursor pyrolysis [68–75].

Employing conventional solid-state reaction technology, $LaAlO_3$ samples can be directly obtained from $La_2O_3$ and $Al_2O_3$. In [61], these initial reactants were ground down, homogenized in a water media, desiccated and pressed into pellets annealed at a temperature range of 780–1100 °C. Such a temperature regime allows for single-phase $LaAlO_3$ samples to be prepared. A similar technology was used in work [62] to synthesize $LaAl_{1-x}Zn_xO_{3-\delta}$ (here, $\delta$ is the oxygen nonstoichiometry; $\delta = x/2$ in the case of oxidation-state stable cations and one charge state difference between the host and impurity cations). As initial reagents, stoichiometric amounts of aluminium and zinc oxides were milled in ethanol. The heat treatment included five 24-h stages at a temperature range of 700–1100 °C. Single-phased $LaAlO_3$ and $LaAl_{0.95}Zn_{0.05}O_{3-\delta}$ were obtained at 1250 and 1200 °C, respectively.

Fabian et al. [65] synthesised Ca-doped $LaAlO_3$ powders using the mechanochemical method. Oxide powders of $La_2O_3$, $\gamma$-$Al_2O_3$ and CaO in appropriate proportions were milled in a planetary mill at 600 rpm. The prepared powders were pressed into disks with polyethylene glycol as a plasticizer. The $LaAlO_3$ and $La_{1-x}Ca_xAlO_{3-\delta}$ pellets were sintered at 1700 and 1450 °C, respectively, to achieve a desirable ceramic densification.

$LaAlO_3$ complex oxides were prepared starting from water solutions of aluminium and lanthanum chlorides with a molar ratio for the metal components of 1:1 [66]. Solutions with high and low concentrations of starting reagents were mixed with an ammonium solution serving as a precipitation agent. The obtained gels were filtered, washed with distilled water and dried twice, at 25 °C for 24 h and at 100 °C for 2 h. The prepared powders were calcined at a temperature range of 600–900 °C for 1 h. The powder obtained from the high-concentration solution was annealed at 900 °C for 2 h in air, then ground in a rotary mill with zirconia balls in dry ethanol, pressed and calcined at 1300–1500 °C for 2 h.

The most widely used technology for the preparation of $LaAlO_3$ and its doped derivatives is the pyrolysis of organic-nitrate compositions, known as the sol-gel [68,69,74] or autocombustion methods (or self-propagating high-temperature synthesis, and the Pechini method) [70–73,75]. Utilizing different fuels during the pyrolysis process coupled with various annealing temperatures affects the crystallinity, powder dispersity, and ceramics density, determining the functional properties of the obtained $LaAlO_3$-based ceramic materials [74,76,77].

$LaAlO_3$ powders were prepared by Zhang et al. [68] from $La(NO_3)_3 \cdot 6H_2O$ and $Al(NO_3)_3 \cdot 9H_2O$: they were dissolved in 2-methoxyethanol and then mixed with citric acid at a molar ratio of 1:1 to the total content of metal ions. The obtained solutions were heated and dried at 80 °C until gelatinous $LaAlO_3$ precursors were obtained, which were then calcined at 600–900 °C for 2 h.

To obtain $La_{0.9}Sr_{0.1}Al_{0.97}Mg_{0.03}O_{3-\delta}$ powder, $La(NO_3)_3{\cdot}6H_2O$, $Al(NO_3)_3{\cdot}9H_2O$, $Mg(NO_3)_2{\cdot}6H_2O$, $Sr(NO_3)_2$, EDTA, $C_2H_5NO_2$ and $NH_3{\cdot}H_2O$ were used in [69]. The molar ratio of glycine and EDTA to overall metal-ion content was 1.2:1:1; the ratio of $NH_3{\cdot}H_2O$ to EDTA was adjusted to 1.15:1. The aqueous solution of metal nitrates was prepared and heated at 80 °C, and then the EDTA-ammonia solution and glycine were added. The colourless solution was dried, and the obtained brown resin was calcined at 350 °C; it was then ground down and calcined at 600–1000 °C for 3 h. The obtained powders were finally pressed into disks followed by sintering at 1600–1700 °C for 5 h.

According to Adak and Pramanik [70], $LaAlO_3$ was prepared from a 10% aqueous polyvinyl alcohol precursor that was added to a solution obtained from $La_2O_3$ (99%) dissolved in nitric acid and $Al(NO_3)_3{\cdot}9H_2O$. The organic-nitrate mixture was evaporated at 200 °C until dehydration; then, spontaneous decomposition and the formation of a voluminous black fluffy powder occurred. The obtained powders were ground down and annealed at 600–800 °C for 2 h to form a pure phase.

Verma et al. [71] synthesized $LaAlO_3$ and $La_{0.9-x}Sr_{0.1}Ba_xAl_{0.9}Mg_{0.1}O_{3-\delta}$ ($x$ = 0.00, 0.01 and 0.03) samples from initial reagents composed of $La(NO_3)_3{\cdot}H_2O$, $Sr(NO_3)_2$, $Ba(NO_3)_2$, $Al(NO_3)_3{\cdot}6H_2O$ and $Mg(NO_3)_2{\cdot}6H_2O$ initial reagents. $C_6H_8O_7{\cdot}H_2O$ was used as an organic fuel. The metal nitrates and citric acid were dissolved in distilled water, resulting in the formation of a transparent solution. The pH value required for proper combustion was achieved by the addition of ammonia solution. The self-propagating synthesis method is shown in Figure 2a. The obtained powders were calcined at 700 °C for 4 h, then pressed into pellets and sintered at 1300 °C for 8 h to achieve 92-to-96% relative density, depending on the aluminate composition.

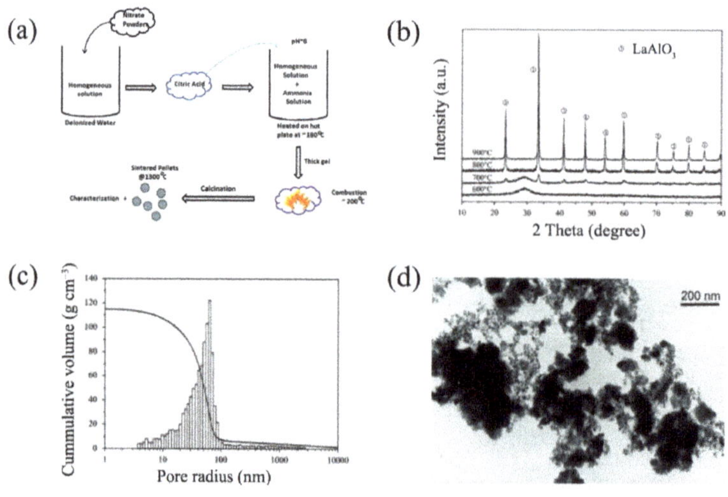


**Figure 2.** Preparation features of $LaAlO_3$-based materials: (**a**) diagram of auto-combustion synthesis. Reproduced from [71] with permission from Springer Nature (Berlin/Heidelberg, Germany), 2021; (**b**) XRD patterns for $LaAlO_3$ powders prepared and calcined at a temperature range of 600–900 °C for 1 h on each stage. Reproduced from [66] with permission by Elsevier Ltd., 2013; (**c**) pore size distributions of $LaAlO_3$ powder bodies calcined at 900 °C for 2 h. Reproduced from [66] with permission by Elsevier Ltd., 2013; (**d**) TEM image of $LaAlO_3$ powder calcined at 900 °C for 2 h. Reproduced from [66] with permission by Elsevier Ltd., 2013.

The literature shows that the annealing temperature of the precursor powders plays a significant role in complex oxide synthesis: this regulates the density of the final polycrystalline ceramic samples [78]. For practical applications, it is important to obtain

$LaAlO_3$-based samples with a narrow distribution of fine-grained particles. These requirements were fulfilled in [66], where a fully converted $LaAlO_3$ phase was formed at relatively low temperatures. In more detail, the authors developed a co-precipitation technique enabling the formation of single-phase $LaAlO_3$ powders after its calcination in air at 900 °C for 2 h (Figure 2b). A narrow particle size distribution for $LaAlO_3$ powder was achieved in [66], where milling in an ethanol medium was conducted. As shown in Figure 2c, the milled $LaAlO_3$ powder exhibited mono-modal pore size distribution. The TEM image (Figure 2d) demonstrates that the calcined powder consisted of isometric particles of up to 15 nm in size. The use of a precursor solution with a high concentration of metal chlorides and ammonia allowed for the researchers to realize gel homogeneity and the direct synthesis of $LaAlO_3$.

A Rietveld analysis of the XRD pattern confirmed the presence of a pure perovskite phase with a rhombohedral structure, referring to the *R-3c* space group. Reference [66] calculated unit cell parameters for the $LaAlO_3$ sample ($a$ = 5.3556(1) Å and $c$ = 13.1518(2) Å) agreed well with results from neutron powder diffraction [79]. The primitive $LaAlO_3$ cell consists of two formula units, as shown in Figure 1b. The rotation of $AlO_6$ octahedra is caused by changes to the $\theta$ angle (Al–O–Al). Above 540 °C, a phase transition from the rhombohedral to cubic structure was observed for $LaAlO_3$ [79]. The cubic lattice of $LaAlO_3$ with a unit cell parameter of $a$ = 3.8106(1) Å corresponds to the *Pm3m* space group [79] (see Figure 1a).

Concluding the chapter about the synthesis methods of doped $LaAlO_3$ oxides, from the perspective of their use in SOFCs, the co-precipitation method should be noted as the most optimal synthetic method. The co-precipitation method with a subsequent sintering of samples at 900 °C is well-approved and allows for both single-phase powders with a narrow nano-size particle distribution and ceramic samples with high relative densities to be obtained.

### 2.2. Functional Properties

$LaAlO_3$, a basic (undoped) lanthanum aluminate, has very low electrical conductivity, equal to around $1 \times 10^{-6}$ S cm$^{-1}$ at 900 °C [75]. La-site doping of $LaAlO_3$ with strontium enhances electrical conductivity because it improves the oxygen vacancy concentration responsible for oxygen-ion transport (Equation (2), [80]). Al-site modification of $LaAlO_3$ with acceptor dopants (for example, magnesium) can also increase the total and ionic conductivities (see Figure 3a).

$$2SrO \overset{'La_2O_3'}{\rightarrow} 2Sr'_{La} + V_O^{\bullet\bullet} + 2O_O^{x}. \quad (2)$$

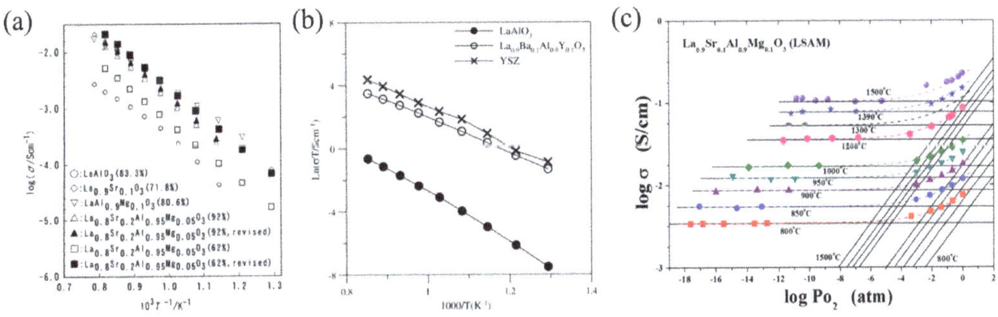

**Figure 3.** Functional properties of $LaAlO_3$-doped materials: (**a**) electrical conductivity of $LaAlO_3$, $La_{0.9}Sr_{0.1}AlO_{3-\delta}$, $LaAl_{0.9}Mg_{0.1}O_{3-\delta}$, $La_{0.8}Sr_{0.2}Al_{0.95}Mg_{0.05}O_{3-\delta}$ samples. Reproduced from [80] with

permission by Elsevier Ltd., 2000; (**b**) electrical conductivity of $LaAlO_3$, $La_{0.9}Ba_{0.1}Al_{0.9}Y_{0.1}O_{3-\delta}$, and YSZ as a reference sample. Reproduced from [53] with permission by Elsevier Ltd., 2011; (**c**) total conductivity of the $La_{0.9}Ba_{0.1}Al_{0.9}Y_{0.1}O_{3-\delta}$ ceramic as function of oxygen partial pressures [81].

The possibility of forming good oxygen-ionic conductivity by doping $LaAlO_3$ oxides has promoted studies on their potential application in SOFCs [53,65,71,82–90]. The co-doping strategy is a beneficial way to further increase ionic conductivity [80,82,83,87]; this is due to the fact that, along with Equation (2), an additional quantity of oxygen vacancies can be formed according to the following mechanism [80]:

$$2MgO \overset{'Al_2O_3'}{\rightarrow} 2Mg'_{Al} + V_O^{\bullet\bullet} + 2O_O^{x}. \quad (3)$$

According to the results of [53], the simultaneous doping of $LaAlO_3$ with barium and yttrium drastically enhanced ionic transport. For example, the total conductivity of $La_{0.9}Ba_{0.1}Al_{0.9}Y_{0.1}O_{3-\delta}$ at 800 °C was close to that of YSZ ($2 \times 10^{-2}$ S $cm^{-1}$), as shown in Figure 3b. There are various ways to tailor the transport properties of $LaAlO_3$-based materials. For example, the doping of (La,Sr)$AlO_3$ with manganese resulted in total conductivity rising due to the substitution of $Mn^{3+}$ ions, which were transformed into $Mn^{2+}$ and $Mn^{4+}$ ions at the $Al^{3+}$ position, enhancing an electronic contribution [75,84]. Therefore, co-doped (La,Sr)(Al,Mn)$O_3$ is attributed to mixed ionic-electronic conductors (MIEC). The Pr-doping of (La,Sr)$AlO_3$ had a positive influence on transport properties due to the suppression of grain boundary resistivity [85], and the isovalent substitution of $La^{3+}$-ions with $Sm^{3+}$-ions in (La,Sr)$AlO_{3-\delta}$ resulted in the formation of a pronounced mixed ion-electron conduction [88] due to the generation of more electrons than in the case of the aliovalent substitution of $La^{3+}$ ions with $Ba^{2+}$ ions.

The electrical conductivity values of $LaAlO_3$-based ceramic materials are summarized in Table 1. Analysis of these data confirms that the simultaneous modification of both sublattices of $LaAlO_3$ results in improved conductivity compared to those reached using single doping approaches (see Figure A1). However, it should be noted that the Sr- and Mg-co-doped $LaAlO_3$ materials exhibit mixed ionic-electronic conduction in air atmospheres over a wide temperature range (800–1400 °C, see Figure 3c), while predominant ionic transport occurs for more reduced atmospheres (for example, wet hydrogen). This is typical behaviour for various La-based perovskites [49] as well as for other perovskite-related ion-conducting electrolytes [91].

Thermal expansion coefficients (TECs) play an important role in material selection when seeking to avoid thermal incompatibilities between various parts of SOFCs. According to da Silva and de Miranda [75], the average TEC values for $LaAlO_3$ and $La_{0.8}Sr_{0.2}AlO_3$ were equal to around $11.4 \times 10^{-6}$ and $9.9 \times 10^{-6}$ $K^{-1}$, respectively. These data confirm that the TEC values of $LaAlO_3$-based materials were close to those of the conventional YSZ electrolyte, i.e., $10.9 \times 10^{-6}$ $K^{-1}$ [92].

The chemical compatibility of $La_{0.9}Sr_{0.1}Al_{0.97}Mg_{0.03}O_{3-\delta}$ as an electrolyte material with NiO-$Ce_{0.9}Gd_{0.1}O_{2-\delta}$, $Sr_{0.88}Y_{0.08}TiO_{3-\delta}$ and $La_{0.75}Sr_{0.25}Cr_{0.5}Mn_{0.5}O_{3-\delta}$ as anode SOFC materials was thoroughly investigated in [87] using XRD analysis and scanning electron microscopy with energy-dispersive X-ray spectroscopy. The obtained results demonstrated that $Sr_{0.88}Y_{0.08}TiO_{3-\delta}$ and $La_{0.75}Sr_{0.25}Cr_{0.5}Mn_{0.5}O_{3-\delta}$ interacted with $La_{0.9}Sr_{0.1}Al_{0.97}Mg_{0.03}O_{3-\delta}$ due to the interdiffusion of $Sr^{2+}$, $Ti^{4+}$, $Mn^{3+}$ and $Cr^{3+}$ cations into the $La_{0.9}Sr_{0.1}Al_{0.97}Mg_{0.03}O_{3-\delta}$ lattice. An interaction between $La_{0.9}Sr_{0.1}Al_{0.97}Mg_{0.03}O_{3-\delta}$ and NiO-$Ce_{0.9}Gd_{0.1}O_{2-\delta}$ at 1300 °C was not detected, which means that joint utilization is possible.

The XRD patterns of two mixtures, $La_{0.8}Sr_{0.2}Ga_{0.85}Mg_{0.15}O_{3-\delta}/La_{0.9}Sr_{0.1}AlO_{3-\delta}$ and $NiO/La_{0.9}Sr_{0.1}AlO_{3-\delta}$ (annealed at 1450 °C), confirmed that there were no chemical interactions between these components [93]. The authors noted that doped $LaAlO_3$ materials can serve as additives to the composite electrolytes and the anode-protective layers [93]. In addition, Mn-doped $LaAlO_3$ phases are considered a constituent part of the composite electrolytes, providing for the effective electrochemical oxidation of methane via ethylene and ethane [94].

**Table 1.** Total conductivity and activation energy values for $LaAlO_3$ ceramic materials. Figure A1 (see the Appendix A) provides a visualization of these data.

| Sample | $T$ (°C) | $\sigma$ (S cm$^{-1}$) | $E_a$ (eV) | Ref. |
|---|---|---|---|---|
| $LaAlO_3$ | 900 | $6 \times 10^{-4}$ | 1.30 | [53] |
| $LaAlO_3$ | 700 | $6.7 \times 10^{-4}$ | 0.99 | [71] |
| $LaAlO_3$ | 900 | $1.1 \times 10^{-6}$ | 1.83 | [75] |
| $LaAlO_3$ | 900 | $1.4 \times 10^{-3}$ | 1.88 | [80] |
| $LaAlO_3$ | 800 | $2.0 \times 10^{-4}$ | 1.30 | [83] |
| $La_{0.9}Ca_{0.1}AlO_{3-\delta}$ | 900 | $6.0 \times 10^{-3}$ | 1.08 | [65] |
| $La_{0.9}Sr_{0.1}AlO_{3-\delta}$ | 900 | $1.1 \times 10^{-2}$ | 1.14 | [80] |
| $La_{0.9}Sr_{0.1}AlO_{3-\delta}$ | 800 | $9.0\times10^{-3}$ | 0.93 | [85] |
| $La_{0.8}Sr_{0.2}AlO_{3-\delta}$ | 800 | $6.2 \times 10^{-3}$ | 1.06 | [75] |
| $La_{0.8}Sr_{0.2}AlO_{3-\delta}$ | 900 | $1.5 \times 10^{-2}$ | 1.06 | [75] |
| $La_{0.8}Sr_{0.2}AlO_{3-\delta}$ | 900 | $1.1 \times 10^{-2}$ | 1.16 | [80] |
| $La_{0.8}Sr_{0.2}AlO_{3-\delta}$ | 810 | $4.3 \times 10^{-3}$ | 1.06 | [84] |
| $La_{0.7}Pr_{0.2}Sr_{0.1}AlO_{3-\delta}$ | 800 | $2.3 \times 10^{-2}$ | 0.84 | [85] |
| $LaAl_{0.95}Zn_{0.05}O_{3-\delta}$ | 700 | $8.5 \times 10^{-4}$ | 1.05 | [62] |
| $LaAl_{0.95}Zn_{0.05}O_{3-\delta}$ | 900 | $1.1 \times 10^{-3}$ | 1.05 | [62] |
| $LaAl_{0.9}Mg_{0.1}O_{3-\delta}$ | 900 | $9.6 \times 10^{-3}$ | 1.05 | [80] |
| $LaAl_{0.5}Mn_{0.5}O_{3-\delta}$ | 800 | 4.7(2) | 0.22 | [75] |
| $LaAl_{0.5}Mn_{0.5}O_{3-\delta}$ | 900 | 5.8(2) | 0.22 | [75] |
| $La_{0.9}Sr_{0.1}Al_{0.9}Mg_{0.1}O_{3-\delta}$ | 700 | $2.6 \times 10^{-3}$ | 1.56 | [71] |
| $La_{0.9}Sr_{0.1}Al_{0.9}Mg_{0.1}O_{3-\delta}$ | 700 | $5.3 \times 10^{-4}$ | 1.38 | [88] |
| $La_{0.9}Sr_{0.1}Al_{0.9}Mg_{0.1}O_{3-\delta}$ | 900 | $2.0 \times 10^{-2}$ | 0.90 | [82] |
| $La_{0.8}Sr_{0.2}Al_{0.95}Mg_{0.05}O_{3-\delta}$ | 900 | $1.3 \times 10^{-2}$ | 1.15 | [80] |
| $La_{0.89}Sr_{0.1}Ba_{0.01}Al_{0.9}Mg_{0.1}O_{3-\delta}$ | 700 | $2.6 \times 10^{-3}$ | 1.48 | [71] |
| $La_{0.89}Sr_{0.1}Ba_{0.01}Al_{0.9}Mg_{0.1}O_{3-\delta}$ tape | 700 | $6.0 \times 10^{-4}$ | 0.60 | [86] |
| $La_{0.89}Sr_{0.1}Ba_{0.01}Al_{0.9}Mg_{0.1}O_{3-\delta}$ pellet | 700 | $4.6 \times 10^{-2}$ | 0.75 | [86] |
| $La_{0.87}Sr_{0.1}Ba_{0.03}Al_{0.9}Mg_{0.1}O_{3-\delta}$ | 700 | $1.7 \times 10^{-3}$ | 1.38 | [71] |
| $La_{0.8}Sr_{0.2}Al_{0.5}Mn_{0.5}O_{3-\delta}$ | 800 | 8.6(3) | 0.15 | [75] |
| $La_{0.8}Sr_{0.2}Al_{0.5}Mn_{0.5}O_{3-\delta}$ | 900 | 9.8(2) | 0.15 | [75] |
| $La_{0.8}Sr_{0.2}Al_{0.7}Mn_{0.3}O_{3-\delta}$ | 810 | 0.75 | 0.29 | [84] |
| $La_{0.8}Sr_{0.2}Al_{0.5}Mn_{0.5}O_{3-\delta}$ | 810 | 10 | 0.17 | [84] |
| $(La_{0.8}Sr_{0.2})_{0.94}Al_{0.5}Mn_{0.5}O_{3-\delta}$ | 810 | 12 | 0.14 | [84] |
| $La_{0.9}Ba_{0.1}Al_{0.9}Y_{0.1}O_{3-\delta}$ | 800 | $1.8 \times 10^{-2}$ | 0.82 | [53] |
| $La_{0.9}Ba_{0.1}Al_{0.9}Y_{0.1}O_{3-\delta}$ | 900 | $3.1 \times 10^{-2}$ | 0.82 | [53] |
| $La_{0.87}Sr_{0.1}Sm_{0.03}Al_{0.9}Mg_{0.1}O_{3-\delta}$ | 700 | $1.2 \times 10^{-3}$ | 1.09 | [88] |
| $La_{0.85}Sr_{0.1}Sm_{0.05}Al_{0.9}Mg_{0.1}O_{3-\delta}$ | 700 | $1.1 \times 10^{-3}$ | 1.10 | [88] |

### 2.3. Applications in SOFCs

There are fragmentary data on the application of lanthanum aluminate electrolytes in SOFCs; see Figure 4.

For example, an SOFC was fabricated with 70% NiO–30% YSZ as an anode, SDC as an interlayer, $La_{0.9}Ba_{0.1}Al_{0.9}Y_{0.1}O_{3-\delta}$ (LBAYO) as an electrolyte and LSM as a cathode, and tested in [53]. LBAYO films with thicknesses of 63 and 74 μm were electrophoretically deposited on the LSM pellets with a diameter of 25 mm and a thickness of 2 mm. The LSM substrates and the deposited LBAYO films were then annealed at 1450 °C for 2 h to achieve full electrolyte densification. The thickness of the LBAYO film varied due to increases in the applied voltage. A NiO/YSZ anode with a thickness of 40 μm was screen-printed on the LBAYO/LSM sample and then sintered at 1500 °C for 6 h. To avoid chemical interactions between the NiO and the LBAYO film, an SDC buffer layer with a thickness of 10 μm was additionally screen-printed on the LBAYO film between the electrolyte and the anode. Humidified hydrogen was used as a fuel, while air was used as an oxidant. Figure 4a presents the SEM micrograph of the NiO–YSZ/SDC/LBAYO/LSM cell, indicating that after the annealing procedure, the LBAYO film was highly densified without cracks with a uniform thickness and a strong adhesion to the LSM substrate. The open-circuit voltage (OCV) values of the fabricated cells were 0.927 and 0.953 V, while the maximum power density values were 0.306 and 0.235 W cm$^{-2}$ for the LBAYO electrolyte layers with thicknesses of 63 and 74 μm, respectively (Figure 4b). The authors of the work attributed the sharp decrease in the cells' voltage at a small current to the slow oxygen reduction reaction kinetics for the LSM cathode.

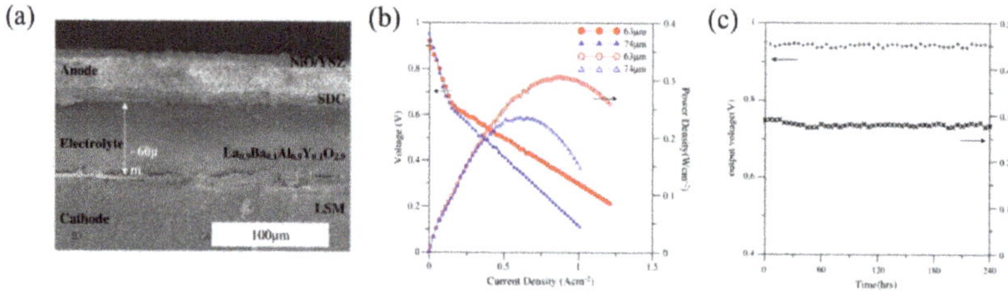


**Figure 4.** Properties of the NiO–YSZ/SDC/LBAYO/LSM SOFC: (**a**) SEM micrograph of a cell sintered at 1500 °C for 6 h; (**b**) current-voltage and current-power dependencies of a cell with an electrolyte thickness of 63 μm tested at different temperatures; (**c**) time dependencies of OCV and $P_{max}$ measured at 800 °C for 10 days. These images were reproduced from [53] with permission from Elsevier Ltd., 2011.

The long-term stability experiments demonstrated negligible degradation of the LBAYO electrolyte over 10 days. Figure 4c illustrates the time dependencies of the obtained open circuit voltage (OCV) and the maximum power density ($P_{max}$) for a cell tested at 800 °C.

Another Ni-GDC/GDC/$La_{0.9}Sr_{0.1}Al_{0.97}Mg_{0.03}O_{3-\delta}$/GDC/$La_{0.75}Sr_{0.25}FeO_{3-\delta}$ electrolyte-supported cell was tested in [87]. For this single cell with a $La_{0.9}Sr_{0.1}Al_{0.97}Mg_{0.03}O_{3-\delta}$ electrolyte thickness of 550 μm, the OCV and $P_{max}$ values at 800 °C were found to be equal to 0.925 V and 19.5 mW cm$^{-2}$, respectively.

## 3. Electrolyte Materials Based on Doped $LaGaO_3$

### *3.1. Synthesis, Structure and Morphology*

Historically, $La_{1-x}Sr_xGa_{1-y}Mg_yO_{3-\delta}$ (LSGM) oxides were the first well-studied doped materials in the $LaGaO_3$ system. In 1998, Huang, Tichy and Goodenough determined the existence of single-phase $La_{1-x}Sr_xGa_{1-y}Mg_yO_{3-0.5(x+y)}$ perovskites while studying a $LaO_{1.5}$-SrO-$GaO_{1.5}$-MgO quasi-quaternary diagram [95] (see Figure 5a). This was possible due to variations in both $x$ and $y$ contents in a composition range of 0.05–0.30 with a step of 0.05. Sr- and Mg- co-doped $LaGaO_3$ samples were prepared from $La_2O_3$, $SrCO_3$, $Ga_2O_3$, and MgO using solid-state reaction technology. The obtained powders were pressed into pellets and calcined at 1250 °C for 12 h. After remilling and repressing, the final pellets were finally sintered in air at 1470 °C for 24 h and quenched in a furnace at 500 °C.

Similar conventional techniques for synthesizing $La_{1-x}Sr_xGa_{1-y}Mg_yO_{3-\delta}$ were used in other studies [96,97]. $La_{0.9}Sr_{0.1}Ga_{0.8}Mg_{0.2}O_{3-\delta}$ samples were obtained from $La_2O_3$, $SrCO_3$, $Ga_2O_3$ and MgO sources, which were mixed and sintered in a platinum crucible at 1350 °C for 12 h [96]. The annealed powder was milled with zirconia balls and dried. Then, the powder was pressed into disks and sintered at 1350 °C in air, nitrogen or oxygen atmospheres for various times ranging from 20 min to 5 h. Moure et al. [97] obtained $La_{0.8}Sr_{0.2}Ga_{0.85}Mg_{0.15}O_{3-\delta}$ and $La_{0.8}Sr_{0.15}Ga_{0.85}Mg_{0.2}O_{3-\delta}$ samples from $La_2O_3$, $SrCO_3$, $Ga_2O_3$ and MgO, which were mechanochemically activated in a Pulverizette 6 Fritsch planetary mill with stainless steel balls. The mixtures were synthesized at 1300 °C for 16 h; then after milling for 2 h and sieving with a 100-μm sieve, the powders were pressed into pellets and finally sintered at 1550 °C to form the desired ceramic samples.

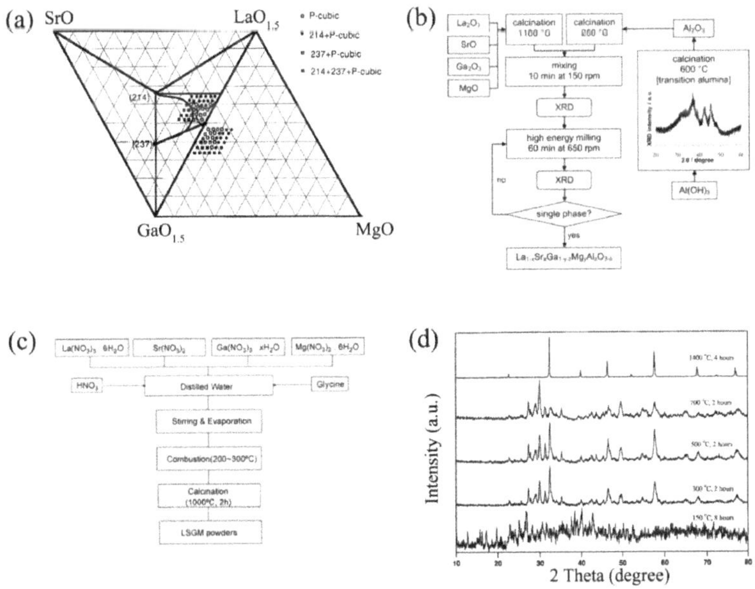


**Figure 5.** The phase and structure features of $LaGaO_3$-based materials: (**a**) phase diagram of a $LaO_{1.5}$–SrO–$GaO_{1.5}$–MgO system up to 800 °C, P-cubic = single-phase $La_{1-x}Sr_xGa_{1-y}Mg_yO_{3-\delta}$, 214 = $LaSrGaO_4$, 237 = $LaSrGa_3O_7$. Reproduced from [95] with permission from John Wiley & Sons, Inc. (Hoboken, NJ, USA), 1998; (**b**) the scheme of mechanosynthesis for the preparation of $La_{1-x}Sr_xGa_{1-y-z}Mg_yAl_zO_{3-\delta}$. Reproduced from [98] with permission by Elsevier Masson SAS, 2012; (**c**) the combustion scheme synthesis for the preparation of $La_{1-x}Sr_xGa_{1-y}Mg_yO_{3-\delta}$. Reproduced from [99] with permission by Elsevier Ltd., 2007; (**d**) XRD pattern evaluation of $La_{0.8}Sr_{0.2}Ga_{0.83}Mg_{0.17}O_{3-\delta}$ precursor powders at various calcination temperatures. Reproduced from [100] with permission from Elsevier Ltd., 1998.

For the synthesis of $La_{0.9}Sr_{0.1}Ga_{1-x}Ni_xO_{3-\delta}$, Colomer and Kilner [101] grinded a mixture of $La_2O_3$, $SrCO_3$, $Ga_2O_3$ and NiO in an agate mortar with acetone medium and then calcined them at 1000 °C for 6 h. After sieving with a 65-μm sieve, milling for 1 h, drying and secondary sieving to 65 μm, the finishing powders were pressed into disks and sintered at 1450–1500 °C for 48 h in air. The authors chose nickel as element for gallium substitution in $La_{0.9}Sr_{0.1}GaO_{3-\delta}$ owing to the proposal about achieving a hopping conductivity among the Ni-sites.

Al-substituted $La_{0.95}Sr_{0.05}Ga_{0.9}Mg_{0.1}O_{3-\delta}$ and $La_{0.9}Sr_{0.1}Ga_{0.8}Mg_{0.2}O_{3-\delta}$ derivatives were prepared using $La_2O_3$, $Ga_2O_3$, SrO, MgO and $Al_2O_3$ [98]. Mechanosynthesis was employed in a planetary mill (Retsch PM100, PM200) with tetragonal zirconia balls, according to a scheme presented in Figure 5b. The powders were pressed into disks that were sintered at 1300–1450 °C for 2–24 h.

As can be seen, the aforementioned methods (solid-state reaction synthesis and the mechanochemical route) that were conventionally used for the preparation of $La_{1-x}Sr_xGa_{1-y}Mg_yO_{3-\delta}$ and its derivatives have two considerable disadvantages. First, high sintering temperatures (above 1450–1500 °C) are required for full densification of the pressed pellets [51]. This can influence the production cost of the final electrolyte materials. Second, the appearance of $Sr_3La_4O_9$, $SrLaGa_3O_7$ and/or $SrLaGaO_4$ impurity phases in $La_{1-x}Sr_xGa_{1-y}Mg_yO_{3-\delta}$ samples was frequently observed. This was due to gallium evaporation [102], which resulted in the deterioration of the gallate material's ionic conductivity [51]. To solve the problems that arise during $La_{1-x}Sr_xGa_{1-y}Mg_yO_{3-\delta}$ preparation, techniques based on co-precipitation [103,104], organic-nitrate precursors combustion [96,99,100,105–109], self-propagating, high-temperature synthesis [110,111] and spray-pyrolysis [112] were developed.

For example, $La_{0.8}Sr_{0.2}Ga_{0.8}Mg_{0.2}O_{3-\delta}$ samples were prepared with carbonate co-precipitation from $La(NO_3)_3{\cdot}6H_2O$, $Sr(NO_3)_2$, $Ga(NO_3)_3{\cdot}xH_2O$ and $Mg(NO_3)_2{\cdot}6H_2O$ starting reagents [103]. The resulting aqueous solution containing $La^{3+}$, $Sr^{2+}$, $Ga^{3+}$ and $Mg^{2+}$ cations was gradually dropped into an aqueous $(NH_4)_2CO_3$ solution with heating at 70 °C. After 2 h of homogenization with continuous stirring, the formed sediments were washed, dried at 25 °C for 24 h in a $N_2$ atmosphere, and finally calcined in air at 900–1300 °C for 12 h.

Huang and Goodenough [100] have reported the use of wet synthesis techniques (the sol-gel technique and the Pechini method) for forming single-phase $La_{0.8}Sr_{0.2}Ga_{0.83}Mg_{0.17}O_{3-\delta}$ materials. Solutions of $La(CH_3COO)_3$, $Sr(CH_3COO)_2$ and $Mg(CH_3COO)_2$ acetates and $La(NO_3)_3$, $Sr(NO_3)_2$, $Ga(NO_3)_3$ and $Mg(NO_3)_2$ nitrates were used in these preparation methods. During synthesis with sol-gel technology, the required amounts of metal acetates and gallium nitrate solutions were mixed by stirring. An ammonia solution was then added, forming a white gel. This was aged at 25 °C for 72 h and heated at 150 °C for 8 h upon full water evaporation. The resulting product was fired at 300, 500 and 700 °C at varying times. Using the Pechini method, $La_{0.8}Sr_{0.2}Ga_{0.83}Mg_{0.17}O_{3-\delta}$ samples were prepared from a mixture of the necessary amounts of metal nitrate solutions at 25 °C: citric acid was then added. The citric acid was used to fulfil a mole ratio of citric acid/total cations around 1.5/1. After stirring the precursor solution, ethylene glycol was added in an equal amount to the citric acid. The obtained solution was heated at 150 °C for 12 h and resulted in a polymer-like solid material. This resin was slowly heated to 300 °C and, after several sintering stages, it was finally calcined at 1400 °C for 4 h [100]. The pressed $La_{0.85}Sr_{0.15}Ga_{0.8}Mg_{0.2}O_{3-\delta}$ samples were found to be single-phase after they were obtained via the Pechini method and annealed at 1400 °C for 6 h [105].

A $La_{0.8}Sr_{0.2}Ga_{0.85}Mg_{0.15}O_{3-\delta}$ sample was also obtained via the glycine-nitrate combustion method [106]. Ga, $La_2O_3$, MgO and $SrCO_3$ powders were dissolved in strong $HNO_3$ and mixed with water. Glycine was then added with a molar ratio of glycine/nitrate ions equal to 1:1. The glass beaker with the precursor glycine–nitrate solution was heated on a hot plate with spontaneous burning, which resulted in a white powder. Dense samples were formed at a temperature range of 1400–1550 °C for 6 h at each stage [106]. A similar method was used in [107] for the synthesis of $La_{0.9}Sr_{0.1}Ga_{0.8}Mg_{0.2}O_{3-\delta}$. The experimental procedure included the heating of the precursor glycine–nitrate solution at 550 °C upon combustion, initial calcination of voluminous oxide powders at 800 °C for 3 h, annealing the powders at 1000 °C and final annealing at 1300 °C for 2 h. It should be noted that the authors of [107] could not achieve single-phase sample. Huang and Goodenough also concluded that a $La_{0.8}Sr_{0.2}Ga_{0.83}Mg_{0.17}O_{3-\delta}$ single-phase material cannot be formed via hydrothermal treatment synthesis [100]. A typical diagram of $La_{1-x}Sr_xGa_{1-y}Mg_yO_{3-\delta}$ synthesis via the glycine–nitrate combustion method described in [99] is presented in Figure 5c.

In [110], Ishikawa et al., prepared $La_{0.9}Sr_{0.1}Ga_{0.8}Mg_{0.2}O_{3-\delta}$ and $La_{0.9}Sr_{0.1}Ga_{0.7}Mg_{0.3}O_{3-\delta}$ samples via self-propagating high-temperature synthesis from $La_2O_3$, $SrCO_3$, $Ga_2O_3$, Mg and $NaClO_4$. An initial powder mixture was supplied to a self-propagating synthesis reactor: it was then ignited with a disposable carbon foil in contact with the sample. The obtained powders were washed with water to remove NaCl. The samples were pressed into disks in vacuum and then sintered at a temperature range of 1000–1500 °C for 6 h in air. An alternative process for $La_{0.9}Sr_{0.1}Ga_{0.8}Mg_{0.2}O_{3-\delta}$ synthesis based on a preliminarily mechanically activated powder mixture was proposed by Ishikawa et al. [111]. The initial mixture was grinded in a planetary mill with stainless steel balls. The powder sample was pressed into a disk, which was placed in a self-propagating synthesis reactor: the aforementioned algorithm [110] was then used.

The literature points out that temperature of about 1400 °C (or more) is required for the synthesis of single-phase LSGM samples. Figure 5d presents the thermal evolution of the

XRD pattern for a $La_{0.8}Sr_{0.2}Ga_{0.83}Mg_{0.17}O_{3-\delta}$ precursor powder [100]. The powders calcined at the intermediate temperatures were multiphase, containing $La_{0.8}Sr_{0.2}Ga_{0.83}Mg_{0.17}O_{3-\delta}$ and $La_2O_3$, $LaSrGa_3O_7$ and $La_2O_2CO_3$ impurities. A single-phase $La_{0.8}Sr_{0.2}Ga_{0.83}Mg_{0.17}O_{3-\delta}$ sample with a cubic structure was formed during calcination at 1400 °C.

It is worth noting that the crystal structure of the obtained LSGM samples depends on the strontium and manganese dopant contents. Basic $LaGaO_3$ at room temperature has an orthorhombic structure [113] but varying the doping contents can change the crystal structure symmetry [100,114]. Generally, the substitution of $La^{3+}$-ions with $Sr^{2+}$-ions increases the tolerance factor $t$ (Equation (1)), while Ga-with-Mg substitution decreases it. Therefore, the $t$ factor for $La_{1-x}Sr_xGa_{1-y}Mg_yO_{3-\delta}$ is nearly equal to that calculated for undoped $LaGaO_3$.

The $t$ factor is equal to 1 for $La_{0.8}Sr_{0.2}Ga_{0.8}Mg_{0.2}O_{3-\delta}$, which exhibits an ideal *Pm-3m* cubic structure with a unit cell parameter of $a$ = 3.9146(1) Å [114] (Figure 6a). According to [114], the crystal structure of $La_{0.9}Sr_{0.1}Ga_{0.8}Mg_{0.2}O_{3-\delta}$ and $La_{0.9}Sr_{0.1}Ga_{0.9}Mg_{0.1}O_{3-\delta}$ samples (Figure 5a) was refined in a *I2/a* monoclinic space group.

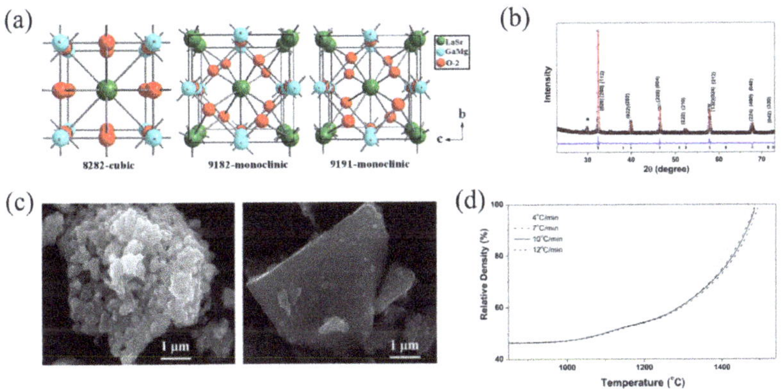


**Figure 6.** Properties of $LaGaO_3$-based phases: (**a**) the crystal structure of $La_{0.8}Sr_{0.2}Ga_{0.8}Mg_{0.2}O_{3-\delta}$ (8282), $La_{0.9}Sr_{0.1}Ga_{0.8}Mg_{0.2}O_{3-\delta}$ (9182) and $La_{0.9}Sr_{0.1}Ga_{0.9}Mg_{0.1}O_{3-\delta}$ (9191). Reproduced from [114] with permission from John Wiley & Sons, Inc., 2021; (**b**) observed and Rietveld-refined XRD patterns of $La_{0.9}Sr_{0.1}Ga_{0.8}Mg_{0.2}O_{3-\delta}$. Reproduced from [115] with permission by Elsevier Ltd., 2018; (**c**) an SEM micrograph of a $La_{0.9}Sr_{0.1}Ga_{0.8}Mg_{0.2}O_{3-\delta}$ ceramic obtained via mechanically activated and conventional self-propagating synthesis. Reproduced from [111] with permission by Elsevier Ltd., 2009; (**d**) the temperature dependencies of the relative density of a $La_{0.9}Sr_{0.1}Ga_{0.8}Mg_{0.2}O_{3-\delta}$ ceramic material. Reproduced from [115] with permission from Elsevier Ltd., 2018.

The crystal structure of $LaGaO_3$ and $La_{0.9}Sr_{0.1}Ga_{0.8}Mg_{0.2}O_{3-\delta}$ samples was investigated via powder neutron diffraction at 25, 800 and 1000 °C in [116]. According to the Rietveld refinement analysis of the diffraction data collected at 25 °C, an orthorhombic structure was observed for both samples: fitting was provided in the *Pnma* space group for $LaGaO_3$ (unit cell parameters were equal to $a$ = 5.4908(1), $b$ = 7.7925(1) and $c$ = 5.5227(1) Å) and in the *Imma* space group for $La_{0.9}Sr_{0.1}Ga_{0.8}Mg_{0.2}O_{3-\delta}$ (unit cell parameters were equal to $a$ = 5.5179(1), $b$ = 7.8200(1) and $c$ = 5.5394(1) Å). The high temperature measurements [116] show that the $LaGaO_3$ sample possessed a rhombohedral structure in the *R-3c* space group (unit cell parameters were equal to $a$ = 5.5899(1) Å and $a$ = 5.5987(1) Å at 800 and 1000 °C, correspondingly), whereas $La_{0.9}Sr_{0.1}Ga_{0.8}Mg_{0.2}O_{3-\delta}$ exhibits a cubic structure in the *Pm3m* space group (unit cell parameters were equal to $a$ = 3.9760(1) Å and $a$ = 3.9866(1) Å at 800 and 1000 °C, correspondingly). Similar data at 25 °C (the *Imma* space group, $a$ = 5.5056(9), $b$ = 7.8241(7), $c$ = 5.5387(5) Å) for a $La_{0.9}Sr_{0.1}Ga_{0.8}Mg_{0.2}O_{3-\delta}$ sample obtained via solid-state route and sintered at 1350 °C for 2 h was reported in [115]. However, this sample consisted of an LSGM phase and a $LaSrGa_3O_7$ impurity phase, as indicated by '*' in Figure 6b. This fact proves the necessity of sintering temperatures of 1400 °C for obtaining single-phase LSGM samples.

Comparative analysis of the microstructural parameters for $La_{0.9}Sr_{0.1}Ga_{0.8}Mg_{0.2}O_{3-\delta}$ disks sintered at 1400 °C for 6 h obtained via the self-propagating high-temperature and solid-reaction synthesis techniques showed that the first sample was denser [110]. The relative densities of the samples were 98 and 92%, respectively, despite the fact that the sintering temperature for the first disk was 100 °C lower than that for the second one. Images in Figure 6c show the SEM micrographs of $La_{0.9}Sr_{0.1}Ga_{0.8}Mg_{0.2}O_{3-\delta}$ samples obtained via self-propagating synthesis with and without mechanical activation of the starting mixture for 24 h [111]. These SEM images testify that mechanically activated self-propagating synthesis provided the high-grade powders with nano-size particles. The specific surface areas of the samples were 3.36 and 2.06 $m^2\ g^{-1}$, respectively. Based on both studies, Ishikawa et al. [110,111] concluded that this proved the advantages of using self-propagating high-temperature synthesis (especially with mechanical activation of the starting mixture) in comparison with the solid-reaction method.

The evolution of a $La_{0.9}Sr_{0.1}Ga_{0.8}Mg_{0.2}O_{3-\delta}$ sample's density against temperature was provided in by Batista et al. [115]. Based on dilatometry experimental results (Figure 6d), the authors separated the process into three steps: an insignificant increase of relative density at 25–1000 °C; gradual densification at 1000–1300 °C; and, finally, a fast densification above 1300 °C. According to [117], a relative density of over 99% was achieved after calcination at 1450 °C for 6 h.

Summing up the review section, which was devoted to the synthesis methods of Sr, Mg-doped $LaGaO_3$ oxides as electrolyte materials, the self-propagating high-temperature synthesis with mechanical activation of the starting mixtures can be identified as one of the most optimal techniques. The above-mentioned method can obtain the single-phase $La_{0.9}Sr_{0.1}Ga_{0.8}Mg_{0.2}O_{3-\delta}$ powders with high specific surface areas, a narrow distribution of nano-size particles, and high relative densities for the sintered ceramic samples.

### 3.2. Functional Properties

In 1994, Ishihara et al. [51] were the first to show that the La-substitution of $LaGaO_3$ with strontium and gallium with magnesium increased the electrical conductivity of doped materials (Figure 7a,b) owing to the formation of oxygen vacancies in $La_{1-x}Sr_xGa_{1-y}Mg_yO_{3-\delta}$ [118].

The measurements of Ishihara [51], Stevenson [119] and Goodenough [95] demonstrate that the $La_{1-x}Sr_xGa_{1-y}Mg_yO_{3-\delta}$ samples possess maximal electrical conductivity values at $x = 0.15/0.2$ and $y = 0.2$, as can be seen in Table 2. It should be also noted that conductivity of nominally similar materials can be varied over a wide range (see Figure A2). This confirms that the microstructural parameters of ceramics, as well as the presence of insulating impurity phases, considerably affect the transport properties of gallates, encouraging the continuous search for their new synthesis and fabricating techniques.

Hayashi et al. [120] concluded that the electrical conductivity of $La_{1-x}Sr_xGa_{1-y}Mg_yO_{3-\delta}$ becomes greater when approaching the tolerance factor of the doped sample to $t$ for $LaGaO_3$ and decreases when the tolerance factor for the doped samples differed from $t$ for $LaGaO_3$. It was established that increasing the Sr, Mg-doping levels led to the association of oxygen vacancies [51,119,120]; for this reason, further electrical investigations of the doped-$LaGaO_3$ oxides were performed on $La_{1-x}Sr_xGa_{1-y}Mg_yO_{3-\delta}$ samples with a fixed content of Sr and Mg dopants (nearly 20 mol.%, i.e., $x = y = 0.2$). The literature on the transport properties of $La_{1-x}Sr_xGa_{1-y}Mg_yO_{3-\delta}$ ceramic samples is summarised in Table 2. Figure 7c presents the temperature dependencies of conductivity for the $La_{0.9}Sr_{0.1}Ga_{0.9}Mg_{0.1}O_{3-\delta}$ (LSGM9191), $La_{0.9}Sr_{0.1}Ga_{0.8}Mg_{0.2}O_{3-\delta}$ (LSGM9182) and $La_{0.8}Sr_{0.2}Ga_{0.8}Mg_{0.2}O_{3-\delta}$ (LSGM8282) samples obtained in [114]. These data agree with the conclusion that the maximal conductivity for LSGM is achieved at $x = y = 0.2$.

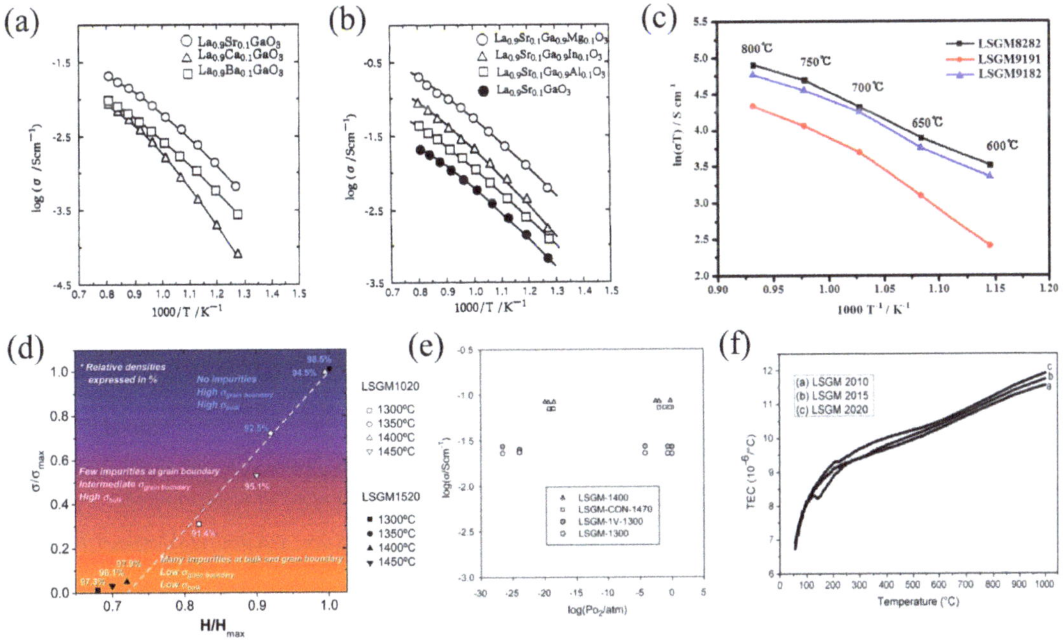

**Figure 7.** Properties of $LaGaO_3$-based phases: (**a**) the crystal structure of $La_{0.8}Sr_{0.2}Ga_{0.8}Mg_{0.2}O_{3-\delta}$ (8282), $La_{0.9}Sr_{0.1}Ga_{0.8}Mg_{0.2}O_{3-\delta}$ (9182) and $La_{0.9}Sr_{0.1}Ga_{0.9}Mg_{0.1}O_{3-\delta}$ (9191). Reproduced from [114] with permission from John Wiley & Sons, Inc., 2021; (**b**) observed and Rietveld-refined XRD patterns of $La_{0.9}Sr_{0.1}Ga_{0.8}Mg_{0.2}O_{3-\delta}$. Reproduced from [115] with permission form Elsevier Ltd., 2018; (**c**) an SEM micrograph of a $La_{0.9}Sr_{0.1}Ga_{0.8}Mg_{0.2}O_{3-\delta}$ ceramic obtained via mechanically activated and conventional self-propagating synthesis. Reproduced from [111] with permission by Elsevier Ltd., 2009; (**d**) the temperature dependencies of the relative density of a $La_{0.9}Sr_{0.1}Ga_{0.8}Mg_{0.2}O_{3-\delta}$ ceramic material. Reproduced from [115] with permission from Elsevier Ltd., 2018; (**e**) conductivity of $La_{0.8}Sr_{0.2}Ga_{0.8}Mg_{0.2}O_{3-\delta}$ as a function of oxygen partial pressure. Reproduced from [109] with permission from Elsevier Ltd., 2011; (**f**) the temperature dependencies of TEC for $La_{0.8}Sr_{0.2}Ga_{0.9}Mg_{0.1}O_{3-\delta}$ (LSGM2010), $La_{0.8}Sr_{0.2}Ga_{0.85}Mg_{0.15}O_{3-\delta}$ (LSGM2015) and $La_{0.8}Sr_{0.2}Ga_{0.8}Mg_{0.2}O_{3-\delta}$ (LSGM2020). Reproduced from [121] with permission from Elsevier Ltd., 2009.

**Table 2.** Total conductivities of $LaGaO_3$-based materials depending on their compositions, preparation methods and temperatures. Figure A2 provides a visualization of these data.

| Sample | Samples Obtaining Method; Annealing Temperature (°C) | $T$ (°C) | $\sigma$ (S cm$^{-1}$) | Ref. |
|---|---|---|---|---|
| $LaGaO_3$ | Solid-state route; 1500 | 950 | 0.02 | [51] |
| $La_{0.9}Sr_{0.1}Ga_{0.9}Mg_{0.1}O_{3-\delta}$ | Solid-state route; 1500 | 950 | 0.20 | [51] |
| $La_{0.9}Sr_{0.1}Ga_{0.85}Mg_{0.15}O_{3-\delta}$ | Solid-state route; 1500 | 950 | 0.27 | [51] |
| $La_{0.9}Sr_{0.1}Ga_{0.8}Mg_{0.2}O_{3-\delta}$ | Solid-state route; 1500 | 950 | 0.29 | [51] |
| $La_{0.9}Sr_{0.1}Ga_{0.7}Mg_{0.3}O_{3-\delta}$ | Solid-state route; 1500 | 950 | 0.28 | [51] |
| $La_{0.9}Sr_{0.1}Ga_{0.6}Mg_{0.4}O_{3-\delta}$ | Solid-state route; 1500 | 950 | 0.10 | [51] |
| $La_{0.9}Sr_{0.1}Ga_{0.8}Mg_{0.2}O_{3-\delta}$ | Glycine-combustion method; 1400 | 1000 | 0.26 | [51] |
| $La_{0.85}Sr_{0.15}Ga_{0.8}Mg_{0.2}O_{3-\delta}$ | Glycine-combustion method; 1400 | 1000 | 0.36 | [51] |
| $La_{0.8}Sr_{0.2}Ga_{0.85}Mg_{0.15}O_{3-\delta}$ | Glycine-combustion method; 1400 | 1000 | 0.31 | [51] |
| $La_{0.8}Sr_{0.2}Ga_{0.8}Mg_{0.2}O_{3-\delta}$ | Glycine-combustion method; 1400 | 1000 | 0.40 | [51] |

**Table 2.** *Cont.*

| Sample | Samples Obtaining Method; Annealing Temperature (°C) | *T* (°C) | $\sigma$ (S cm$^{-1}$) | Ref. |
|---|---|---|---|---|
| $La_{0.9}Sr_{0.1}Ga_{0.9}Mg_{0.1}O_{3-\delta}$ | Solid-state route; 1470 | 800 | 0.116 | [95] |
| $La_{0.9}Sr_{0.1}Ga_{0.85}Mg_{0.15}O_{3-\delta}$ | Solid-state route; 1470 | 800 | 0.127 | [95] |
| $La_{0.9}Sr_{0.1}Ga_{0.8}Mg_{0.2}O_{3-\delta}$ | Solid-state route; 1470 | 800 | 0.132 | [95] |
| $La_{0.9}Sr_{0.1}Ga_{0.7}Mg_{0.3}O_{3-\delta}$ | Solid-state route; 1470 | 800 | 0.096 | [95] |
| $La_{0.85}Sr_{0.15}Ga_{0.8}Mg_{0.2}O_{3-\delta}$ | Solid-state route; 1470 | 800 | 0.150 | [95] |
| $La_{0.8}Sr_{0.2}Ga_{0.85}Mg_{0.15}O_{3-\delta}$ | Solid-state route; 1470 | 800 | 0.149 | [95] |
| $La_{0.8}Sr_{0.2}Ga_{0.83}Mg_{0.17}O_{3-\delta}$ | Solid-state route; 1470 | 800 | 0.17 | [95] |
| $La_{0.8}Sr_{0.2}Ga_{0.8}Mg_{0.2}O_{3-\delta}$ | Solid-state route; 1470 | 800 | 0.14 | [95] |
| $La_{0.7}Sr_{0.3}Ga_{0.8}Mg_{0.2}O_{3-\delta}$ | Solid-state route; 1470 | 800 | 0.109 | [95] |
| $La_{0.9}Sr_{0.1}Ga_{0.8}Mg_{0.2}O_{3-\delta}$ | Self-propagating high-temperature synthesis; 1500 | 800 | 0.11 | [110] |
| $La_{0.9}Sr_{0.1}Ga_{0.8}Mg_{0.2}O_{3-\delta}$ | Carbonate co-precipitation; 1400 | 800 | 0.045 | [104] |
| $La_{0.9}Sr_{0.1}Ga_{0.9}Mg_{0.1}O_{3-\delta}$ | Solid-state route; 1450 | 800 | 0.071 | [114] |
| $La_{0.9}Sr_{0.1}Ga_{0.8}Mg_{0.2}O_{3-\delta}$ | Solid-state route; 1450 | 800 | 0.1095 | [114] |
| $La_{0.9}Sr_{0.1}Ga_{0.8}Mg_{0.2}O_{3-\delta}$ | Glycine-combustion method; 1500 | 800 | 0.092 | [122] |
| $La_{0.9}Sr_{0.1}Ga_{0.8}Mg_{0.2}O_{3-\delta}$ | Glycine-combustion method; 1400 | 800 | 0.0395 | [123] |
| $La_{0.85}Sr_{0.15}Ga_{0.85}Mg_{0.15}O_{3-\delta}$ | Acrylamide polymerization technique; 1432 | 800 | 0.093 | [124] |
| $La_{0.85}Sr_{0.15}Ga_{0.8}Mg_{0.2}O_{3-\delta}$ | Mechanochemical route; 1380 | 600 | 0.016 | [97] |
| $La_{0.85}Sr_{0.15}Ga_{0.8}Mg_{0.2}O_{3-\delta}$ | Glycine-combustion method; 1300 | 800 | 0.053 | [125] |
| $La_{0.85}Sr_{0.15}Ga_{0.8}Mg_{0.2}O_{3-\delta}$ | EDTA-combustion method; 1300 | 800 | 0.06 | [125] |
| $La_{0.85}Sr_{0.15}Ga_{0.8}Mg_{0.2}O_{3-\delta}$ | Glycine-combustion method; 1400 | 800 | 0.096 | [105] |
| $La_{0.85}Sr_{0.15}Ga_{0.8}Mg_{0.2}O_{3-\delta}$ | Pechini method; 1400 | 800 | 0.135 | [126] |
| $La_{0.8}Sr_{0.2}Ga_{0.8}Mg_{0.2}O_{3-\delta}$ | Carbonate co-precipitation; 1300 | 600 | 0.014 | [103] |
| $La_{0.8}Sr_{0.2}Ga_{0.8}Mg_{0.2}O_{3-\delta}$ | Glycine-combustion method; 1300 | 700 | 0.022 | [109] |
| $La_{0.8}Sr_{0.2}Ga_{0.8}Mg_{0.2}O_{3-\delta}$ | Glycine-combustion method; 1400 | 700 | 0.085 | [109] |
| $La_{0.8}Sr_{0.2}Ga_{0.8}Mg_{0.2}O_{3-\delta}$ | Spray pyrolysis; 1400 | 500 | 0.0029 | [112] |
| $La_{0.8}Sr_{0.2}Ga_{0.8}Mg_{0.2}O_{3-\delta}$ | Solid-state route; 1450 | 800 | 0.126 | [127] |
| $La_{0.8}Sr_{0.2}Ga_{0.8}Mg_{0.2}O_{3-\delta}$ | Solid-state route; 1400 | 800 | 0.035 | [127] |
| $La_{0.8}Sr_{0.2}Ga_{0.8}Mg_{0.2}O_{3-\delta}$ | Hydrothermal urea hydrolysis precipitation; 1400 | 800 | 0.056 | [127] |
| $La_{0.8}Sr_{0.2}Ga_{0.8}Mg_{0.2}O_{3-\delta}$ | Carbonate co-precipitation; 1400 | 800 | 0.137 | [128] |
| $La_{0.8}Sr_{0.2}Ga_{0.8}Mg_{0.2}O_{3-\delta}$ | Solid-state route; 1250 | 727 | 0.019 | [129] |
| $La_{0.8}Sr_{0.2}Ga_{0.8}Mg_{0.2}O_{3-\delta}$ | Sol-gel technique; 1300 | 450 | $2.9 \times 10^{-4}$ | [130] |
| $La_{0.8}Sr_{0.2}Ga_{0.8}Mg_{0.2}O_{3-\delta}$ | Solid-state route; 1400 | 800 | 0.132 | [131] |
| $La_{0.8}Sr_{0.2}Ga_{0.8}Mg_{0.2}O_{3-\delta}$ | Thin film deposited by vacuum cold spray; 200 | 750 | 0.043 | [132] |
| $La_{0.8}Sr_{0.2}Ga_{0.8}Mg_{0.2}O_{3-\delta}$ | Step-wise current-limiting flash sintering process; 690 | 850 | 0.072 | [133] |

It was shown in [119] that the ion-transfer numbers were nearly equal to 1. For $La_{0.9}Sr_{0.1}Ga_{0.8}Mg_{0.2}O_{3-\delta}$ and $La_{0.8}Sr_{0.2}Ga_{0.8}Mg_{0.2}O_{3-\delta}$ ceramic samples, the oxygen-ion transference numbers were found to be equal 1 at 700–1000 °C [107], confirming the presence of electrolyte-type behaviour. Savioli and Watson [134] studied the defect structure of $LaGaO_3$ upon the use of various doping strategies using DFT calculations. They confirmed that Sr-, Ba-, and Mg-doping should result in the greatest improvements to the ionic conductivity of the $LaGaO_3$ parent phase, while the $Ni^{2+}$-, $Co^{2+}$-, $Fe^{2+}$-, and $Zn^{2+}$-doping is responsible for the generation of a mixed ionic-electronic conducting behaviour. Sr- and Mg- co-doped $LaGaO_3$ complex oxides are predominantly oxygen-ionic conductors, for which the electronic conductivity levels are 3–4 magnitudes lower compared to the oxygen-ionic conductivity levels [135].

According to [125], the dependence $\ln(\sigma T)$ vs. $1/T$ had a break at 700 °C for $La_{0.85}Sr_{0.15}Ga_{0.8}Mg_{0.2}O_{3-\delta}$, which indicates that the activation energy value of oxygen-ion conductivity at a low-temperature range was higher than that at a high-temperature range.

A linear correlation between hardness and total ionic conductivity was revealed in [126] for $La_{0.9}Sr_{0.1}Ga_{0.8}Mg_{0.2}O_{3-\delta}$ and $La_{0.85}Sr_{0.15}Ga_{0.8}Mg_{0.2}O_{3-\delta}$ samples. It was shown that the electrical and mechanical properties of $La_{1-x}Sr_xGa_{1-y}Mg_yO_{3-\delta}$ are strongly defined by microstructural peculiarities and the presence of low-conductive $LaSrGaO_4$ and $LaSrGa_3O_7$ impurity phases [123]. The $LaSrGaO_4$ phase exhibits a tetragonal structure $K_2NiF_4$-type and crystalizes in the *I4/mmm* space group; its conductivity is found to be around $2 \cdot 10^{-7}$ S cm$^{-1}$ at 900 °C [136]. The $LaSrGa_3O_7$phase belongs to a melilitestructure described in the $P421m$ space group; its ionic conductivity level is around $2 \cdot 10^{-6}$ S $cm^{-1}$ at 800 °C [137]. The maximum values of ionic conductivity and hardness were achieved for single-phase $La_{0.9}Sr_{0.1}Ga_{0.8}Mg_{0.2}O_{3-\delta}$ (LSGM1020) and $La_{0.85}Sr_{0.15}Ga_{0.8}Mg_{0.2}O_{3-\delta}$ (LSGM1520) samples with a high relative density, as shown in Figure 7d. With a significant amount of impurity phases at the grain boundaries, the samples exhibited a gradual decrease in hardness and the grain boundary conductivity, which resulted in a decreasing total conductivity. The data in Table 2 may also be analysed from the aforementioned perspective.

The electrical conductivity of $La_{0.8}Sr_{0.2}Ga_{0.8}Mg_{0.2}O_{3-\delta}$ was investigated over a $Po_2$ range of $10^{-27}$–1 atm at 700 °C [109]. The results of the measurements are presented in Figure 7e for $La_{0.8}Sr_{0.2}Ga_{0.8}Mg_{0.2}O_{3-\delta}$ samples, sintered at 1470 °C (LSGM-CON-1400), 1400 °C (LSGM-1400) and 1300 °C (LSGM-1300), and an LSGM sample with 1 wt% $V_2O_5$ sintered at 1300 °C (LSGM-1V-1300). All these samples show an approximately constant conductivity over the measuring $pO_2$ range, implying a realization of the electrolytic conduction behaviour.

The thermal expansion of $La_{1-x}Sr_xGa_{1-y}Mg_yO_{3-\delta}$ was studied by Baskaran et al. [138]. The TEC values measured for the $La_{0.9}Sr_{0.1}Ga_{0.8}Mg_{0.2}O_{3-\delta}$ sample were equal to $10 \times 10^{-6}$ $K^{-1}$ over a low-temperature range and 13.5–14.0 $\times$ $10^{-6}$ $K^{-1}$ above 600 °C. Lee et al. [99] reported about an average TEC of $12.1 \times 10^{-6}$ $K^{-1}$ for $La_{0.8}Sr_{0.2}Ga_{0.8}Mg_{0.2}O_{3-\delta}$ at a temperature range of 25–1000 °C, which is close to $12.3 \times 10^{-6}$ $K^{-1}$ for a $La_{0.65}Sr_{0.3}MnO_{3-\delta}$ electrode at the same temperatures [92].

The expansion behaviour for $La_{1-x}Sr_xGa_{1-y}Mg_yO_{3-\delta}$ is correlated with its crystal structure in the observed temperature range. Therefore, the presence of a phase transition from an orthorhombic phase to a cubic one for $La_{0.9}Sr_{0.1}Ga_{0.8}Mg_{0.2}O_{3-\delta}$ [116] and the existence of an ideal perovskite cubic structure for $La_{0.8}Sr_{0.2}Ga_{0.8}Mg_{0.2}O_{3-\delta}$ [114] are responsible for the aforementioned variations in their thermal expansion behaviour.

Datta et al. [121] observed that the temperature of phase transition from an orthorhombic to a rhombohedral structure for $La_{1-x}Sr_xGa_{1-y}Mg_yO_{3-\delta}$ increased as Mg content increased at a fixed Sr content, as shown in Figure 7f, and decreased with increasing Sr content at a fixed Mg content. The effect of Sr and Mg co-doping on TEC values was explained for $La_{1-x}Sr_xGa_{1-y}Mg_yO_{3-\delta}$ in terms of the amount of generated oxygen vacancies. It was concluded that TEC values increased as oxygen vacancies increase, regardless of the dopant type. This was the result of the binding energy weakening as a result of oxygen vacancy formation.

Shkerin et al. [139] analysed the structure and phase transitions of $La_{0.88}Sr_{0.12}Ga_{0.82}Mg_{0.18}O_{3-\delta}$ using dilatometry, XRD and Raman spectroscopy. According to the obtained data, $La_{0.88}Sr_{0.12}Ga_{0.82}Mg_{0.18}O_{3-\delta}$ exhibited two phase transitions of the second order at 502 and 607 °C. The first transition was attributed to a phase transition from an orthorhombic phase to a cubic one, while the second phase transition was attributed to the ordering of the oxygen vacancies.

Wu et al. [140] studied transport properties of $La_{0.85}Sr_{0.15}Ga_{0.8}Mg_{0.2}O_{3-\delta}$ upon the partial or full Sr-substitution with calcium or barium. Their analyses have shown that both types of substitution result in a decrease in ionic conductivity by 20–30%. However, at the same time, the Ca-substituted ceramic materials showed higher conductivities compared to the Ba-substituted analogues. This confirms that strontium is an ideal dopant (from the steric and energetic viewpoints) to be introduced into the La-sublattice of $LaGaO_3$-based phases.

The chemical compatibility of $La_{1-x}Sr_xGa_{1-y}Mg_yO_{3-\delta}$ was investigated with oxide materials used in SOFCs, cathodes [141–152] and anodes [153–167]: this is presented in the corresponding reviews [28,56,153].

Chemical interactions between a $La_{0.9}Sr_{0.1}Ga_{0.8}Mg_{0.2}O_{3-\delta}$ electrolyte and cathode materials such as $La_{0.65}Sr_{0.3}MnO_{3-\delta}$, $La_{0.7}Sr_{0.3}CoO_{3-\delta}$, $La_{0.65}Sr_{0.3}FeO_{3-\delta}$, $La_{0.65}Sr_{0.3}NiO_{3-\delta}$ and $La_{0.6}Sr_{0.4}Co_{0.2}Fe_{0.8}O_{3-\delta}$ are demonstrated in [141]. The LSGM/cathode powders were mixed at a weight ratio of 1:1, pressed into disks and annealed at 1300 °C for 3 h in air. The XRD data revealed that impurity phases were not formed in the LSGM mixed with

$La_{0.65}Sr_{0.3}MnO_{3-\delta}$, $La_{0.7}Sr_{0.3}CoO_{3-\delta}$, and $La_{0.65}Sr_{0.3}FeO_{3-\delta}$, but appear in the calcined mixtures with $La_{0.65}Sr_{0.3}NiO_{3-\delta}$ and $La_{0.6}Sr_{0.4}Co_{0.2}Fe_{0.8}O_{3-\delta}$. The absence of reactivity between $La_{0.8}Sr_{0.2}Ga_{0.8}Mg_{0.2}O_{3-\delta}$ and $La_{0.8}Sr_{0.2}MnO_{3-\delta}$ was also confirmed during calcination at 800 °C [142].

Sydyknazar et al. [143] showed that $La_{0.83}Sr_{0.17}Ga_{0.8}Mg_{0.2}O_{3-\delta}$ exhibited good chemical compatibility with a novel cathode material, $Sr_{0.9}Ba_{0.1}Co_{0.95}Ru_{0.05}O_{3-\delta}$, after joint calcination at 1100 °C for 12 h. According to the literature, $La_{0.9}Sr_{0.1}Ga_{0.8}Mg_{0.2}O_{3-\delta}$ does not react with the following cathodes: $La_{0.4}Sr_{0.6}Co_{0.9}Sb_{0.1}O_{3-\delta}$ after heat treatment at 1150 °C for 6 h [144], $SrCo_{0.8}Fe_{0.1}Nb_{0.1}O_{3-\delta}$ at 950 °C for 10 h [145], $BaCo_{0.7}Fe_{0.2}Ta_{0.1}O_{3-\delta}$ at 950 °C for 10 h [146] and $Sr_2Ti_{0.8}Co_{0.2}FeO_{6-\delta}$ after at 950 °C for 10 h [147]. According to Tarancón et al. [148], $La_{0.8}Sr_{0.2}Ga_{0.8}Mg_{0.2}O_{3-\delta}$ interacted with a $GdBaCo_2O_{5+\delta}$ cathode at temperatures above 900 °C, forming $BaLaGa_3O_4$ and $BaLaGa_3O_7$ secondary phases.

An analysis of works devoted to Ruddlesden–Popper phases demonstrates that $La_{0.9}Sr_{0.1}Ga_{0.8}Mg_{0.2}O_{3-\delta}$ and $Pr_{2-x}La_xNi_{0.85}Cu_{0.1}Al_{0.05}O_{4+\delta}$ ($x = 0, 0.2, 0.5, 1.0$) have no interactions at 1000 °C for 5 h [149], but $La_{0.95}Sr_{0.05}Ga_{0.9}Mg_{0.1}O_{3-\delta}$ reacted with $Nd_2NiO_{4+\delta}$ after annealing at 1000 °C for 5 h [150]. Equally, $La_{0.85}Sr_{0.15}Ga_{0.85}Mg_{0.15}O_{3-\delta}$ reacted with $Pr_{2-x}Ca_xNiO_{4+\delta}$ after annealing at 900 °C for 10 h ($x = 0, 0.5$) [151] and at 1200 °C for 1 h ($x = 0, 0.3$) [152].

Zhang et al. [154] showed that a $La_{0.9}Sr_{0.1}Ga_{0.8}Mg_{0.2}O_{3-\delta}$ electrolyte reacted with the nickel component in a Ni-SDC anode. The chemical interaction between LSGM and the composite was due to the interface diffusion of nickel from the anode to the LSGM electrolyte; this led to the formation of La-based poor-conductive secondary phases, which block oxygen-ion transport. The unit cell design with a buffer layer of SDC was suggested as an effective way of avoiding the problem of interface diffusion [155]. However, chemical reactivity was observed between $La_{1-x}Sr_xGa_{1-y}Mg_yO_{3-\delta}$ and buffer layers of $Gd_{0.1}Ce_{0.9}O_{1.95}$, scandia-doped zirconia [156] and $Gd_{0.8}Ce_{0.2}O_{1.9}$ [157].

An alternate solution to the problem of nickel interface diffusion from a Ni-based anode is to find novel anode materials. A study of the chemical compatibility between $La_{0.9}Sr_{0.1}Ga_{0.8}Mg_{0.2}O_{3-\delta}$ and $Fe_2O_3$, $Co_2O_3$, NiO as anode materials is provided in [158]. Powder mixtures of LSGM with metal oxides at a weight ratio of 1:1 were mixed in ethanol, pressed into pellets and annealed at 1150, 1250 and 1350 °C for 2 h. The obtained XRD data showed that the LSGM reacted with NiO and $Co_2O_3$ at 1150 °C, while a detectable reaction with $Fe_2O_3$ occurred only after calcination at 1350 °C.

Du and Sammes [159] reported good chemical compatibility between $La_{0.8}Sr_{0.2}Ga_{0.8}Mg_{0.2}O_{3-\delta}$ and an alternative $La_{0.75}Sr_{0.25}Cr_{0.5}Mn_{0.5}O_3$ anode at a temperature range of 1100–1500 °C. However, the authors note that a low-conductivity phase formed if the annealing time was more than 6 h or the annealing temperature was greater than 1500 °C.

Good chemical compatibility between LSGM and anodes with a double perovskite structure was shown for: $La_{0.9}Sr_{0.1}Ga_{0.8}Mg_{0.2}O_{3-\delta}$ and $Sr_2TiMoO_{6-\delta}$ after calcining the samples at 1000 °C for 10 h in an atmosphere of 5% $H_2$/Ar [160], $La_{0.8}Sr_{0.2}Ga_{0.8}Mg_{0.2}O_{3-\delta}$ and $Sr_2Fe_{1.5}Mo_{0.5}O_{6-\delta}$ after heat treatment at 1200 °C for 24 h in air [161], $La_{0.88}Sr_{0.12}Ga_{0.82}Mg_{0.18}O_{3-\delta}$ with $Sr_2NiMoO_{6-\delta}$ at 1000 °C for 20 h [162,163] and $Sr_2Ni_{0.75}Mg_{0.25}MoO_{6-\delta}$ at 1100 °C for 20 h [164] and at 1250 °C for 2 h [163]. The formation of secondary phases between LSGM and double perovskite anodes was observed for $La_{0.9}Sr_{0.1}Ga_{0.8}Mg_{0.2}O_{3-\delta}$ and $Sr_2MgMoO_{6-\delta}$ after calcining at 1100 °C [165], for $La_{0.88}Sr_{0.12}Ga_{0.82}Mg_{0.18}O_{3-\delta}$ and $Sr_2ZnMoO_6$ at 1000 °C for 20 h [166] and for $La_{0.8}Sr_{0.2}Ga_{0.8}Mg_{0.2}O_{3-\delta}$ at 1300 °C for 10 h with $Sr_2Ni_{0.7}Mg_{0.3}MoO_{6-\delta}$ [167] and, after heat treatment at 1200 °C for 24 h, with $Sr_2CoMoO_{6-\delta}$ [161], $Sr_2NiMoO_{6-\delta}$ [161] and $Sr_2MgMoO_{6-\delta}$ [168].

According to Takano et al. [165], $La_{0.9}Sr_{0.1}Ga_{0.8}Mg_{0.2}O_{3-\delta}$ did not react with $Ce_{0.8}La_{0.2}O_{1.8}$ after annealing at 1300 °C for 1 h; therefore, it was concluded that $La_{0.9}Sr_{0.1}Ga_{0.8}Mg_{0.2}O_{3-\delta}$ and $Ce_{0.8}La_{0.2}O_{2-\delta}$ might be recommended as SOFC electrolyte and buffer materials, respectively, with $Sr_2MgMoO_{6-\delta}$ used as the anode material. However, a comprehensive investigation of the chemical compatibility between various compositions of $La_{1-x}Sr_xGa_{1-y}Mg_yO_{3-\delta}$ and lanthanum-doped $CeO_2$, provided in [169], showed that only a $La_{0.9}Sr_{0.1}Ga_{0.8}Mg_{0.2}O_{3-\delta}$/

$Ce_{0.6}La_{0.4}O_{2-\delta}$ mixture did not result in additional phases after being annealed twice at 1350 °C for 2 h at each stage.

### 3.3. Applications in SOFCs

The problem of reactivity between the LSGM and SOFC electrode materials during sintering can be solved by reducing sintering temperatures or/and using the SDC buffer layer as a barrier, eliminating lanthanum- and nickel-cation diffusion. Several unit cell designs have been proposed in the literature. Table 3 presents a summary of electrochemical performances for different types of hydrogen-fuelled SOFCs with LSGM-based electrolytes. These data testify that enhanced power densities were achieved for electrolyte-supported SOFCs when the LSGM electrolyte thickness was in a range of 100–300 μm. Buffer layers of doped ceria were used between the electrolyte and anode: $Ce_{0.8}Sm_{0.2}O_{2-\delta}$ [144,145,149,155,160,167], $Ce_{0.8}Gd_{0.2}O_{2-\delta}$ [170] and $Ce_{0.6}La_{0.4}O_{2-\delta}$ [171,172].

**Table 3.** The performances of SOFCs with $La_{1-x}Sr_xGa_{1-y}Mg_yO_{3-\delta}$ electrolytes. Figure A3 provides a visualization of these data.

| Anode | Buffer Layer/ Electrolyte (Thickness, μm)/ Buffer Layer | Cathode | *T* (°C) | Power Density (mW cm$^{-2}$) | Ref. |
|---|---|---|---|---|---|
| Ni-$Ce_{0.8}Sm_{0.2}O_{2-\delta}$ | $La_{0.8}Sr_{0.2}Ga_{0.83}Mg_{0.17}O_{3-\delta}$ (265) | $La_{0.6}Sr_{0.4}O_{3-\delta}$ | 800 | 290 | [52] |
| Ni-$La_{0.8}Sr_{0.2}Ga_{0.83}Mg_{0.17}O_{2.815}$ | $La_{0.8}Sr_{0.2}Ga_{0.83}Mg_{0.17}O_{3-\delta}$ (395) | $La_{0.6}Sr_{0.4}O_{3-\delta}$ | 800 | 363 | [52] |
| Ni-$Ce_{0.8}Sm_{0.2}O_{2-\delta}$ | $Ce_{0.8}Sm_{0.2}O_{2-\delta}$/$La_{0.9}Sr_{0.1}Ga_{0.8}Mg_{0.2}O_{3-\delta}$ (300) | $La_{0.4}Sr_{0.6}Co_{0.9}Sb_{0.1}O_{3-\delta}$-$Ce_{0.8}Sm_{0.2}O_{2-\delta}$ | 700 | 432 | [144] |
| Ni-$Ce_{0.8}Sm_{0.2}O_{2-\delta}$ | $Ce_{0.8}Sm_{0.2}O_{2-\delta}$/$La_{0.9}Sr_{0.1}Ga_{0.8}Mg_{0.2}O_{3-\delta}$ (100) | $SrCo_{0.8}Fe_{0.1}Nb_{0.1}O_{3-\delta}$ | 800 | 756 | [145] |
| Ni-$Ce_{0.8}Sm_{0.2}O_{2-\delta}$ | $Ce_{0.8}Sm_{0.2}O_{2-\delta}$/$La_{0.9}Sr_{0.1}Ga_{0.8}Mg_{0.2}O_{3-\delta}$ (100) | $SrCo_{0.8}Fe_{0.1}Nb_{0.1}O_{3-\delta}$–$Ce_{0.9}Gd_{0.1}O_{2-\delta}$ | 800 | 829 | [145] |
| Ni-$Ce_{0.8}Sm_{0.2}O_{2-\delta}$ | $La_{0.9}Sr_{0.1}Ga_{0.8}Mg_{0.2}O_{3-\delta}$ (300) | $BaCo_{0.7}Fe_{0.2}Ta_{0.1}O_{3-\delta}$ | 800 | 460 | [146] |
| Ni-$Ce_{0.8}Sm_{0.2}O_{2-\delta}$ | $Ce_{0.8}Sm_{0.2}O_{2-\delta}$/$La_{0.9}Sr_{0.1}Ga_{0.8}Mg_{0.2}O_{3-\delta}$ (300) | $Pr_2Ni_{0.85}Cu_{0.1}Al_{0.05}O_{4+\delta}$ | 700 | 392 | [149] |
| Ni-$Ce_{0.8}Sm_{0.2}O_{2-\delta}$ | $La_{0.8}Sr_{0.2}Ga_{0.83}Mg_{0.17}O_{3-\delta}$ (500) | $La_{0.6}Sr_{0.4}O_{3-\delta}$ | 800 | 270 | [155, 173] |
| Ni-$Ce_{0.8}Sm_{0.2}O_{2-\delta}$ | $Ce_{0.8}Sm_{0.2}O_{2-\delta}$/$La_{0.8}Sr_{0.2}Ga_{0.83}Mg_{0.17}O_{3-\delta}$ (500) | $La_{0.6}Sr_{0.4}O_{3-\delta}$ | 800 | 550 | [155, 173] |
| Ni-$Ce_{0.8}Sm_{0.2}O_{2-\delta}$ | $La_{0.87}Sr_{0.13}Ga_{0.85}Mg_{0.15}O_{3-\delta}$ (3.8) | $La_{0.87}Sr_{0.13}Ga_{0.85}Mg_{0.15}O_{3-\delta}$-$La_{0.6}Sr_{0.4}Fe_{0.8}Co_{0.2}O_{3-\delta}$ | 750 | 1420 | [174] |
| Ni-$Ce_{0.8}Y_{0.2}O_{2-\delta}$ | $La_{0.9}Sr_{0.1}Ga_{0.8}Mg_{0.2}O_{3-\delta}$ (45) | $La_{0.6}Sr_{0.4}O_{3-\delta}$ | 700 | 500 | [175] |
| Ni-$Ce_{0.6}La_{0.4}O_{2-\delta}$ | $La_{0.8}Sr_{0.2}Ga_{0.8}Mg_{0.2}O_{3-\delta}$ (30) | $La_{0.8}Sr_{0.2}Fe_{0.8}Co_{0.2}O_{3-\delta}$ | 700 | 780 | [176] |
| Ni-$Ce_{0.6}La_{0.4}O_{2-\delta}$ | $Ce_{0.6}La_{0.4}O_{2-\delta}$/$La_{0.8}Sr_{0.2}Ga_{0.83}Mg_{0.17}O_{3-\delta}$ (500) | $SrCo_{0.8}Fe_{0.2}O_{3-\delta}$ | 800 | 900 | [177] |
| Ni-$Ce_{0.9}Gd_{0.1}O_{2-\delta}$ | $Ce_{0.55}La_{0.45}O_{2-\delta}$/$La_{0.9}Sr_{0.1}Ga_{0.8}Mg_{0.2}O_{3-\delta}$ (75) | $La_{0.9}Sr_{0.1}O_{3-\delta}$-$Ce_{0.55}La_{0.45}O_{2-\delta}$ | 800 | 1100 | [178] |
| Ni-$Ce_{0.9}Gd_{0.1}O_{2-\delta}$ | $Ce_{0.55}La_{0.45}O_{2-\delta}$/$La_{0.9}Sr_{0.1}Ga_{0.8}Mg_{0.2}O_{3-\delta}$ (50) | $La_{0.6}Sr_{0.4}O_{3-\delta}$ | 800 | 1565 | [179] |
| Ni-$Ce_{0.9}Gd_{0.1}O_{2-\delta}$ | $Ce_{0.55}La_{0.45}O_{2-\delta}$/$La_{0.9}Sr_{0.1}Ga_{0.8}Mg_{0.2}O_{3-\delta}$ (50)/$Ce_{0.55}La_{0.45}O_{2-\delta}$ | $La_{0.6}Sr_{0.4}O_{3-\delta}$ | 800 | 871 | [179] |
| Ni-$Ce_{0.8}Gd_{0.2}O_{2-\delta}$ | $Ce_{0.8}Gd_{0.2}O_{2-\delta}$/$La_{0.9}Sr_{0.1}Ga_{0.8}Mg_{0.2}O_{3-\delta}$ (75) | $Ba_{0.5}Sr_{0.5}Co_{0.8}Fe_{0.2}O_{3-\delta}$ | 700 | 760 | [180] |
| Ni-Fe | $Ce_{0.8}Sm_{0.2}O_{2-\delta}$/$La_{0.9}Sr_{0.1}Ga_{0.8}Mg_{0.2}O_{3-\delta}$ (6) | $Sm_{0.5}Sr_{0.5}O_{3-\delta}$ | 700 | 1790 | [181] |
| Ni-$Ce_{0.6}La_{0.4}O_{2-\delta}$ | $Ce_{0.6}La_{0.4}O_{2-\delta}$/$La_{0.9}Sr_{0.1}Ga_{0.8}Mg_{0.2}O_{3-\delta}$ (9)/$Ce_{0.6}La_{0.4}O_{1.8}$ | $La_{0.9}Sr_{0.1}Ga_{0.8}Mg_{0.2}O_{3-\delta}$-$La_{0.6}Sr_{0.4}Fe_{0.8}Co_{0.2}O_{3-\delta}$ | 700 | 910 | [182] |
| Ni-$Ce_{0.8}Sm_{0.2}O_{2-\delta}$ | $Ce_{0.6}La_{0.4}O_{2-\delta}$/$La_{0.9}Sr_{0.1}Ga_{0.8}Mg_{0.2}O_{3-\delta}$ (11)/$Ce_{0.6}La_{0.4}O_{1.8}$ | $La_{0.9}Sr_{0.1}Ga_{0.8}Mg_{0.2}O_{3-\delta}$-$La_{0.6}Sr_{0.4}Fe_{0.8}Co_{0.2}O_{3-\delta}$ | 800 | 1230 | [183] |
| Ni-$Ce_{0.8}Gd_{0.2}O_{2-\delta}$ | $Ce_{0.8}Gd_{0.2}O_{2-\delta}$/$(La_{0.9}Sr_{0.1})_{0.97}Ga_{0.9}Mg_{0.1}O_{3-\delta}$ (120) | $La_{0.6}Sr_{0.4}Fe_{0.8}Co_{0.2}O_{3-\delta}$ | 800 | 540 | [170] |
| Ni-$Ce_{0.8}Sm_{0.2}O_{2-\delta}$ | $La_{0.9}Sr_{0.1}Ga_{0.8}Mg_{0.2}O_{3-\delta}$ (3.4) | $La_{0.9}Sr_{0.1}Ga_{0.8}Mg_{0.2}O_{3-\delta}$-$La_{0.6}Sr_{0.4}Fe_{0.8}Co_{0.2}O_{3-\delta}$ | 750 | 736 | [184] |
| Ni-$Ce_{0.8}Gd_{0.2}O_{2-\delta}$ | $La_{0.8}Sr_{0.2}Ga_{0.8}Mg_{0.2}O_{3-\delta}$ (50) | $La_{0.6}Sr_{0.4}Fe_{0.8}Co_{0.2}O_{3-\delta}$ | 700 | 831 | [185] |
| Ni-Fe | $Ce_{0.6}La_{0.4}O_{2-\delta}$/$La_{0.9}Sr_{0.1}Ga_{0.8}Mg_{0.2}O_{3-\delta}$ (200) | $Sm_{0.5}Sr_{0.5}O_{3-\delta}$ | 800 | 1350 | [171] |
| Pd-$Sr_2TiMoO_{6-\delta}$ | $Ce_{0.8}Sm_{0.2}O_{2-\delta}$/$La_{0.9}Sr_{0.1}Ga_{0.8}Mg_{0.2}O_{3-\delta}$ (300) | $NdBaCo_{0.67}Fe_{0.67}Cu_{0.67}O_{5+\delta}$ | 850 | 1009 | [160] |
| $Sr_2NiMoO_{6-\delta}$ | $La_{0.88}Sr_{0.12}Ga_{0.82}Mg_{0.18}O_{3-\delta}$ (700)/$Ce_{0.8}Sm_{0.2}O_{2-\delta}$ | $La_{0.7}Sr_{0.3}Fe_{0.9}Co_{0.1}O_{3-\delta}$ | 800 | 61 | [163] |
| $Sr_2NiMoO_{6-\delta}$ | $La_{0.9}Sr_{0.1}Ga_{0.8}Mg_{0.2}O_{3-\delta}$ (300) | $Ba_{0.5}Sr_{0.5}Co_{0.8}Fe_{0.2}O_{3-\delta}$ | 800 | 595 | [186] |
| $Sr_2MgMoO_{6-\delta}$ | $Ce_{0.8}Sm_{0.2}O_{2-\delta}$/$La_{0.8}Sr_{0.2}Ga_{0.8}Mg_{0.2}O_{3-\delta}$ (700) | $SmBaCo_2O_{5+\delta}$ | 800 | 39 | [167] |

**Table 3.** *Cont.*

| Anode | Buffer Layer/ Electrolyte (Thickness, μm)/ Buffer Layer | Cathode | $T$ (°C) | Power Density (mW cm$^{-2}$) | Ref. |
|---|---|---|---|---|---|
| $Sr_2MgMoO_{6-\delta}$ | $Ce_{0.8}Gd_{0.2}O_{2-\delta}/La_{0.8}Sr_{0.2}Ga_{0.8}Mg_{0.2}O_{3-\delta}$ (600) | $La_{0.6}Sr_{0.4}Fe_{0.8}Co_{0.2}O_{3-\delta}$ | 800 | 330 | [168] |
| $Sr_2Ni_{0.75}Mg_{0.25}MoO_{6-\delta}$ | $La_{0.88}Sr_{0.12}Ga_{0.82}Mg_{0.18}O_{3-\delta}$ (700)/$Ce_{0.8}Sm_{0.2}O_{2-\delta}$ | $La_{0.7}Sr_{0.3}Fe_{0.9}Co_{0.1}O_{3-\delta}$ | 800 | 429 | [163] |
| $Sr_2Ni_{0.75}Mg_{0.25}MoO_{6-\delta}$ | $La_{0.88}Sr_{0.12}Ga_{0.82}Mg_{0.18}O_{3-\delta}$ (500)/$Ce_{0.8}Sm_{0.2}O_{2-\delta}$ | $La_2NiO_{4+\delta}$ | 800 | 276 | [187] |
| $Sr_2Ni_{0.75}Mg_{0.25}MoO_{6-\delta}$ | $La_{0.88}Sr_{0.12}Ga_{0.82}Mg_{0.18}O_{3-\delta}$ (500)/$Ce_{0.8}Sm_{0.2}O_{2-\delta}$ | $La_{1.5}Ca_{0.5}Ni_{0.67}Fe_{0.33}O_{4+\delta}$ | 800 | 273 | [187] |
| $Sr_2Ni_{0.7}Mg_{0.3}MoO_{6-\delta}$ | $Ce_{0.8}Sm_{0.2}O_{2-\delta}/La_{0.8}Sr_{0.2}Ga_{0.8}Mg_{0.2}O_{3-\delta}$ (700) | $SmBaCo_2O_{5+\delta}$ | 800 | 160 | [167] |
| $Sr_2Ni_{0.3}Mg_{0.7}MoO_{6-\delta}$ | $Ce_{0.8}Sm_{0.2}O_{2-\delta}/La_{0.8}Sr_{0.2}Ga_{0.8}Mg_{0.2}O_{3-\delta}$ (700) | $SmBaCo_2O_{5+\delta}$ | 800 | 119 | [167] |
| $Ba_{0.5}Sr_{0.5}Mo_{0.1}Fe_{0.9}O_{3-\delta}$ | $La_{0.8}Sr_{0.2}Ga_{0.8}Mg_{0.2}O_{3-\delta}$ (150) | $Ba_{0.5}Sr_{0.5}Mo_{0.1}Fe_{0.9}O_{3-\delta}$ | 800 | 2280 | [188] |
| $SrFe_{0.75}Mo_{0.25}O_{3-\delta}$ | $La_{0.9}Sr_{0.1}Ga_{0.8}Mg_{0.2}O_{3-\delta}$ (30) | $SrFe_{0.75}Mo_{0.25}O_{3-\delta}$ | 800 | 703 | [189] |
| $PrBa(Fe_{0.8}Sc_{0.2})_2O_{5+\delta}$ | $La_{0.9}Sr_{0.1}Ga_{0.8}Mg_{0.2}O_{3-\delta}$ (275) | $PrBa(Fe_{0.8}Sc_{0.2})_2O_{5+\delta}$ | 800 | 713 | [190] |
| $Sr_2Fe_{1.5}Mo_{0.5}O_{6-\delta}$-$La_{0.9}Sr_{0.1}Ga_{0.8}Mg_{0.2}O_{2.85}$ | $La_{0.9}Sr_{0.1}Ga_{0.8}Mg_{0.2}O_{3-\delta}$ (10) | $Sr_2Fe_{1.5}Mo_{0.5}O_{6-\delta}$-$La_{0.9}Sr_{0.1}Ga_{0.8}Mg_{0.2}O_{3-\delta}$ | 700 | 880 | [191] |
| $Pr_{0.6}Sr_{0.4}Fe_{0.8}Ni_{0.2}O_{3-\delta}$ | $Ce_{0.8}Gd_{0.2}O_{2-\delta}/La_{0.9}Sr_{0.1}Ga_{0.8}Mg_{0.2}O_{3-\delta}$ (320)/$Ce_{0.8}Gd_{0.2}O_{1.9}$ | $Pr_{0.6}Sr_{0.4}Fe_{0.8}Ni_{0.2}O_{3-\delta}$ | 800 | 500 | [192] |
| $PrBaMn_{1.5}Fe_{0.5}O_{5+\delta}$ | $La_{0.8}Sr_{0.2}Ga_{0.8}Mg_{0.2}O_{3-\delta}$ (520) | $PrBaMn_{1.5}Fe_{0.5}O_{5+\delta}$ | 800 | 540 | [193] |
| $La_{0.5}Sr_{0.5}Fe_{0.9}Nb_{0.1}O_{3-\delta}$ | $La_{0.82}Sr_{0.18}Ga_{0.83}Mg_{0.17}O_{3-\delta}$ (300) | $La_{0.5}Sr_{0.5}Fe_{0.9}Nb_{0.1}O_{3-\delta}$ | 750 | 630 | [194] |
| $La_{0.54}Sr_{0.36}Co_{0.2}Fe_{0.6}Nb_{0.2}O_{3-\delta}$ | $Ce_{0.8}Sm_{0.2}O_{2-\delta}/La_{0.9}Sr_{0.1}Ga_{0.8}Mg_{0.2}O_{3-\delta}$ (200)/$Ce_{0.8}Sm_{0.2}O_{1.9}$ | $La_{0.54}Sr_{0.36}Co_{0.2}Fe_{0.6}Nb_{0.2}O_{3-\delta}$ | 800 | 539 | [195] |
| $Sr_2TiFe_{0.9}Mo_{0.1}O_{6-\delta}$ | $Ce_{0.8}Sm_{0.2}O_{2-\delta}/La_{0.9}Sr_{0.1}Ga_{0.8}Mg_{0.2}O_{3-\delta}$ (200)/$Ce_{0.8}Sm_{0.2}O_{1.9}$ | $Sr_2TiFe_{0.9}Mo_{0.1}O_{6-\delta}$ | 800 | 444 | [196] |
| $Sr_2Fe_{1.4}Nb_{0.1}Mo_{0.5}O_{6-\delta}$ | $La_{0.8}Sr_{0.2}Ga_{0.83}Mg_{0.17}O_{3-\delta}$ (243) | $Sr_2Fe_{1.4}Nb_{0.1}Mo_{0.5}O_{6-\delta}$ | 800 | 531 | [197] |
| $Sr_{0.95}Ti_{0.3}Fe_{0.63}Ni_{0.07}O_{3-\delta}$ | $Ce_{0.6}La_{0.4}O_{2-\delta}/La_{0.8}Sr_{0.2}Ga_{0.83}Mg_{0.17}O_{3-\delta}$ (300) | $La_{0.6S}Sr_{0.4}Co_{0.2}Fe_{0.8}O_{3-\delta}$-$Gd_{0.1}Ce_{0.9}O_{2-\delta}$ | 800 | 1000 | [172] |

Considering the details in Figure A3, one can see that the SOFCs' power density tends to increase with a decrease in the electrolyte's thickness (due to a corresponding decline in the ohmic resistance) despite the existence/absence of $CeO_2$-based buffer layers. Nevertheless, the performance of the compared SOFCs varies greatly, even for close electrolyte thicknesses, indicating that other functional components (cermets, oxygen electrodes) have a significant effect on the achievable output characteristics.

A diagram of a typical LSGM-supported cell with a barrier layer between the anode and the electrolyte, using a Ni-Fe/$Ce_{0.6}La_{0.4}O_{2-\delta}/La_{0.9}Sr_{0.1}Ga_{0.8}Mg_{0.2}O_{3-\delta}/Sm_{0.5}Sr_{0.5}O_{3-\delta}$ cell, is presented in Figure 8a. In [171], it was shown that the OCV values were equal to 1.07 and 1.15 V at 800 °C and 700 °C, respectively, and there was no significant difference in the thickness of the $Ce_{0.6}La_{0.4}O_{1.8}$ interlayer. This LSGM-supported cell yielded up to 2200 and 1350 mW cm$^{-2}$ at 850 and 800 °C, respectively. The typical $I$–$V$ curve and power densities at different temperatures for the LSGM-supported cell are shown in Figure 8b, which is based on the Ni-$Ce_{0.8}Gd_{0.2}O_{2-\delta}/Ce_{0.8}Gd_{0.2}O_{2-\delta}/(La_{0.9}Sr_{0.1})_{0.97}Ga_{0.9}Mg_{0.1}O_{3-\delta}/La_{0.6}Sr_{0.4}Fe_{0.8}Co_{0.2}O_{3-\delta}$ cell tested in [170]. The maximum power density of the aforementioned cell reached 540 mW cm$^{-2}$ at 800 °C, while the maximum power density of a cell containing a $La_{0.9}Sr_{0.1}Ga_{0.9}Mg_{0.1}O_{2.9}$ electrolyte reached 450 mW cm$^{-2}$ at 800 °C. The electrode polarization resistance values of the $La_{0.9}Sr_{0.1}Ga_{0.9}Mg_{0.1}O_{3-\delta}$ and $(La_{0.9}Sr_{0.1})_{0.97}Ga_{0.9}Mg_{0.1}O_{3-\delta}$ based cells were equal to 0.34 and 0.30 Ω cm$^2$ at 800 °C, respectively.

Table 3 shows that, for electrode-supported SOFCs with thin-film LSGM electrolytes, a barrier layer between the electrolyte and the electrodes is not necessary [174–176,184,185]. An anode-supported cell containing a $La_{0.9}Sr_{0.1}Ga_{0.8}Mg_{0.2}O_{3-\delta}$ film deposited on an anode supported substrate using radio-frequency magnetron sputtering was fabricated in [174]. The anode substrate was composed of a Ni-$Sm_{0.2}Ce_{0.8}O_{2-\delta}$ functional layer and a Ni collector layer; an LSGM-$La_{0.6}Sr_{0.4}Co_{0.2}Fe_{0.8}O_{3-\delta}$ composite layer was used as a cathode. The obtained SOFC revealed no cracking, delamination or discontinuity, as shown in Figure 8c. The polarization resistance of an anode-supported cell containing a $La_{0.9}Sr_{0.1}Ga_{0.8}Mg_{0.2}O_{3-\delta}$ film decreased from 0.41 to 0.05 Ω cm$^2$ as the temperature increased from 600 to 800 °C. The OCV and $P_{max}$ values were in the range of 0.85–0.95 V and 650-1420 mW cm$^{-2}$, respectively, at a temperature range of 600–750 °C.

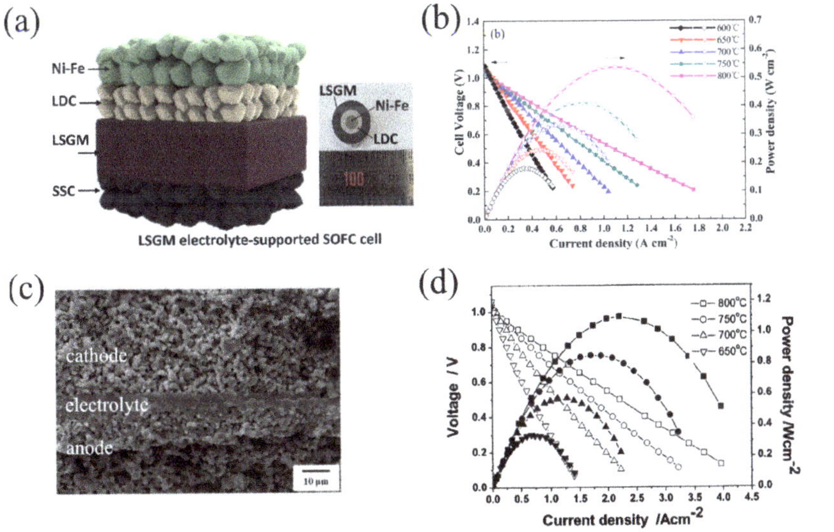


**Figure 8.** Design and performances of $LaGaO_3$-based SOFCs: (**a**) schematic illustration of Ni–Fe/$Ce_{0.6}La_{0.4}O_{2-\delta}$/$La_{0.9}Sr_{0.1}Ga_{0.8}Mg_{0.2}O_{3-\delta}$/$Sm_{0.5}Sr_{0.5}O_{3-\delta}$. Reproduced from [171] with permission from Elsevier Ltd., 2021; (**b**) *I–V* and power density curves of the electrolyte-supported cell with an LSGM electrolyte at different temperatures. Reproduced from [170] with permission from John Wiley & Sons, Inc., 2018; (**c**) SEM micrograph of an anode-supported cell with an LSGM electrolyte. Reproduced from [174] with permission from Elsevier Ltd., 2002; (**d**) *I–V* and power density curves of an anode-supported cell with a $Ce_{0.6}La_{0.4}O_{1.8}$-LSGM bi-layered electrolyte at different temperatures. Reproduced from [178] with permission from The Electrochemical Society, 2004.

Combining the two approaches for SOFC design can be found in [178–181]. Bi et al. deposited a $Ce_{0.6}La_{0.4}O_{2-\delta}$/LSGM bi-layer film on a Ni-$Ce_{0.9}Gd_{0.1}O_{2-\delta}$ anode. Therefore, the cell design allowed for high OCVs (1.02 and 1.043 V at 800 °C) and high power density values (1100 and 1565 mW cm$^{-2}$ at 800 °C) to be achieved at a LDC/LSGM bi-layer thickness of 100 and 65 μm, respectively [178,179]. The *I–V* and power density curves for a Ni-$Ce_{0.6}La_{0.4}O_{2-\delta}$/$Ce_{0.6}La_{0.4}O_{2-\delta}$/LSGM(100 μm)/$La_{0.9}Sr_{0.1}O_{3-\delta}$-$Ce_{0.55}La_{0.45}O_{2-\delta}$ cell at different temperatures, are shown in Figure 8d [178]. Ju et al. [181] reached a paramount performance of 1790 mW cm$^{-2}$ at 700 °C for a SOFC based on an LSGM film with a thickness of 6 μm: this used an SDC buffer layer with a thickness of 500 nm, which was deposited on a Ni–Fe porous anode support. After a thermal cycle going from 700 to 25 °C, the fabricated cell showed an OCV of 1.1 V and $P_{max}$ of 1620 mW cm$^{-2}$, which was almost the same as the first cycles.

According to a number of investigations [179,182,183,198], the most effective design for SOFCs composed of barrier layers is the LDC/LSGM/LDC tri-layered electrolyte. Bi et al. reported [179] that an anode-supported SOFC with an LDC/LSGM/LDC tri-layered electrolyte film significantly increased when using a cell with an LDC/LSGM bi-layered electrolyte film with the same thickness [178]. Guo et al. [183], depositing an LDC/LSGM/LDC tri-layer with thickness of 30 μm on a Ni-$Ce_{0.8}Sm_{0.2}O_{2-\delta}$ anode, fabricated a cell with a 75 mL min$^{-1}$ $H_2$ flow rate that generated 1230 W cm$^{-2}$ at 800 °C. The specific ohmic resistance across the LDC/LSGM/LDC tri-layer electrolyte film was measured to be equal to 0.086 Ω cm$^2$ at 800 °C. The obtained data showed that the polarization resistance was higher than the ohmic resistance at temperatures below 700 °C. A long-term stability experiment was performed on the aforementioned cell with a current density of 1000 mA cm$^{-2}$ and a 30 mL min$^{-1}$ $H_2$ flow rate at 800 °C. The results of 95 h-test demonstrated that the maximum power density values decreased from 1.08 to 0.81

W cm$^{-2}$. The authors of [183] suggest that there was little diffusion of the transition metal from the electrodes to the electrolyte during the test.

Serious efforts have been made to replace traditional cermet anodes with single-phase oxide materials: this is in an attempt to avoid chemical interactions. Complex oxides with double perovskite ($Sr_2MMoO_{6-\delta}$ (M = Mg, Ti, Ni, Fe) [160,162,167,168,186,187,191,196]), layered [190,193] and perovskite [172,189] structures were successfully tested as alternative anode materials for SOFCs with LSGM electrolytes. A buffer layer of doped ceria was used to avoid chemical interactions between an LSGM electrolyte and double perovskites [160,167,168], as well as between an LSGM electrolyte and an oxide cathode [163,187,199]. The composite electrodes $Sr_2Fe_{1.5}Mo_{0.5}O_{6-\delta}$-$La_{0.9}Sr_{0.1}Ga_{0.8}Mg_{0.2}O_{3-\delta}$ [191], $Sr_2CoMoO_{6-\delta}$-$La_{0.9}Sr_{0.1}Ga_{0.8}Mg_{0.2}O_{3-\delta}$ and $Sr_2Co_{0.9}Mn_{0.1}NbO_{6-\delta}$-$La_{0.9}Sr_{0.1}Ga_{0.8}Mg_{0.2}O_{3-\delta}$ [200] have been proposed to solve the thermo-mechanical incompatibility between an electrolyte and an electrode due to a mismatch in the materials' thermal expansion [174,182–184,191,200–203].

An analysis of recent studies illustrates that LSGM can be used as a base matrix for the formation of both composite electrodes and new composite electrolytes [200,204–210]. Xu et al. [200] fabricated a cell based on a $La_{0.9}Sr_{0.1}Ga_{0.8}Mg_{0.2}O_{3-\delta}$-$Ce_{0.8}Gd_{0.2}O_{1.9}$ electrolyte, with $Sr_2CoMoO_{6-\delta}$-$La_{0.9}Sr_{0.1}Ga_{0.8}Mg_{0.2}O_{3-\delta}$ as the anode and $Sr_2Co_{0.9}Mn_{0.1}NbO_{6-\delta}$-$La_{0.9}Sr_{0.1}Ga_{0.8}Mg_{0.2}O_{3-\delta}$ as the cathode. For this cell, obtained with a 95 wt.% $La_{0.9}Sr_{0.1}Ga_{0.8}Mg_{0.2}O_{3-\delta}$-5 wt.% $Ce_{0.8}Gd_{0.2}O_{2-\delta}$ electrolyte, the OCV, $P_{max}$ and current density values at 800 °C were equal to 1.08 V, 192 mW cm$^{-2}$, and 720 mA cm$^{-2}$, respectively [200].

The electrochemical investigations in [211–215] for LSGM-based SOFCs confirm that these cells can operate in both fuel cell and electrolysis cell modes. Reversible cells were fabricated in [215] with NiO–YSZ-substrate as an anode, $La_{0.9}Sr_{0.1}Ga_{0.8}Mg_{0.2}O_{3-\delta}$ film as an electrolyte and $Sm_{0.5}Sr_{0.5}CoO_{3-\delta}$ as an air electrode. It was established that the infiltration of cerium nitrate into the substrate was an effective means of increasing cell performance. The maximum power density of this cell at 3 M Ce nitrate infiltration achieved 950 mW cm$^{-2}$ at 600 °C.

## 4. Conclusions

Complex oxides based on $LaGaO_3$ offer a convenient basis for the design of oxygen-conducting electrolytes that can be employed in intermediate-temperature solid oxide fuel cells (SOFCs). A rational combination of appropriate dopants incorporated at various sublattices of $LaGaO_3$ allows superior transport properties to be achieved for co-doped derivatives ($La_{1-x}Sr_xGa_{1-y}Mg_yO_{3-\delta}$, LSGM). LSGM materials are considered one of the most conductive oxygen-ionic electrolytes, enabling a decrease in SOFC operation temperatures by 100–300 °C compared to YSZ-based SOFCs. As a result, very high SOFC performances (from 0.5 to 1.5 W cm$^{-2}$ at 700 °C) were reported for lab-type electrochemical cells. However, to efficiently place laboratory studies on a manufacturing scale, several issues remain, including the development of simple and low-cost technologies for electrolyte preparation (including thin-film forms), searching for strategies to improve the chemical stability of LSGM with other SOFC components (especially with nickel) and the design of new electrochemically active electrodes. In this regard, the present review serves as the starting point for further research in fields such as solid-state chemistry, physical chemistry, electrochemistry and the technology of $LaGaO_3$-based materials and electrochemical cells.

**Author Contributions:** Conceptualization, E.F. and D.M.; investigation, E.F. and D.M.; writing—original draft preparation, E.F. and D.M.; writing—review and editing, E.F. and D.M.; visualization, E.F. and D.M.; project administration, E.F.; funding acquisition, D.M. All authors have read and agreed to the published version of the manuscript.

**Funding:** This research was funded by to the Ministry of Education and Science of the Russian Federation, contract no. 075-03-2021-051/5.

**Data Availability Statement:** Not applicable.

**Conflicts of Interest:** The authors declare no conflict of interest.

## Appendix A

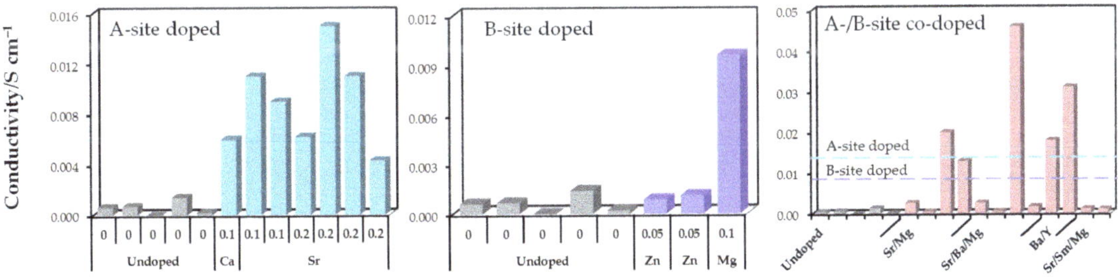


**Figure A1.** Total conductivity of the $LaAlO_3$ ceramic materials at 700–900 °C depending on doping strategies. These data are taken from Table 1.

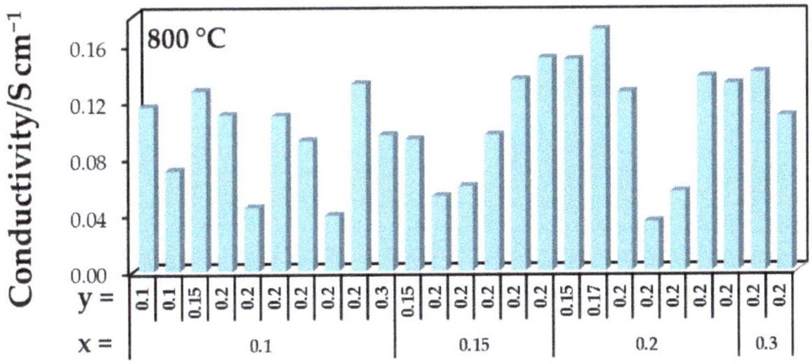


**Figure A2.** Total conductivity of the$La_{1-x}Sr_xGa_{1-y}Mg_yO_{3-\delta}$ ceramic materials at 800 °C. These data are taken from Table 2.

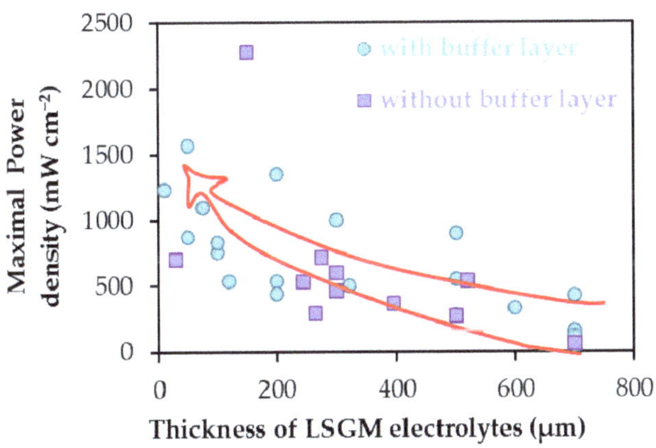


**Figure A3.** Maximum power densities of SOFCs based on the LSGM-based electrolytes at 800 °C. These data are taken from Table 3.

## References


1. Holden, E.; Linnerud, K.; Rygg, B.J. A review of dominant sustainable energy narratives. *Renew. Sustain. Energy Rev.* **2021** *144*, 110955. [CrossRef]
2. Chu, W.; Calise, F.; Duić, N.; Østergaard, P.A.; Vicidomini, M.; Wang, Q. Recent advances in technology, strategy and application of sustainable energy systems. *Energies* **2020**, *13*, 5229. [CrossRef]
3. Østergaard, P.A.; Duic, N.; Noorollahi, Y.; Mikulcic, H.; Kalogirou, S. Sustainable development using renewable energy technology *Renew. Energy* **2020**, *146*, 2430–2437. [CrossRef]
4. Kolosok, S.; Bilan, Y.; Vasylieva, T.; Wojciechowski, A.; Morawski, M. A scoping review of renewable energy, sustainability and the environment. *Energies* **2021**, *14*, 4490. [CrossRef]
5. Erixno, O.; Rahim, N.A.; Ramadhani, F.; Adzman, N.N. Energy management of renewable energy-based combined heat and power systems: A review. *Sustain. Energy Technol. Assess.* **2022**, *51*, 101944. [CrossRef]
6. Martínez, M.L.; Vázquez, G.; Pérez-Maqueo, O.; Silva, R.; Moreno-Casasola, P.; Mendoza-González, G.; López-Portillo, J. MacGregor-Fors, I.; Heckel, G.; Hernández-Santana, J.R.; et al. A systemic view of potential environmental impacts of ocean energy production. *Renew. Sustain. Energy Rev.* **2021**, *149*, 111332. [CrossRef]
7. Pradhan, S.; Chakraborty, R.; Mandal, D.K.; Barman, A.; Bose, P. Design and performance analysis of solar chimney power plant (SCPP): A review. *Sustain. Energy Technol. Assess.* **2021**, *47*, 101411. [CrossRef]
8. Nazir, M.S.; Ali, N.; Bilal, M.; Iqbal, H.M.N. Potential environmental impacts of wind energy development: A global perspective *Curr. Opin. Environ. Sci. Health.* **2020**, *13*, 85–90. [CrossRef]
9. Soltani, M.; Moradi Kashkooli, F.; Souri, M.; Rafiei, B.; Jabarifar, M.; Gharali, K.; Nathwani, J.S. Environmental, economic, and social impacts of geothermal energy systems. *Renew. Sustain. Energy Rev.* **2021**, *140*, 110750. [CrossRef]
10. Yuan, X.; Su, C.-W.; Umar, M.; Shao, X.; Lobont, O.-R. The race to zero emissions: Can renewable energy be the path to carbon neutrality? *J. Environ. Manag.* **2022**, *308*, 114648. [CrossRef]
11. Tomkins, P.; Müller, T.E. Evaluating the carbon inventory, carbon fluxes and carbon cycles for a long-term sustainable world. *Green Chem.* **2019**, *21*, 3994–4013. [CrossRef]
12. Nurdiawati, A.; Urban, F. Towards deep decarbonisation of energy-intensive industries: A review of current status, technologies and policies. *Energies* **2021**, *14*, 2408. [CrossRef]
13. Bolatkhan, K.; Kossalbayev, B.D.; Zayadan, B.K.; Tomo, T.; Veziroglu, T.N.; Allakhverdiev, S.I. Hydrogen production from phototrophic microorganisms: Reality and perspectives. *Int. J. Hydrogen Energy* **2019**, *44*, 5799–5811. [CrossRef]
14. Barba, F.J.; Gavahian, M.; Es, I.; Zhu, Z.; Chemat, F.; Lorenzo, J.M.; Mousavi Khaneghah, A. Solar radiation as a prospective energy source for green and economic processes in the food industry: From waste biomass valorization to dehydration, cooking, and baking. *J. Clean. Prod.* **2019**, *220*, 1121–1130. [CrossRef]
15. Wijayasekera, S.C.; Hewage, K.; Siddiqui, O.; Hettiaratchi, P.; Sadiq, R. Waste-to-hydrogen technologies: A critical review of techno-economic and socio-environmental sustainability. *Int. J. Hydrogen Energy* **2022**, *47*, 5842–5870. [CrossRef]
16. Testoni, R.; Bersano, A.; Segantin, S. Review of nuclear microreactors: Status, potentialities and challenges. *Prog. Nucl. Energy* **2021**, *138*, 103822. [CrossRef]
17. Timmer, M.A.G.; De Blok, K.; Van Der Meer, T.H. Review on the conversion of thermoacoustic power into electricity. *J. Acoustic. Soc. Am.* **2018**, *143*, 841–857. [CrossRef]
18. Selvan, K.V.; Hasan, M.N.; Mohamed Ali, M.S. Methodological reviews and analyses on the emerging research trends and progresses of thermoelectric generators. *Int. J. Energy Res.* **2019**, *43*, 113–140. [CrossRef]
19. Cigolotti, V.; Genovese, M.; Fragiacomo, P. Comprehensive review on fuel cell technology for stationary applications as sustainable and efficient poly-generation energy systems. *Energies* **2021**, *14*, 4963. [CrossRef]
20. Mishra, P.; Saravanan, P.; Packirisamy, G.; Jang, M.; Wang, C. A subtle review on the challenges of photocatalytic fuel cell for sustainable power production. *Int. J. Hydrogen Energy* **2021**, *46*, 22877–22906. [CrossRef]
21. Minh, N.Q. Solid oxide fuel cell technology-features and applications. *Solid State Ion.* **2004**, *174*, 271–277. [CrossRef]
22. Bilal Hanif, M.; Motola, M.; Qayyum, S.; Rauf, S.; Khalid, A.; Li, C.-J.; Li, C.-X. Recent advancements, doping strategies and the future perspective of perovskite-based solid oxide fuel cells for energy conversion. *Chem. Engin. J.* **2022**, *428*, 132603. [CrossRef]
23. Peng, J.; Huang, J.; Wu, X.-L.; Xu, Y.-W.; Chen, H.; Li, X. Solid oxide fuel cell (SOFC) performance evaluation, fault diagnosis and health control: A review. *J. Power Sources* **2021**, *505*, 230058. [CrossRef]
24. Zarabi Golkhatmi, S.; Asghar, M.I.; Lund, P.D. A review on solid oxide fuel cell durability: Latest progress, mechanisms, and study tools. *Renew. Sustain. Energy Rev.* **2022**, *161*, 112339. [CrossRef]
25. Jacobson, A.J. Materials for solid oxide fuel cells. *Chem. Mater.* **2010**, *22*, 660–674. [CrossRef]
26. Saadabadi, S.A.; Thallam Thattai, A.; Fan, L.; Lindeboom, R.E.F.; Spanjers, H.; Aravind, P.V. Solid oxide fuel cells fuelled with biogas: Potential and constraints. *Renew. Energy* **2019**, *134*, 194–214. [CrossRef]
27. Yang, B.C.; Koo, J.; Shin, J.W.; Go, D.; Shim, J.H.; An, J. Direct alcohol-fueled low-temperature solid oxide fuel cells: A review. *Energy Technol.* **2019**, *7*, 5–19. [CrossRef]
28. Zhang, L.; Chen, G.; Dai, R.; Lv, X.; Yang, D.; Geng, S. A review of the chemical compatibility between oxide electrodes and electrolytes in solid oxide fuel cells. *J. Power Sources* **2021**, *492*, 229630. [CrossRef]
29. Brandon, N.P.; Skinner, S.; Steele, B.C.H. Recent advances in materials for fuel cells. *Ann. Rev. Mater. Res.* **2003**, *33*, 182–213. [CrossRef]

30. Zhou, Z.; Nadimpalli, V.K.; Pedersen, D.B.; Esposito, V. Degradation mechanisms of metal-supported solid oxide cells and countermeasures: A review. *Materials* **2021**, *14*, 3139. [CrossRef]
31. Sreedhar, I.; Agarwal, B.; Goyal, P.; Agarwal, A. An overview of degradation in solid oxide fuel cells-potential clean power sources. *J. Solid State Electrochem.* **2020**, *24*, 1239–1270. [CrossRef]
32. Yang, Z.; Guo, M.; Wang, N.; Ma, C.; Wang, J.; Han, M. A short review of cathode poisoning and corrosion in solid oxide fuel cell. *Int. J. Hydrogen Energy* **2017**, *42*, 24948–24959. [CrossRef]
33. Zakaria, Z.; Abu Hassan, S.H.; Shaari, N.; Yahaya, A.Z.; Boon Kar, Y. A review on recent status and challenges of yttria stabilized zirconia modification to lowering the temperature of solid oxide fuel cells operation. *Int. J. Energy Res.* **2020**, *44*, 631–650. [CrossRef]
34. Hanif, M.B.; Rauf, S.; Motola, M.; Babar, Z.U.D.; Li, C.-J.; Li, C.-X. Recent progress of perovskite-based electrolyte materials for solid oxide fuel cells and performance optimizing strategies for energy storage applications. *Mat. Res. Bull.* **2022**, *146*, 111612. [CrossRef]
35. Atkinson, A.; Sun, B. Residual stress and thermal cycling of planar solid oxide fuel cells. *Mater. Sci. Technol.* **2007**, *23*, 1135–1143. [CrossRef]
36. Brett, D.J.L.; Atkinson, A.; Brandon, N.P.; Skinner, S.J. Intermediate temperature solid oxide fuel cells. *Chem. Soc. Rev.* **2008**, *37*, 1568–1578. [CrossRef]
37. Tarancón, A. Strategies for lowering solid oxide fuel cells operating temperature. *Energies* **2009**, *2*, 1130–1150. [CrossRef]
38. Kilner, J.A.; Burriel, M. Materials for intermediate-temperature solid-oxide fuel cells. *Ann. Rev. Mater. Res.* **2014**, *44*, 365–393. [CrossRef]
39. Yang, D.; Chen, G.; Zhang, L.; Chen, Z.; Zhang, R.; Asghar, M.I.; Lund, P.D. Low temperature ceramic fuel cells employing lithium compounds: A review. *J. Power Sources* **2021**, *503*, 230070. [CrossRef]
40. Su, H.; Hu, Y.H. Progress in low-temperature solid oxide fuel cells with hydrocarbon fuels. *Chem. Eng. J.* **2020**, *402*, 126235. [CrossRef]
41. Kharton, V.V.; Marques, F.M.B.; Atkinson, A. Transport properties of solid oxide electrolyte ceramics: A brief review. *Solid State Ion.* **2004**, *174*, 135–149. [CrossRef]
42. Mahato, N.; Banerjee, A.; Gupta, A.; Omar, S.; Balani, K. Progress in material selection for solid oxide fuel cell technology: A review. *Prog. Mater. Sci.* **2015**, *72*, 141–331. [CrossRef]
43. Wang, F.; Lyu, Y.; Chu, D.; Jin, Z.; Zhang, G.; Wang, D. The electrolyte materials for SOFCs of low-intermediate temperature: Review. *Mater. Sci. Technol.* **2019**, *35*, 1551–1562. [CrossRef]
44. Shi, H.; Su, C.; Ran, R.; Cao, J.; Shao, Z. Electrolyte materials for intermediate-temperature solid oxide fuel cells. *Prog. Nat. Sci. Mater. Int.* **2020**, *30*, 764–774. [CrossRef]
45. Abdalla, A.M.; Hossain, S.; Azad, A.T.; Petra, P.M.I.; Begum, F.; Eriksson, S.G.; Azad, A.K. Nanomaterials for solid oxide fuel cells: A review. *Renew. Sustain. Energy Rev.* **2018**, *82*, 353–368. [CrossRef]
46. Fan, L.; Zhu, B.; Su, P.-C.; He, C. Nanomaterials and technologies for low temperature solid oxide fuel cells: Recent advances, challenges and opportunities. *Nano Energy* **2018**, *45*, 148–176. [CrossRef]
47. Ellingsen, L.A.W.; Hung, C.R.; Bettez, G.M.; Singh, B.; Chen, Z.; Whittingham, M.S.; Strømman, A.H. Nanotechnology for environmentally sustainable electromobility. *Nat. Nanotechnol.* **2016**, *11*, 1039–1051. [CrossRef]
48. Zhigachev, A.O.; Rodaev, V.V.; Zhigacheva, D.V.; Lyskov, N.V.; Shchukina, M.A. Doping of scandia-stabilized zirconia electrolytes for intermediate-temperature solid oxide fuel cell: A review. *Ceram. Int.* **2021**, *47*, 32490–32504. [CrossRef]
49. Kasyanova, A.V.; Rudenko, A.O.; Lyagaeva, Y.G.; Medvedev, D.A. Lanthanum-containing proton-conducting electrolytes with perovskite structures. *Membr. Membr. Technol.* **2021**, *3*, 73–97. [CrossRef]
50. Artini, C. Crystal chemistry, stability and properties of interlanthanide perovskites: A review. *J. Eur. Ceram. Soc.* **2017**, *37*, 427–440. [CrossRef]
51. Ishihara, T.; Matsuda, H.; Takita, Y. Doped $LaGaO_3$ perovskite type oxide as a new oxide ionic conductor. *J. Am. Chem. Soc.* **1994**, *116*, 3801–3803. [CrossRef]
52. Feng, M.; Goodenough, J.B.; Huang, K.; Milliken, C. Fuel cells with doped lanthanum gallate electrolyte. *J. Power Sources* **1996**, *63*, 47–51. [CrossRef]
53. Fung, K.Z.; Chen, T.Y. Cathode-supported SOFC using a highly conductive lanthanum aluminate-based electrolyte. *Solid State Ion.* **2011**, *188*, 64–68. [CrossRef]
54. Kaur, J.; Singh, D.; Dubey, V.; Suryanarayana, N.S.; Parganiha, Y.; Jha, P. Review of the synthesis, characterization, and properties of $LaAlO_3$ phosphors. *Res. Chem. Intermed.* **2013**, *40*, 2737–2771. [CrossRef]
55. Rizwan, M.; Gul, S.; Iqbal, T.; Mushtaq, U.; Farooq, M.H.; Farman, M.; Bibi, R.; Ijaz, M. A review on perovskite lanthanum aluminate ($LaAlO_3$), its properties and applications. *Mat. Res. Express* **2019**, *6*, 112001. [CrossRef]
56. Morales, M.; Roa, J.J.; Tartaj, J.; Segarra, M. A review of doped lanthanum gallates as electrolytes for intermediate temperature solid oxides fuel cells: From materials processing to electrical and thermo-mechanical properties. *J. Eur. Ceram. Soc.* **2016**, *36*, 1–16. [CrossRef]
57. Fu, D.; Itoh, M. Ferroelectricity in silver perovskite oxides. In *Ferroelectrics-Material Aspects*; Lallart, M., Ed.; InTech: Houston, TX, USA, 2011; ISBN 978-953-307-332-3. [CrossRef]

58. Shen, J.X.; Schleife, A.; Janotti, A.; Van de Walle, C.G. Effects of La 5d and 4f states on the electronic and optical properties c $LaAlO_3$. *Phys. Rev. B* **2016**, *94*, 205203. [CrossRef]
59. Chen, T.Y.; Fung, K.Z. A and B-site substitution of the solid electrolyte $LaGaO_3$ and $LaAlO_3$ with the alkaline-earth oxides Mg and SrO. *J. Alloys Compd.* **2004**, *368*, 106–115. [CrossRef]
60. Goldschmidt, V.M. Die Gesetze der Krystallochemie. *Die Nat.* **1926**, *14*, 477–485. [CrossRef]
61. Popova, V.F.; Tugova, E.A.; Zvereva, I.A.; Gusarov, V.V. Phase equilibria in the $LaAlO_3$-$LaSrAlO_4$ system. *Glass Phys. Chem.* **200** *30*, 564–567. [CrossRef]
62. Egorova, A.V.; Belova, K.G.; Animitsa, I.E.; Morkhova, Y.A.; Kabanov, A.A. Effect of zinc doping on electrical properties c $LaAlO_3$ perovskite. *Chim. Technol. Acta* **2021**, *8*, 20218103. [CrossRef]
63. Azaiz, A.; Kadari, A.; Alves, N.; Faria, L.O. Influence of carbon doping on the thermoluminescence properties of LaAlO3 cryst grown by solid state reaction method. *Int. J. Microstruct. Mater. Prop.* **2020**, *15*, 156–167. [CrossRef]
64. Beheshti, M.; Malekfar, R. A novel approach for the synthesis of lanthanum aluminate nanoparticles using thermal shock assiste solid-state method as a microwave absorber layer. *Mater. Chem. Phys.* **2021**, *270*, 124848. [CrossRef]
65. Fabián, M.; Arias-Serrano, B.I.; Yaremchenko, A.A.; Kolev, H.; Kaňuchová, M.; Briančin, J. Ionic and electronic transport i calcium-substituted $LaAlO_3$ perovskites prepared via mechanochemical route. *J. Eur. Ceram. Soc.* **2019**, *39*, 5298–5308. [CrossRe
66. Brylewski, T.; Bućko, M.M. Low-temperature synthesis of lanthanum monoaluminate powders using the co-precipitation calcination technique. *Ceram. Int.* **2013**, *39*, 5667–5674. [CrossRef]
67. Jin, X.; Zhang, L.; Luo, H.; Fan, X.; Jin, L.; Liu, B.; Li, D.; Qiu, Z.; Gan, Y. Preparation of $Eu^{3+}$ doped $LaAlO_3$ phosphors b coprecipitation-molten salt synthesis. *Integr. Ferroelectr.* **2018**, *188*, 1–11. [CrossRef]
68. Zhang, X.Y.; Zhang, T.; Qi, X.W.; Qi, J.Q.; Sun, G.F.; Chen, H.H.; Zhong, R.X. Preparation and characterization of $LaAlO_3$ vi sol-gel process. *Adv. Mater. Res.* **2013**, *624*, 26–29. [CrossRef]
69. Qin, G.; Huang, X.; Chen, J.; He, Z. Synthesis of Sr and Mg double-doped $LaAlO_3$ nanopowders via EDTA-glycine combine process. *Powder Technol.* **2013**, *235*, 880–885. [CrossRef]
70. Adak, A.K.; Pramanik, P. Synthesis and characterization of lanthanum aluminate powder at relatively low temperature. *Mate Lett.* **1997**, *30*, 269–273. [CrossRef]
71. Verma, O.N.; Singh, S.; Singh, V.K.; Najim, M.; Pandey, R.; Singh, P. Influence of Ba doping on the electrical behaviour o $La_{0.9}Sr_{0.1}Al_{0.9}Mg_{0.1}O_{3-\delta}$ system for a solid electrolyte. *J. Electron. Mater.* **2021**, *50*, 1010–1021. [CrossRef]
72. Rivera-Montalvo, T.; Morales-Hernandez, A.; Barrera-Angeles, A.A.; Alvarez-Romero, R.; Falcony, C.; Zarate-Medina, J. Modifie Pechini's method to prepare $LaAlO_3$:RE thermoluminescent. *Mater. Radiat. Phys. Chem.* **2017**, *140*, 68–73. [CrossRef]
73. Garcia, A.B.S.; Bispo-Jr, A.G.; Lima, S.A.M.; Pires, A.M. Effects of the Pechini's modified synthetic route on structural an photophysical properties of $Eu^{3+}$ or $Tb^{3+}$-doped $LaAlO_3$. *Mat. Res. Bull.* **2021**, *143*, 111462. [CrossRef]
74. Silveira, I.S.; Ferreira, N.S.; Souza, D.N. Structural, morphological and vibrational properties of $LaAlO_3$ nanocrystals produce by four different methods. *Ceram. Int.* **2021**, *47*, 27748–27758. [CrossRef]
75. da Silva, C.A.; de Miranda, P.E.V. Synthesis of $LaAlO_3$ based materials for potential use as methane-fueled solid oxide fuel cel anodes. *Int. J. Hydrogen Energy* **2015**, *40*, 10002–10015. [CrossRef]
76. Lee, G.; Kim, I.; Yang, I.; Ha, J.-M.; Na, H.B.; Jung, J.C. Effects of the preparation method on the crystallinity and catalytic activit of $LaAlO_3$ perovskites for oxidative coupling of methane. *Appl. Surf. Sci.* **2018**, *429*, 55–61. [CrossRef]
77. Stathopoulos, V.N.; Kuznetsova, T.; Lapina, O.; Khabibulin, D.; Pandis, P.K.; Krieger, T.; Chesalov, Y.; Gulyalev, R.; Krivensov, V Larina, T.; et al. Evolution of bulk and surface structures in stoichiometric $LaAlO_3$ mixed oxide prepared by using starch a template. *Mater. Chem. Phys.* **2018**, *207*, 423–434. [CrossRef]
78. Lessing, P.A. Mixed-cation oxide powders via polymeric precursors. *Am. Ceram. Soc. Bull.* **1989**, *68*, 1002–1007.
79. Howard, C.J.; Kennedy, B.J.; Chakoumakos, B.C. Neutron powder diffraction study of rhombohedral rare-earth aluminates an the rhombohedral to cubic phase transition. *J. Phys. Condens. Matter* **2000**, *12*, 349–365. [CrossRef]
80. Nguyen, T.L.; Dokiya, M.; Wang, S.; Tagawa, H.; Hashimoto, T. The effect of oxygen vacancy on the oxide ion mobility ir $LaAlO_3$-based oxides. *Solid State Ion.* **2000**, *130*, 229–241. [CrossRef]
81. Park, J.Y.; Park, H.J. High temperature electrical properties of Sr- and Mg-doped $LaAlO_3$. *J. Korean Cryst. Growth. Cryst. Techno* **2019**, *29*, 187–191. [CrossRef]
82. Park, J.Y.; Choi, G.M. Electrical conductivity of Sr and Mg doped $LaAlO_3$. *Solid State Ion.* **2002**, *154–155*, 535–540. [CrossRef]
83. Chen, T.Y.; Fung, K.Z. Comparison of dissolution behavior and ionic conduction between Sr and/or Mg doped $LaGaO_3$ an $LaAlO_3$. *J. Power Sources* **2004**, *132*, 1–10. [CrossRef]
84. Fu, Q.X.; Tietz, F.; Lersch, P.; Stöver, D. Evaluation of Sr- and Mn-substituted $LaAlO_3$ as potential SOFC anode materials. *Soli State Ion.* **2006**, *177*, 1059–1069. [CrossRef]
85. Villas-Boas, L.A.; De Souza, D.P.F. The effect of Pr co-doping on the densification and electrical properties of Sr-$LaAlO_3$. *Mater Res.* **2013**, *16*, 982–989. [CrossRef]
86. Verma, O.N.; Jha, P.A.; Melkeri, A.; Singh, P. A comparative study of aqueous tape and pellet of $(La_{0.89}Ba_{0.01})Sr_{0.1}Al_{0.9}Mg_{0.1}O_{3-\delta}$ electrolyte material. *Phys. B Condens. Matter* **2017**, *521*, 230–238. [CrossRef]
87. Guo-Heng, Q.; Xiao-Wei, H.; Zh, H.U. Chemical compatibility and electrochemical performance between $LaAlO_3$-based electrolyt and selected anode materials. *Wuli Huaxue Xuebao/Acta Phys.-Chim. Sin.* **2013**, *29*, 311–318. [CrossRef]

88. Verma, O.N.; Jha, P.A.; Singh, P.; Jha, P.K.; Singh, P. Influence of iso-valent "Sm" double substitution on the ionic conductivity of $La_{0.9}Sr_{0.1}Al_{0.9}Mg_{0.1}O_{3-\delta}$ ceramic system. *Mater. Chem. Phys.* **2020**, *241*, 122345. [CrossRef]
89. Villas-Boas, L.A.; Goulart, C.A.; De Souza, D.P.F. Desenvolvimento microestrutural e mobilidade de ions oxigênio em perovskitas do tipo $LaAlO_3$ dopadas com Sr, Ba e Ca. *Rev. Mater.* **2020**, *25*, e-12801. [CrossRef]
90. Villas-Boas, L.A.; Goulart, C.A.; De Souza, D.P.F. Effects of Sr and Mn co-doping on microstructural evolution and electrical properties of $LaAlO_3$. *Process. Appl. Ceram.* **2019**, *13*, 333–341. [CrossRef]
91. Zvonareva, I.; Fu, X.-Z.; Medvedev, D.; Shao, Z. Electrochemistry and energy conversion features of protonic ceramic cells with mixed ionic-electronic electrolytes. *Energy Environ. Sci.* **2022**, *15*, 439–465. [CrossRef]
92. Tietz, F. Thermal expansion of SOFC materials. *Ionics* **1999**, *5*, 129–139. [CrossRef]
93. Nguyen, T.L.; Dokiya, M. Electrical conductivity, thermal expansion and reaction of (La, Sr)(Ga, Mg)$O_3$ and (La, Sr)$AlO_3$ system. *Solid State Ion.* **2000**, *132*, 217–226. [CrossRef]
94. Venâncio, S.A.; de Miranda, P.E.V. Direct utilization of carbonaceous fuels in multifunctional SOFC anodes for the electrosynthesis of chemicals or the generation of electricity. *Int. J. Hydrogen Energy* **2017**, *42*, 13927–13938. [CrossRef]
95. Huang, K.; Tichy, R.S.; Goodenough, J.B. Superior perovskite oxide-ion conductor; strontium- and magnesium-doped $LaGaO_3$: I, phase relationships and electrical properties. *J. Am. Ceram. Soc.* **1998**, *81*, 2565–2575. [CrossRef]
96. Chang, J.; Lee, H.-W.; Kang, S.-J.L. Low-temperature pressureless sintering of Sr- and Mg-doped lanthanum gallate ceramics by sintering atmosphere control. *J. Am. Ceram. Soc.* **2009**, *92*, 927–930. [CrossRef]
97. Moure, A.; Castro, A.; Tartaj, J.; Moure, C. Single-phase ceramics with $La_{1-x}Sr_xGa_{1-y}Mg_yO_{3-\delta}$ composition from precursors obtained by mechanosynthesis. *J. Power Sources* **2009**, *188*, 489–497. [CrossRef]
98. Domingues, E.M.; Gonçalves, P.; Figueiredo, F.M. Synthesis of nanopowders of the aluminum-substituted lanthanum gallate solid electrolyte by mechanochemical route. *Solid State Sci.* **2012**, *14*, 820–827. [CrossRef]
99. Lee, D.; Han, J.-H.; Chun, Y.; Song, R.-H.; Shin, D.R. Preparation and characterization of strontium and magnesium doped lanthanum gallates as the electrolyte for IT-SOFC. *J. Power Sources* **2007**, *166*, 35–40. [CrossRef]
100. Huang, K.; Goodenough, J.B. Wet chemical synthesis of Sr- and Mg-doped $LaGaO_3$, a perovskite-type oxide-ion conductor. *J. Solid State Chem.* **1998**, *136*, 274–283. [CrossRef]
101. Colomer, M.T.; Kilner, J.A. Ni-doped lanthanum gallate perovskites: Synthesis and structural, microstructural, and electrical characterization. *Solid State Ion.* **2011**, *182*, 76–81. [CrossRef]
102. Kuncewicz-Kupczyk, W.; Kobertz, D.; Miller, M.; Singheiser, L.; Hilpert, K. Vaporization of Sr- and Mg-doped lanthanum gallate and implications for solid oxide fuel cells. *J. Electrochem. Soc.* **2001**, *148*, E276–E281. [CrossRef]
103. Cho, P.-S.; Park, S.-Y.; Cho, Y.H.; Kim, S.-J.; Kang, Y.C.; Mori, T.; Lee, J.-H. Preparation of LSGM powders for low temperature sintering. *Solid State Ion.* **2009**, *180*, 788–791. [CrossRef]
104. Chae, N.S.; Park, K.S.; Yoon, Y.S.; Yoo, I.S.; Kim, J.S.; Yoon, H.H. Sr- and Mg-doped $LaGaO_3$ powder synthesis by carbonate coprecipitation. *Colloids Surf. A Physicochem. Eng. Asp.* **2008**, *313–314*, 154–157. [CrossRef]
105. Shi, M.; Xu, Y.; Liu, A.; Liu, N.; Wang, C.; Majewski, P.; Aldinger, F. Synthesis and characterization of Sr- and Mg-doped lanthanum gallate electrolyte materials prepared via the Pechini method. *Mater. Chem. Phys.* **2009**, *114*, 43–46. [CrossRef]
106. Cong, L.; He, T.; Ji, Y.; Guan, P.; Huang, Y.; Su, W. Synthesis and characterization of IT-electrolyte with perovskite structure $La_{0.8}Sr_{0.2}Ga_{0.85}Mg_{0.15}O_{3-\delta}$ by glycine-nitrate combustion method. *J. Alloys Compd.* **2003**, *348*, 325–331. [CrossRef]
107. Kumar, M.; Nesaraj, A.S.; Raj, I.A.; Pattabiraman, R. Synthesis and characterization of $La_{0.9}Sr_{0.1}Ga_{0.8}Mg_{0.2}O_{3-\delta}$ electrolyte for intermediate temperature solid oxide fuel cells (IT SOFC). *Ionics* **2004**, *10*, 93–98. [CrossRef]
108. Ishikawa, H.; Enoki, M.; Ishihara, T.; Akiyama, T. Combustion synthesis of doped lanthanum gallate as an electrolyte for solid oxide fuel cells. *Mater. Trans.* **2006**, *47*, 149–155. [CrossRef]
109. Ha, S.B.; Cho, Y.H.; Ji, H.-I.; Lee, J.-H.; Kang, Y.C.; Lee, J.-H. Low-temperature sintering and electrical properties of strontium- and magnesium-doped lanthanum gallate with $V_2O_5$ additive. *J. Power Sources* **2011**, *196*, 2971–2978. [CrossRef]
110. Ishikawa, H.; Enoki, M.; Ishihara, T.; Akiyama, T. Self-propagating high-temperature synthesis of La(Sr)Ga(Mg)$O_{3-\delta}$ for electrolyte of solid oxide fuel cells. *J. Alloys Compd.* **2007**, *430*, 246–251. [CrossRef]
111. Ishikawa, H.; Enoki, M.; Ishihara, T.; Akiyama, T. Mechanically activated self-propagating high-temperature synthesis of $La_{0.9}Sr_{0.1}Ga_{0.8}Mg_{0.2}O_{3-\delta}$ as an electrolyte for SOFC. *J. Alloys Compd.* **2009**, *488*, 238–242. [CrossRef]
112. Jung, D.S.; Koo, H.Y.; Jang, H.C.; Kim, J.H.; Cho, Y.H.; Lee, J.-H.; Kang, Y.C. Firing characteristics of $La_{0.8}Sr_{0.2}Ga_{0.8}Mg_{0.2}O_{3-\delta}$ electrolyte powders prepared by spray pyrolysis. *J. Alloys Compd.* **2009**, *487*, 693–697. [CrossRef]
113. Sammes, N.M.; Tompsett, G.A.; Phillips, R.J.; Cartner, A.M. Characterisation of doped-lanthanum gallates by X-ray diffraction and Raman spectroscopy. *Solid State Ion.* **1998**, *111*, 1–7. [CrossRef]
114. Yu, J.; Liu, H.; Chen, X.; Xing, J.; Yuan, B.; Wang, M.; Ma, W. Ionic conductivity and crystal structure of LSGM with different element mole ratios. *Fuel Cells* **2021**, *21*, 149–154. [CrossRef]
115. Batista, R.M.; Reis, S.L.; Muccillo, R.; Muccillo, E.N.S. Sintering evaluation of doped lanthanum gallate based on thermodilatometry. *Ceram. Int.* **2018**, *45*, 5218–5222. [CrossRef]
116. Lerch, M.; Boysen, H.; Hansen, T. High-temperature neutron scattering investigation of pure and doped lanthanum gallate. *J. Phys. Chem. Solids* **2001**, *62*, 445–455. [CrossRef]
117. Traina, K.; Henrist, C.; Vertruyen, B.; Cloots, R. Dense $La_{0.9}Sr_{0.1}Ga_{0.8}Mg_{0.2}O_{2.85}$ electrolyte for IT-SOFC's: Sintering study and electrochemical characterization. *J. Alloys Compd.* **2011**, *509*, 1493–1500. [CrossRef]

118. Biswal, R.C.; Biswas, K. Novel way of phase stability of LSGM and its conductivity enhancement. *Int. J. Hydrogen Energy* **2015**, *40*, 509–518. [CrossRef]
119. Stevenson, J.W.; Armstrong, R.; McCready, D.E.; Pederson, L.R.; Weber, W.J. Processing and electrical properties of alkaline earth-doped lanthanum gallate. *J. Electrochem. Soc.* **1997**, *144*, 3613–3620. [CrossRef]
120. Hayashi, H.; Inaba, H.; Matsuyama, M.; Lan, N.G.; Dokiya, M.; Tagawa, H. Structural consideration on the ionic conductivity of perovskite-type oxides. *Solid State Ion.* **1999**, *122*, 1–15. [CrossRef]
121. Datta, P.; Majewski, P.; Aldinger, F. Thermal expansion behaviour of Sr- and Mg-doped $LaGaO_3$ solid electrolyte. *J. Eur. Ceram. Soc.* **2009**, *29*, 1463–1468. [CrossRef]
122. Yu, S.; Bi, H.; Sun, J.; Zhu, L.; Yu, H.; Lu, C.; Liu, X. Effect of grain size on the electrical properties of strontium and magnesium doped lanthanum gallate electrolytes. *J. Alloys Compd.* **2019**, *777*, 244–251. [CrossRef]
123. Zhang, Q.; Liu, W.J.; Wang, J.; Liu, D.; Sun, Z.H.C. Processing of perovskite $La_{0.9}Sr_{0.1}Ga_{0.8}Mg_{0.2}O_{3-\delta}$ electrolyte by glycine-nitrate combustion method. *Int. J. Hydrogen Energy* **2021**, *46*, 31362–31369. [CrossRef]
124. Liu, N.; Yuan, Y.P.; Majewski, P.; Aldinger, F. Synthesis of $La_{0.85}Sr_{0.15}Ga_{0.85}Mg_{0.15}O_{2.85}$ materials for SOFC applications by acrylamide polymerization. *Mat. Res. Bull.* **2006**, *41*, 461–468. [CrossRef]
125. Shi, M.; Chen, M.; Zuo, R.; Xu, Y.; Su, H.; Wang, L.; Yu, T. Synthesis and characterization of $La_{0.85}Sr_{0.15}Ga_{0.8}Mg_{0.2}O_{2.825}$ by glycine combustion method and EDTA combustion method. *Powder Technol.* **2010**, *204*, 188–193. [CrossRef]
126. Morales, M.; Roa, J.J.; Perez-Falcón, J.M.; Moure, A.; Tartaj, J.; Espiell, F.; Segarra, M. Correlation between electrical and mechanical properties in $La_{1-x}Sr_xGa_{1-y}Mg_yO_{3-\delta}$ ceramics used as electrolytes for solid oxide fuel cells. *J. Power Sources* **2014**, *246*, 918–925. [CrossRef]
127. Chen, T.-Y.; Fung, K.-Z. Synthesis of and densification of oxygen-conducting $La_{0.8}Sr_{0.2}Ga_{0.8}Mg_{0.2}O_{2.8}$ nano powder prepared from a low temperature hydrothermal urea precipitation process. *J. Eur. Ceram. Soc.* **2008**, *28*, 803–810. [CrossRef]
128. Cristiani, C.; Zampori, L.; Latorrata, S.; Pelosato, R.; Dotelli, G.; Ruffo, R. Carbonate coprecipitation synthesis of Sr- and Mg-doped $LaGaO_3$. *Mater. Lett.* **2009**, *63*, 1892–1894. [CrossRef]
129. Kaleva, G.M.; Politova, E.D.; Mosunov, A.V.; Sadovskaya, N.V. Modified ion-conducting ceramics based on lanthanum gallate: Synthesis, structure, and properties. *Russ. J. Phys. Chem.* **2018**, *92*, 1138–1144. [CrossRef]
130. Wang, Y.; Zhou, D.F.; Chen, L.; Xie, S.K.; Liu, X.J.; Meng, J. Improvement in the sintering and electrical properties of strontium- and magnesium-doped lanthanum gallate by $MoO_3$ dopant. *J. Alloys Compd.* **2017**, *710*, 748–755. [CrossRef]
131. Ishii, K.; Matsunaga, C.; Munakata, F.; Uchikoshi, T. Effect of A-site ion non-stoichiometry on the chemical stability and electric conductivity of strontium and magnesium-doped lanthanum gallate. *J. Am. Ceram. Soc.* **2019**, *103*, 790–799. [CrossRef]
132. Wang, L.-S.; Li, C.-X.; Ma, K.; Zhang, S.-L.; Yang, G.-J.; Li, C.-J. Microstructure and electrochemical properties of $La_{0.8}Sr_{0.2}Ga_{0.8}Mg_{0.2}O_3$ thin film deposited by vacuum cold spray for solid oxide fuel cells. *ECS Trans.* **2017**, *78*, 405–412. [CrossRef]
133. Zhang, J.; Zhao, Y.; Qiao, J.; Sun, W.; Sun, K.; Wang, Z. An easily controllable flash sintering process for densification of electrolyte for application in solid oxide fuel cells. *Int. J. Hydrogen Energy* **2020**, *45*, 17824–17832. [CrossRef]
134. Savioli, J.; Watson, G.W. Defect chemistry of $LaGaO_3$ doped with divalent cations. *Solid State Ion.* **2022**, *374*, 115828. [CrossRef]
135. Rupp, G.M.; Glowacki, M.; Fleig, J. Electronic and ionic conductivity of $La_{0.95}Sr_{0.05}Ga_{0.95}Mg_{0.05}O_{3-\delta}$ (LSGM) single crystals. *J. Electrochem. Soc.* **2016**, *163*, F1189–F1197. [CrossRef]
136. Mariño, C.; Basbus, J.; Larralde, A.L.; Alonso, J.A.; Fernández-Díaz, M.T.; Troncoso, L. Structural, electrical characterization and oxygen-diffusion paths in $LaSrGa_{1-x}Mg_xO_{4-\delta}$ (x = 0.0–0.2) layered perovskites: An impedance spectroscopy and neutron diffraction study. *New J. Chem.* **2021**, *45*, 10248–10256. [CrossRef]
137. Li, Y.; Yi, H.; Xu, J.; Kuang, X. High oxide ion conductivity in the $Bi^{3+}$ doped melilite $LaSrGa_3O_7$. *J. Alloys Compd.* **2018**, *740*, 143–147. [CrossRef]
138. Baskaran, S.; Lewinsohn, C.A.; Chou, Y.-S.; Qian, M.; Stevenson, J.W.; Armstrong, T.R. Mechanical properties of alkaline earth-doped lanthanum gallate. *J. Mater. Sci.* **1999**, *34*, 3913–3922. [CrossRef]
139. Shkerin, S.N.; Bronin, D.I.; Kovyazina, S.A.; Gorelov, V.P.; Kuzmin, A.V.; Martemyanova, Z.S.; Beresnev, S.M. Structure and phase transitions of $(La,Sr)(Ga,Mg)O_{3-\alpha}$ solid electrolyte. *Solid State Ion.* **2004**, *171*, 129–134. [CrossRef]
140. Wu, Y.-C.; Rao, C.-R. An investigation on the microstructural and electrical properties of $La_{0.85}D_xSr_{0.15-x}Ga_{0.8}Mg_{0.2}O_{2.825}$ (D = Ba and Ca) electrolytes in solid oxide fuel cells. *Ceram. Int.* **2018**, *44*, 19706–19717. [CrossRef]
141. Nesaraj, A.S.; Kumar, M.; Arul Raj, I.; Radhakrishna, I.; Pattabiraman, R. Investigations on chemical interactions between alternate cathodes and lanthanum gallate electrolyte for intermediate temperature solid oxide fuel cell (ITSOFC). *J. Iran. Chem. Soc.* **2007**, *4*, 89–106. [CrossRef]
142. Garcia-Garcia, F.J.; Tang, Y.; Gotor, F.J.; Sayagués, M.J. Development by mechanochemistry of $La_{0.8}Sr_{0.2}Ga_{0.8}Mg_{0.2}O_{2.8}$ electrolyte for SOFCs. *Materials* **2020**, *13*, 1366. [CrossRef] [PubMed]
143. Sydyknazar, S.; Cascos, V.; Troncoso, L.; Larralde, A.L.; Fernández-Díaz, M.T.; Alonso, J.A. Design, synthesis, structure and properties of Ba-doped derivatives of $SrCo_{0.95}Ru_{0.05}O_{3-\delta}$ perovskite as cathode materials for SOFCs. *Materials* **2019**, *12*, 1957. [CrossRef] [PubMed]
144. Zhang, L.; Li, X.; Zhang, L.; Hongdongcai, H.; Xu, J.; Wang, L.; Long, W. Improved thermal expansion and electrochemical performance of $La_{0.4}Sr_{0.6}Co_{0.9}Sb_{0.1}O_{3-\delta}$-$Ce_{0.8}Sm_{0.2}O_{1.9}$ composite cathode for IT-SOFCs. *Solid State Sci.* **2019**, *91*, 126–132. [CrossRef]

145. Wang, S.; Jin, F.; Li, L.; Li, R.; Qu, B.; He, T. Stability, compatibility and performance improvement of $SrCo_{0.8}Fe_{0.1}Nb_{0.1}O_{3-\delta}$ perovskite as a cathode for intermediate-temperature solid oxide fuel cells. *Int. J. Hydrogen Energy* **2017**, *42*, 4465–4477. [CrossRef]
146. Yang, X.; Han, X.; He, T.; Du, Y. Enhanced stability of $BaCoO_{3-\delta}$ using doping process as a cathode material for IT-SOFCs. *ECS Trans.* **2017**, *78*, 543–550. [CrossRef]
147. Niu, B.; Lu, C.; Yi, W.; Luo, S.; Li, X.; Zhong, X.; Zhao, X.; Xu, B. In-situ growth of nanoparticles-decorated double perovskite electrode materials for symmetrical solid oxide cells. *Appl. Catal. B Environ.* **2020**, *270*, 118842. [CrossRef]
148. Tarancón, A.; Peña-Martínez, J.; Marrero-López, D.; Morata, A.; Ruiz-Morales, J.; Núñez, P. Stability, chemical compatibility and electrochemical performance of $GdBaCo_2O_{5+x}$ layered perovskite as a cathode for intermediate temperature solid oxide fuel cells. *Solid State Ion.* **2008**, *179*, 2372–2378. [CrossRef]
149. Zhou, Q.; Qu, L.; Zhang, T.; He, Y.; Zhao, C.; Wang, M.; Wei, T.; Zhang, Y. Preparation and electrochemical properties of an La-doped $Pr_2Ni_{0.85}Cu_{0.1}Al_{0.05}O_{4+\delta}$ cathode material for an IT-SOFC. *J. Alloys Compd.* **2020**, *824*, 153967. [CrossRef]
150. Chen, T.; Zhou, Y.; Yuan, C.; Liu, M.; Meng, X.; Zhan, Z.; Xia, C.; Wang, S. Impregnated $Nd_2NiO_{4+\delta}$-scandia stabilized zirconia composite cathode for intermediate-temperature solid oxide fuel cells. *J. Power Sources* **2014**, *269*, 812–817. [CrossRef]
151. Pikalova, E.Y.; Medvedev, D.A.; Khasanov, A.F. Structure, stability, and thermomechanical properties of Ca-substituted $Pr_2NiO_{4+\delta}$. *Phys. Solid State* **2017**, *59*, 694–702. [CrossRef]
152. Pikalova, E.; Kolchugin, A.; Bogdanovich, N.; Medvedev, D.; Lyagaeva, J.; Vedmid', L.; Ananyev, M.; Plaksin, S.; Farlenkov, A. Suitability of $Pr_{2-x}Ca_xNiO_{4+\delta}$ as cathode materials for electrochemical devices based on oxygen ion and proton conducting solid state electrolytes. *Int. J. Hydrogen Energy* **2020**, *42*, 4465–4477. [CrossRef]
153. Skutina, L.; Filonova, E.; Medvedev, D.; Maignan, A. Undoped $Sr_2MMoO_6$ double perovskite molybdates (M = Ni, Mg, Fe) as promising anode materials for solid oxide fuel cells. *Materials* **2021**, *14*, 1715. [CrossRef] [PubMed]
154. Zhang, X.; Ohara, S.; Mukai, K.; Fukui, T.; Yoshida, H.; Nishimura, M.; Inagaki, T.; Miura, K. Ni-SDC cermet anode for medium-temperature solid oxide fuel cell with lanthanum gallate electrolyte. *J. Power Sources* **1999**, *83*, 170–177. [CrossRef]
155. Huang, K.; Goodenough, J.B. A solid oxide fuel cell based on Sr- and Mg-doped $LaGaO_3$ electrolyte: The role of a rare-earth oxide buffer. *J. Alloys Compd.* **2000**, *303–304*, 454–464. [CrossRef]
156. Kim, K.; Kim, B.; Son, J.; Kim, J.; Lee, H.; Lee, J.; Moon, J. Characterization of the electrode and electrolyte interfaces of LSGM-based SOFCs. *Solid State Ion.* **2006**, *177*, 2155–2158. [CrossRef]
157. Eba, H.; Anzai, C.; Ootsuka, S. Observation of cation diffusion and phase formation between solid oxide layers of lanthanum gallate-based fuel cells. *Mater. Trans.* **2018**, *59*, 244–250. [CrossRef]
158. Zhang, X.; Ohara, S.; Okawa, H.; Maric, R.; Fukui, T. Interactions of a $La_{0.9}Sr_{0.1}Ga_{0.8}Mg_{0.2}O_{3-\delta}$ electrolyte with $Fe_2O_3$, $Co_2O_3$ and NiO anode materials. *Solid State Ion.* **2001**, *139*, 145–152. [CrossRef]
159. Du, Y.; Sammes, N.M. Interactions and compatibilities of LSGM electrolyte and LSCM anode. *ECS Proc. Vol.* **2005**, *2005–2007*, 1127–1136. [CrossRef]
160. Niu, B.; Jin, F.; Liu, J.; Zhang, Y.; Jiang, P.; Feng, T.; Xu, B.; He, T. Highly carbon– and sulfur–tolerant $Sr_2TiMoO_{6-\delta}$ double perovskite anode for solid oxide fuel cells. *Int. J. Hydrogen Energy* **2019**, *44*, 20404–20415. [CrossRef]
161. Dos Santos-Gómez, L.; León-Reina, L.; Porras-Vázquez, J.M.; Losilla, E.R.; Marrero-López, D. Chemical stability and compatibility of double perovskite anode materials for SOFCs. *Solid State Ion.* **2013**, *239*, 1–7. [CrossRef]
162. Filonova, E.A.; Dmitriev, A.S. Physicochemical properties of potential cathode $La_{1-x}Ba_xMn_{1-y}Cr_yO_3$ and anode $Sr_2NiMoO_6$ materials for solid-oxide fuel cells. *Eurasian Chem.-Technol. J.* **2012**, *14*, 139–145. [CrossRef]
163. Filonova, E.A.; Gilev, A.R.; Skutina, L.S.; Vylkov, A.I.; Kuznetsov, D.K.; Shur, V.Y. Double $Sr_2Ni_{1-x}Mg_xMoO_6$ perovskites (x = 0, 0.25) as perspective anode materials for $LaGaO_3$-based solid oxide fuel cells. *Solid State Ion.* **2018**, *314*, 112–118. [CrossRef]
164. Filonova, E.A.; Dmitriev, A.S.; Pikalov, P.S.; Medvedev, D.A.; Pikalova, E.Y. The structural and electrical properties of $Sr_2Ni_{0.75}Mg_{0.25}MoO_6$ and its compatibility with solid state electrolytes. *Solid State Ion.* **2014**, *262*, 365–369. [CrossRef]
165. Takano, S.; Shin-mura, K.; Niwa, E.; Hashimoto, T.; Sasaku, K. Chemical compatibility of $Sr_2MgMoO_{6-\delta}$ with representative electrolyte materials and interlayer materials for solid oxide fuel cells. *J. Ceram. Soc. Japan* **2018**, *126*, 482–487. [CrossRef]
166. Filonova, E.A.; Dmitriev, A.S. Crystal structure and thermal properties of $Sr_2ZnMoO_6$. *Inorg. Mater.* **2014**, *49*, 602–605. [CrossRef]
167. Xie, Z.; Zhao, H.; Du, Z.; Chen, T.; Chen, N. Electrical, chemical, and electrochemical properties of double perovskite oxides $Sr_2Mg_{1-x}Ni_xMoO_{6-\delta}$ as anode materials for solid oxide fuel cells. *J. Phys. Chem. C* **2014**, *118*, 18853–18860. [CrossRef]
168. Marrero-López, D.; Peña-Martínez, J.; Ruiz-Morales, J.C.; Gabás, M.; Núñez, P.; Aranda, M.A.G.; Ramos-Barrado, J.R. Redox behaviour, chemical compatibility and electrochemical performance of $Sr_2MgMoO_{6-\delta}$ as SOFC anode. *Solid State Ion.* **2010**, *180*, 1672–1682. [CrossRef]
169. Kumar, S.; Chakraborty, A.; Kobi, S.; Gopalan, P.; Prasanna, T.R.S. Phase formation between La(Sr)Ga(Mg)$O_3$ and Ce(La)$O_2$ for solid oxide fuel cell applications. *J. Amer. Ceram. Soc.* **2022**, *105*, 3625–3635. [CrossRef]
170. Wang, J.Q.; Zhou, D.F.; Gao, J.Q.; Sun, H.R.; Zhu, X.F.; Meng, J. Effect of A/B-Site Non-stoichiometry on the structure and properties of $La_{0.9}Sr_{0.1}Ga_{0.9}Mg_{0.1}O_{3-\delta}$ solid electrolyte in intermediate-temperature solid oxide fuel cells. *ChemElectroChem* **2018**, *5*, 665–673. [CrossRef]
171. Hwang, K.-J.; Jang, M.; Kim, M.K.; Lee, S.H.; Shin, T.H. Effective buffer layer thickness of La-doped $CeO_2$ for high durability and performance on $La_{0.9}Sr_{0.1}Ga_{0.8}Mg_{0.2}O_{3-\delta}$ electrolyte supported type solid oxide fuel cells. *J. Eur. Ceram. Soc.* **2021**, *41*, 2674–2681. [CrossRef]

172. Zhu, T.; Troiani, H.E.; Mogni, L.V.; Han, M.; Barnett, S.A. Ni-Substituted Sr(Ti,Fe)$O_3$ SOFC anodes: Achieving high performanc via metal alloy nanoparticle exsolution. *Joule* **2018**, *2*, 478–496. [CrossRef]
173. Huang, K.; Tichy, R.S.; Goodenough, J.B.; Milliken, C. Superior perovskite oxide-ion conductor; strontium- and magnesium-dope $LaGaO_3$: III. Performance tests of single ceramic fuel cells. *J. Am. Ceram. Soc.* **1998**, *81*, 2581–2585. [CrossRef]
174. Fukui, T.; Ohara, S.; Murata, K.; Yoshida, H.; Miura, K.; Inagaki, T. Performance of intermediate temperature solid oxide fuel cel with La(Sr)Ga(Mg)$O_3$ electrolyte film. *J. Power Sources* **2002**, *106*, 142–145. [CrossRef]
175. Matsuda, M.; Ohara, O.; Murata, K.; Ohara, S.; Fukui, T.; Miyake, M. Electrophoretic fabrication and cell performance of dense S and Mg-doped $LaGaO_3$-based electrolyte films. *Electrochem. Solid State Lett.* **2003**, *6*, A140–A143. [CrossRef]
176. Bozza, F.; Polini, R.; Traversa, E. High performance anode-supported intermediate temperature solid oxide fuel cells (IT-SOFCs with $La_{0.8}Sr_{0.2}Ga_{0.8}Mg_{0.2}O_{3-\delta}$ electrolyte films prepared by electrophoretic deposition. *Electrochem. Commun.* **2009**, *11*, 1680–168 [CrossRef]
177. Huang, K.; Wan, J.; Goodenough, J.B. Increasing power density of LSGM-based solid oxide fuel cells using new anode material *J. Electrochem. Soc.* **2001**, *148*, A788–A794. [CrossRef]
178. Bi, Z.; Yi, B.; Wang, Z.; Dong, Y.; Wu, H.; She, Y.; Cheng, M. A high-performance anode-supported SOFC with LDC-LSGM bilaye electrolytes. *Electrochem. Solid State Lett.* **2004**, *7*, A105–A107. [CrossRef]
179. Bi, Z.; Dong, Y.; Cheng, M.; Yi, B. Behavior of lanthanum-doped ceria and Sr-, Mg-doped $LaGaO_3$ electrolytes in an anod supported solid oxide fuel cell with a $La_{0.6}Sr_{0.4}CoO_3$ cathode. *J. Power Sources* **2006**, *161*, 34–39. [CrossRef]
180. Wang, W.; Yang, Z.; Wang, H.; Ma, G.; Gao, W.; Zhou, Z. Desirable performance of intermediate-temperature solid oxide fuel ce with an anode-supported $La_{0.9}Sr_{0.1}Ga_{0.8}Mg_{0.2}O_{3-\delta}$ electrolyte membrane. *J. Power Sources* **2011**, *196*, 3539–3543. [CrossRef]
181. Ju, Y.W.; Eto, H.; Inagaki, T.; Ida, S.; Ishihara, T. Preparation of Ni–Fe bimetallic porous anode support for solid oxide fuel cell using $LaGaO_3$ based electrolyte film with high power density. *J. Power Sources* **2010**, *195*, 6294–6300. [CrossRef]
182. Lin, Y.B.; Barnett, S.A. Co-firing of anode-supported SOFCs with thin $La_{0.9}Sr_{0.1}Ga_{0.8}Mg_{0.2}O_{3-\delta}$ electrolytes. *Electrochem. Soli State Lett.* **2006**, *9*, A285–A288. [CrossRef]
183. Guo, W.; Liu, J.; Zhang, Y. Electrical and stability performance of anode-supported solid oxide fuel cells with strontium- an magnesium-doped lanthanum gallate thin electrolyte. *Electrochim. Acta* **2008**, *53*, 4420–4427. [CrossRef]
184. Liu, Y.-X.; Wang, S.-F.; Hsu, Y.-F.; Yeh, W.-Y. Solid oxide fuel cells incorporating doped lanthanum gallate films deposited b radio-frequency magnetron sputtering at various Ar/$O_2$ ratios and annealing conditions. *Surf. Coat. Technol.* **2018**, *344*, 507–51 [CrossRef]
185. Wang, Y.-P.; Gao, J.-T.; Li, J.-H.; Li, C.-J.; Li, C.-X. Preparation of bulk-like $La_{0.8}Sr_{0.2}Ga_{0.8}Mg_{0.2}O_{3-\delta}$ coatings for porous meta supported solid oxide fuel cells via plasma spraying at increased particle temperatures. *Int. J. Hydrogen Energy* **2021**, *4* 32655–32664. [CrossRef]
186. Wei, T.; Ji, Y.; Meng, X.; Zhang, Y. $Sr_2NiMoO_{6-\delta}$ as anode material for $LaGaO_3$-based solid oxide fuel cell. *Electrochem. Commu* **2008**, *10*, 1369–1372. [CrossRef]
187. Gilev, A.R.; Kiselev, E.A.; Cherepanov, V.A. Performance of the lanthanum gallate based solid oxide fuel cells with th $La_{2-x}Ca_xNi_{1-y}Fe_yO_{4+\delta}$ cathodes and $Sr_2Ni_{0.75}Mg_{0.25}MoO_{6-\delta}$ anode. *Solid State Ion.* **2019**, *339*, 115001. [CrossRef]
188. Ma, L.; Wang, Y.; Li, W.; Guan, B.; Qi, H.; Tian, H.; Zhou, L.; De Santiago, H.A.; Liu, X. Redox-stable symmetrical solid oxide fue cells with exceptionally high performance enabled by electrode/electrolyte diffuse interface. *J. Power Sources* **2021**, *488*, 229458 [CrossRef]
189. Gao, J.; Meng, X.; Luo, T.; Wu, H.; Zhan, Z. Symmetrical solid oxide fuel cells fabricated by phase inversion tape casting wit impregnated $SrFe_{0.75}Mo_{0.25}O_{3-\delta}$ (SFMO) electrodes. *Int. J. Hydrogen Energy* **2017**, *42*, 18499–18503. [CrossRef]
190. He, W.; Wu, X.; Dong, F.; Ni, M. A novel layered perovskite electrode for symmetrical solid oxide fuel cells: $PrBa(Fe_{0.8}Sc_{0.2})_2O_{5+\delta}$ *J. Power Sources* **2017**, *363*, 16–19. [CrossRef]
191. Liu, J.; Lei, Y.; Li, Y.; Gao, J.; Han, D.; Zhan, W.; Huang, F.; Wang, S. Infiltrated $Sr_2Fe_{1.5}Mo_{0.5}O_6$/$La_{0.9}Sr_{0.1}Ga_{0.8}Mg_{0.2}O_3$ electrode towards high performance symmetrical solid oxide fuel cells fabricated by an ultra-fast and time-saving procedure. *Electrochem Commun.* **2017**, *78*, 6–10. [CrossRef]
192. Lu, X.; Yang, Y.; Ding, Y.; Chen, Y.; Gu, Q.; Tian, D.; Yu, W.; Lin, B. Mo-doped $Pr_{0.6}Sr_{0.4}Fe_{0.8}Ni_{0.2}O_{3-\delta}$ as potential electrodes fo intermediate-temperature symmetrical solid oxide fuel cells. *Electrochim. Acta* **2017**, *227*, 33–40. [CrossRef]
193. Zhao, L.; Chen, K.; Liu, Y.; He, B. A novel layered perovskite as symmetric electrode for direct hydrocarbon solid oxide fuel cell *J. Power Sources* **2017**, *342*, 313–319. [CrossRef]
194. Bian, L.; Duan, C.; Wang, L.; Zhu, L.; O'Hayre, R.; Chou, K.-C. Electrochemical performance and stability of $La_{0.5}Sr_{0.5}Fe_{0.9}Nb_{0.1}O_{3-\delta}$ symmetric electrode for solid oxide fuel cells. *J. Power Sources* **2018**, *399*, 398–405. [CrossRef]
195. Niu, B.; Jin, F.; Feng, T.; Zhang, L.; Zhang, Y.; He, T. A-site deficient $(La_{0.6}Sr_{0.4})_{1-x}Co_{0.2}Fe_{0.6}Nb_{0.2}O_{3-\delta}$ symmetrical electrod materials for solid oxide fuel cells. *Electrochim. Acta* **2018**, *270*, 174–182. [CrossRef]
196. Niu, B.; Jin, F.; Zhang, L.; Shen, P.; He, T. Performance of double perovskite symmetrical electrode materials $Sr_2TiFe_{1-x}Mo_xO_{6-\delta}$ (x = 0.1, 0.2) for solid oxide fuel cells. *Electrochim. Acta* **2018**, *263*, 217–227. [CrossRef]
197. Gou, M.; Ren, R.; Sun, W.; Xu, C.; Meng, X.; Wang, Z.; Qiao, J.; Sun, K. Nb-doped $Sr_2Fe_{1.5}Mo_{0.5}O_{6-\delta}$ electrode with enhanced stability and electrochemical performance for symmetrical solid oxide fuel cells. *Ceram. Int.* **2019**, *45*, 15696–15704. [CrossRef]
198. Wu, Y.-C.; Huang, W.-H. Processing improvement and performance analysis of $La_{0.85}Sr_{0.15}Ga_{0.8}Mg_{0.2}O_{2.825}$ electrolyte-supported fuel cells. *Ceram. Int.* **2017**, *43*, S729–S738. [CrossRef]

199. Gao, K.; Liu, X.; Wang, Z.; Xiong, Y. A $Sm_{0.2}Ce_{0.8}O_{1.9}$ (SDC) interlayer method to prevent the elemental interdiffusion between $Sm_{0.5}Sr_{0.5}CoO_{3-\delta}$ (SSC) cathode and $La_{0.8}Sr_{0.2}Ga_{0.8}Mg_{0.2}O_{3-\delta}$ (LSGM) electrolyte. *Int. J. Hydrogen Energy* **2017**, *42*, 19170–19177. [CrossRef]
200. Xu, S.; Lin, X.; Ge, B.; Ai, D.; Ma, J.; Peng, Z. Microstructure and electrical conductivity of $La_{0.9}Sr_{0.1}Ga_{0.8}Mg_{0.2}O_{2.85}$-$Ce_{0.8}Gd_{0.2}O_{1.9}$ composite electrolytes for SOFCs. *Int. J. Appl. Ceram. Technol.* **2019**, *16*, 108–118. [CrossRef]
201. Abubaker, O.A.; Singh, K.; Thangadurai, V. Investigating the effect of Cu-doping on the electrochemical properties of perovskite-type $Ba_{0.5}Sr_{0.5}Fe_{1-x}Cu_xO_{3-\delta}$ ($0 \leq x \leq 0.20$) cathodes. *J. Power Sources* **2020**, *451*, 227777. [CrossRef]
202. Hong, G.; Kim, T.W.; Kwak, M.J.; Song, J.; Choi, Y.; Woo, S.-K.; Han, M.H.; Cho, C.H.; Kim, S.-D. Composite electrodes of Ti-doped $SrFeO_{3-\delta}$ and LSGMZ electrolytes as both the anode and cathode in symmetric solid oxide fuel cells. *J. Alloys Compd.* **2020**, *846*, 156154. [CrossRef]
203. Lee, S.; Kim, S.; Choi, S.; Shin, J.; Kim, G. A Nano-Structured SOFC Composite cathode prepared via infiltration of $La_{0.5}Ba_{0.25}Sr_{0.25}Co_{0.8}Fe_{0.2}O_{3-\delta}$ into $La_{0.9}Sr_{0.1}Ga_{0.8}Mg_{0.2}O_{3-\delta}$ for extended triple-phase boundary area. *J. Electrochem. Soc.* **2019**, *166*, F805–F809. [CrossRef]
204. Fujimoto, T.G.; Reis, S.L.; dos Santos Muccillo, E.N. Influence of yttria-stabilized zirconia on microstructure and electrical properties of doped lanthanum gallate. *Mater. Res.* **2019**, *22*, 20190043. [CrossRef]
205. Malik, Y.T.; Noviyanti, A.R.; Syarif, D.G. Lowered sintering temperature on synthesis of $La_{9.33}Si_6O_{26}$ (LSO)–$La_{0.8}Sr_{0.2}Ga_{0.8}Mg_{0.2}O_{2.55}$ (LSGM) electrolyte composite and the electrical performance on $La_{0.7}Ca_{0.3}MnO_3$ (LCM) cathode. *J. Kim. Sains Apl.* **2018**, *21*, 205–210. [CrossRef]
206. Lin, X.P.; Zhong, H.T.; Chen, X.; Ge, B.; Ai, D.S. Preparation and property of LSGM-carbonate composite electrolyte for low temperature solid oxide fuel cell. *Solid State Phenom.* **2018**, *281*, 754–760. [CrossRef]
207. Wang, S.-F.; Lu, H.-C.; Hsu, Y.-F.; Jasinski, P. High-performance anode-supported solid oxide fuel cells with co-fired $Sm_{0.2}Ce_{0.8}O_{2-\delta}/La_{0.8}Sr_{0.2}Ga_{0.8}Mg_{0.2}O_{3-\delta}/Sm_{0.2}Ce_{0.8}O_{2-\delta}$ sandwiched electrolyte. *Int. J. Hydrogen Energy* **2022**, *47*, 5429–5438. [CrossRef]
208. Noviyanti, A.R.; Malik, Y.T.; Rahayu, I.; Eddy, D.R.; Pratomo, U. Electrochemical properties of $La_{9.33}Si_6O_{26}$(LSO)-$La_{0.8}Sr_{0.2}Ga_{0.8}Mg_{0.2}O_{2.55}$(LSGM) electrolyte over NiO and $La_{0.1}Ca_{0.9}MnO_3$(LCM) electrodes. *Mat. Res. Exp.* **2021**, *8*, 115505. [CrossRef]
209. Pandey, R.; Singh, P.; Singh, A.K.; Singh, P. Polyol-mediated synthesis of $La_{0.9}Sr_{0.1}Ga_{0.8}Mg_{0.2}O_{2.85}$-$Ce_{0.85}Sm_{0.15}O_{1.925}$ composite electrolyte for IT-SOFCs. *Mater. Today Proc.* **2020**, *49*, 3071–3075. [CrossRef]
210. Chen, L.; Zhou, D.F.; Wang, Y.; Zhu, X.F.; Meng, J. Enhanced sintering of $Ce_{0.8}Nd_{0.2}O_{2-\delta}$-$La_{0.8}Sr_{0.2}Ga_{0.8}Mg_{0.2}O_{3-\delta}$ using CoO as a sintering aid. *Ceram. Int.* **2017**, *43*, 3583–3589. [CrossRef]
211. Glisenti, A.; Bedon, A.; Carollo, G.; Savaniu, C.; Irvine, J.T.S. Reversible, all-perovskite SOFCs based on La, Sr gallates. *Int. J. Hydrogen Energy* **2020**, *45*, 29155–29165. [CrossRef]
212. Kwon, Y.; Kang, S.; Bae, J. Development of a $PrBaMn_2O_{5+\delta}$-$La_{0.8}Sr_{0.2}Ga_{0.85}Mg_{0.15}O_{3-\delta}$ composite electrode by scaffold infiltration for reversible solid oxide fuel cell applications. *Int. J. Hydrogen Energy* **2020**, *45*, 1748–1758. [CrossRef]
213. Tan, Z.; Ishihara, T. Reversible operation of tubular type solid oxide fuel cells using $LaGaO_3$ electrolyte porous layer on dense film prepared by dip-coating method. *J. Electrochem. Soc.* **2017**, *164*, F1690–F1696. [CrossRef]
214. Tan, Z.; Ishihara, T. Redox stability of tubular solid oxide cell using $LaGaO_3$ electrolyte film prepared by dip-coating. *ECS Trans.* **2021**, *103*, 1685–1693. [CrossRef]
215. Tan, Z.; Song, J.T.; Takagaki, A.; Ishihara, T. Infiltration of cerium into a NiO–YSZ tubular substrate for solid oxide reversible cells using a LSGM electrolyte film. *J. Mater. Chem. A* **2021**, *9*, 1530–1540. [CrossRef]

*nanomaterials*

Review

# MXene-Based Ink Design for Printed Applications

**Zahra Aghayar [1], Massoud Malaki [2,*] and Yizhou Zhang [3,*]**

1 Metallurgy and Materials Engineering, Iran University of Science and Technology, Narmak, Tehran 16846-11314, Iran
2 Department of Mechanical Engineering, Isfahan University of Technology, Isfahan 84156-83111, Iran
3 Institute of Advanced Materials and Flexible Electronics (IAMFE), School of Chemistry and Materials Science, Nanjing University of Information Science and Technology (NUIST), Nanjing 210044, China
* Correspondence: m.malaki@me.iut.ac.ir (M.M.); yizhou.zhang@nuist.edu.cn (Y.Z.)

**Abstract:** MXenes are a class of two-dimensional nanomaterials with a rich chemistry, hydrophilic surface and mechano-ceramic nature, and have been employed in a wide variety of applications ranging from medical and sensing devises to electronics, supercapacitors, electromagnetic shielding, and environmental applications, to name a few. To date, the main focus has mostly been paid to studying the chemical and physical properties of MXenes and MXene-based hybrids, while relatively less attention has been paid to the optimal application forms of these materials. It has been frequently observed that MXenes show great potential as inks when dispersed in solution. The present paper aims to comprehensively review the recent knowledge about the properties, applications and future horizon of inks based on 2D MXene sheets. In terms of the layout of the current paper, 2D MXenes have briefly been presented and followed by introducing the formulation of MXene inks, the process of turning MAX to MXene, and ink compositions and preparations. The chemical, tribological and rheological properties have been deeply discussed with an eye to the recent developments of the MXene inks in energy, health and sensing applications. The review ends with a summary of research pitfalls, challenges, and future directions in this area.

**Keywords:** MXene; ink; printing; 2D material

## 1. Introduction

In the context of printing electronics, printing is commonly employed to precipitate a suspension/solution called ink containing an electrochemical active ingredient onto a substrate of a different kind. Inks usually contain additives such as adhesives, surfactants, and rheology modifiers. The recent boom in the internet of things (IoT) and portable electronics has strongly stimulated the design of advanced and miniatured devices [1–3]. Recently, nano-inks, i.e., the inks with nano-scale additives, have been utilized in many printing applications owing to great personalization, reduced wastes, digitalization, improved productivity, scalability, and high performance, among others [1,4–6].

To date, many attempts have been performed to produce nano-inks with targeted properties. Graphene [7,8], molybdenum disulfide [9], and black phosphorus [10] inks have been reported as the nano-additives of many inks. The mentioned nano-materials are mostly hydrophobic, and their homogenous dispersions in a given ink may sometimes be problematic; as a result, a third-party agent such as a surfactant or a secondary solvent material is commonly added to most of the printable inks in an attempt to improve their homogeneity and to tune the final rheological property, concentration, conductivity, etc., reaching the desired ink [8,11,12]. Conductive additives like silver nanoparticles [13], graphene sheets [14], gallium [15], etc., have already been used for years to make high-performance conductive inks, and yet efforts in this area are continuing.

A new family of 2D few-atoms-thick layers of transition metal carbides, nitrides, or carbonitrides, called MXenes, has recently drawn heated attention due to special properties like hydrophilicity, mechano-ceramic nature and excellent electrical conductivity, making them an ideal additive for conductive inks, hybrids, or nanocomposites [16,17]. To date, MXenes have successfully been employed in next-generation batteries [18], electromagnetic insulations [19], high-strength composite materials [20], biomaterials [21], and energy and environmental applications [22,23], to name a few. The first member of MXenes, $Ti_3C_2T_x$, is a carbon-based laminate synthesized through a selective etching process by the Gogotsi group [24] at Drexel University in 2011. Surface terminations such as -O, -OH, and -F commonly appear on the MXene surface [25] and endow a hydrophilicity behavior without sacrificing key properties such as the mobility of charge carriers or conductivity of MXenes [26]. Moreover, the aforesaid functional groups have a significant effect on the physical and chemical properties [27].

Using a selective etching approach, Figure 1 demonstrates how element A is corroded from a three-dimensional MAX-phase, and a layered accordion structure is then left. The outer surfaces of each layer are covered with functional groups like O, F, or OH.

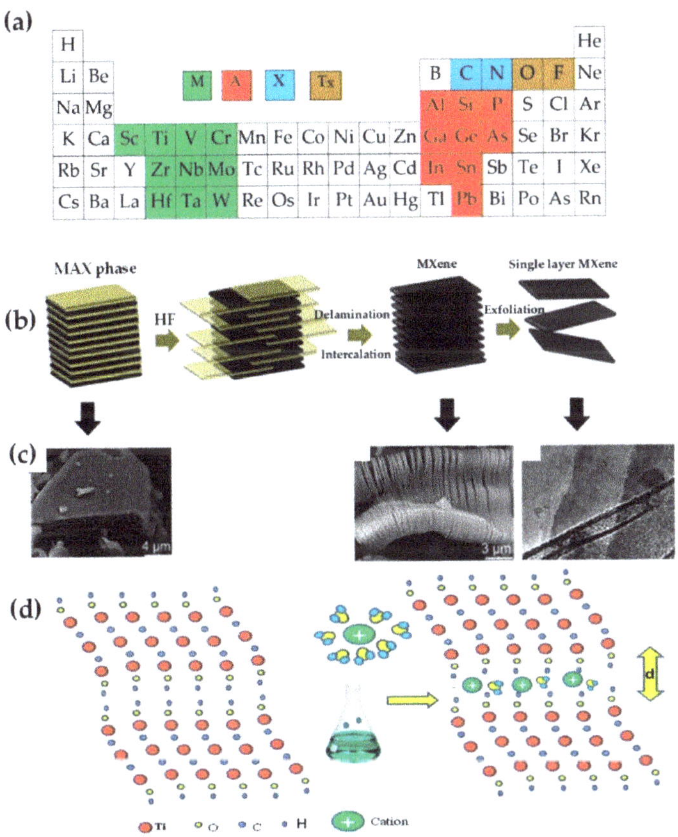


**Figure 1.** (**a**) Different MAX-phase atoms, (**b**) Top-down synthesis steps of MXene nanosheets from the parent MAX-phase, (**c**) Electron microscope images before and after etching, and exfoliated MXene, reprinted with permission from [28], (**d**) Ions intercalation and delamination in $Ti_3C_2T_x$ layers, reprinted with permission from [29].

After etching, the MAX-phase material ($M_{n+1}AX_nT_x$) becomes sheet-structured MXenes ($M_{n+1}X_nT_x$) wherein the number n varies from one to four, and $T_x$ represents the surface termination groups; M a transition metal, A an element of the group of thirteen or fourteen periodic tables, and X carbon or nitrogen (Figure 1). The type and amount of functional groups depend on the etching path and its conditions; owing to the mentioned functionalities, MXene usually has excellent hydrophilic properties. After the completion of etching and weakening the forces between the MXene layers, the layered particles are delaminated and dispersed in their solvent with the help of ultrasonication or even by gentle hand shaking [28].

MXene is a thermodynamically stable nanomaterial in the category of three-component ceramics and usually has dual behaviors of metallic (such as flexibility and high toughness) and ceramic properties (such as high elasticity, high abrasion resistance, low density, and good corrosion resistance). Unlike other hydrophobic nanomaterials, MXene nano-flakes can be easily used in different media without the need for any intermediate material in that environment [24]. The unique combination of hydrophilicity and conductive properties has made MXene nanoflakes a promising candidate for electrical and electronic applications; the conductivity and high concentration of surface functional groups make them particularly promising to be printed on a variety of substrates. Having a high viscosity of about 1–20 mPa $s^{-1}$ and suitable surface tension is required for most printing purposes; further, the amount of surface tension must also correspond to the surface energy and the texture of the substrate to achieve a proper wetting condition. MXenes usually have a zeta potential of about -80 to -30 mV, leading to colloidal stability in water or other polar solvents [30].

Although extensive research efforts have been conducted to figure out different physical or chemical properties, rare studies have been reported about the rheological characteristics as well as the process printing parameters of the 2D MXene sheets. In this work, particular attention is paid to the inks based on MXene nanosheets. In terms of the layout of the present review, the formulation of MXene inks is briefly elaborated first, then the properties derived by both theoretical efforts and experimentations are presented and discussed in detail, followed by their applications in many different fields. Finally, the current research pitfalls, challenges and future opportunities will be provided at the end.

## 2. Formulation of MXene Inks

### *2.1. Turning MAX into MXene*

MXene is generally produced from a MAX phase through an acid etching, and then it is delaminated, intercalated and finally exfoliated by mechanical agitation like simple hand shaking or ultrasonication [28,31]. In an effort made by the Gogotsi group [24] to synthesize the first MXene (i.e., $Ti_3C_2$), $Ti_3AIC_2$ powder was immersed in a hydrofluoric acid (HF) of 50% concentration for 24 h; the resulting suspension was then washed in a distilled water, and then the compound was centrifuged to prepare a precipitate. It is worth noting that Al-Ti bonding to Ti-C is easily detachable [32]; a chemical etching with HF acid could remove Al from the original MAX material through the following reactions:

$$Ti_3AIC_2 + 3HF \rightarrow AlF_3 + 1.5H_2 + Ti_3C_2 \quad (1)$$

$$Ti_3C_2 + 2H_2O \rightarrow Ti_3C_2(OH)_2 + H_2 \quad (2)$$

$$Ti_3C_2 + 2HF \rightarrow Ti_3C_2F_2 + H_2 \quad (3)$$

Reactions 2 and 3 indicate the formation of surface functional groups, i.e., OH and F, on $Ti_3C_2T_x$ [33,34]. Selective etching by HF has a few major drawbacks, like etchant-induce defects, based on which the use of alternative mild etchants is almost always recommended to etch off with safer and faster strategies. Further, the quality of the final MXene strongly depends on the raw materials as well as the processing conditions. For instance, the M-A bond energy determines the time and concentration required for the hydrofluoric acid [29]; reducing the particle size of the MAX phase by a simple ball milling process may greatly reduce the synthesis time with a milder etchant type.

Ghidiu et al. [35] removed Al from the $Ti_3AIC_2$ MAX phase in one step using HCI acid and LiF salts to increase the efficiency, speed, and flexibility of the MXene production process wherein the hydrofluoric acid is in-situ produced locally; a certain amount of LiF is dissolved in HCl at a concentration of 6 M; the $Ti_3AIC_2$ powder was then slowly added to the solution with the temperature kept in 40 °C for 45 h. Upon cooling down to ambient temperature, the product was then washed, centrifuged several times, and then dried. $Ti_3C_2T_x$ was obtained with superior hydrophilicity and excellent electrical conductivity. In an effort to prepare $Ti_3C_2T_x$ by a top-down technique, Shen et al. [36] used a facile hydrothermal method through immersing $Ti_3AIC_2$ powder in $NH_4F$ solution and heating in a stainless steel autoclave at 100 °C for 24 h to produce the final MXene. Liu et al. [37] conducted a study on the effects of various fluoride salts such as LiF, NaF, KF, and $NH_4F$ in HCl in etching $Ti_3AlC_2$ and $Ti_2AlC$. According to Halim et al. [38], when $NH_4HF_2$ is used as an etchant, the etching time increases as compared to HF; however, $NH_4HF_2$ is able to put $NH_4^+$ ions between the MXene layers without the need for another layering agent; further, sodium, potassium, magnesium, and aluminum ions among the MXene layers not only help ion intercalation and delamination but also increase the volumetric capacitance in supercapacitors as shown in Figure 1d.

Surface terminations usually play a key role in the final properties, especially when dealing with the MXene inks and their flow parameters. For example, the bare/pristine MXenes ($M_{n+1}X_n$) usually have quite different features than those terminated MXenes ($M_{n+1}X_nT_x$) [31]; almost all MXenes produced by top-down methods have surface functional groups, and rare studies have focused on the development of pristine MXenes (without end groups) being highly demanded in special applications like printing electronic devices.

### 2.2. Composition of Inks

Ink is typically defined as a liquid of pigments and dyes for writing and printing. The ink composition is usually regulated in a manner to have suitable rheological properties for the desired printing process. Inks based on 2D nanomaterials are usually designed to serve particular functions. Thus, the appearance, or the gloss and color, are not given much attention like graphical inks. Over the past years, interest in composite materials/hybrids, particularly those conductive, semi-conductive, and dielectric materials typically used as active pigments in inks, has been growing. Metal nanoparticles [39], dielectric and organic semiconductors [40,41], carbon allotropes [42,43], and two-dimensional materials [12] are widely used materials having a great potential to tune the ink properties. As discussed below, the most common ink formulations are pigments, adhesives, solvents, and additives.

1. Adhesives mainly consist of polymer nanocomposites to act as films to link the pigments and thereby adhere to a given substrate. There are various polymeric resins, including acrylics, alkyds, cellulose derivatives, and plastic resins. The adhesives used in the ink formulation may affect, to some extent, the printing properties, such as ink gloss, air resistance, chemical resistance, etc. [44];
2. As volatile active ingredients, solvents maintain the ink in a liquid state during printing until a bottom layer is applied. Solvent type and amount vastly influence the final features, among which viscosity and rheology rank highly. A wide variety of organic and inorganic solvents can be utilized as the solvent. Finally, the choice of solvent type may considerably depend on printing technology, substrate, and application [44];
3. Inks are formulated with additives to change their properties; however, the additives are often filled with inks in very small amounts due to their substantial influence. In addition to surfactants being used to make pigments wettable, there are also emulsifiers that facilitate combining pigments with other substances. To make water-based inks more water-resistant, alkalis can be added to create a slight change in pH. Silicone-based additives are typically used to prevent bubbles during printing, making the printed film more resistant to moisture. Those ink containing 2D materials are quite similar to those graphical ink systems [45]. Reduced graphene oxide (rGO) [46], molybdenum disulfide [47], hexagonal boron nitride [48], black phosphorus [49], etc., have already been employed to produce advanced nano-inks. However, the formulations based on 2D materials are different from traditional systems owing to

their special nature. For example, many two-dimensional materials are pre-dispersed in a liquid form, or they may require exfoliation from their bulk condition, while most materials are powdered in traditional systems [50,51]. In the next section, the preparation of inks filled with 2D additives is elaborated on and discussed further.

### 2.3. 2D Inks Preparation

Unlike remarkable scientific achievements in preparing inks with 2D materials, they have not yet been widely utilized in industrial applications. The formula of conductive inks well suited with diverse patterning strategies is crucial for fabricating low-cost, flexible electronics and electrochemical devices [52,53]. Owing to high surface charge and hydrophilicity, MXene inks could simply be applied on a variety of substrates using different processes like writing, printing, stamping, and painting; however, MXene inks up to now are seldom composed of solely delaminated-MXene flakes (i.e., the frequent presence of bulky structures in the deposited flakes), and hence the films made from the inks do not usually exhibit laterally stacked morphologies. The uneven stacking of MXene nanosheets may lead to a reduction in conductivity, therefore restricting the potential applications in electrical circuits or electrodes for energy storage devices. It is believed that formulating MXene ink with ideal elastic properties, permitting the 2D MXene flakes to be deposited uniformly and laterally, is urgently needed at the moment.

Since preparation methods usually consist of multi-stage processing and complex instruments [5,54], developing simple and cost-effective strategies to simulate conductive materials on various substrates or platforms seems to be highly demanded. Ranging from electrochemical capacitors to electromagnetic shielding, MXenes have shown promise in many fields [55,56]. By imparting hydrophilic surface functional groups to facilitate the synthesis of dispersion solvent [57,58], MXenes can be transformed into independent films through different processes like vacuum filtration, thin films on rigid substrates and flexible films using solution spray/spin coating [59,60], screen printing on paper or metal foil with a squeegee [61].

Seok et al. [62] used a vacuum filtering approach followed by an optional equalization procedure to produce a high-purity MAX phase precursor and MXene, wherein a high conductivity and enhanced EMI shielding behavior could be obtained without using any additives. Based on their rheological examinations, it was also found that MXene ink of lower than 45 mg $mL^{-1}$ concentration may be printed on a variety of surfaces with ease, as demonstrated in Figure 2. In other words, the inks with less than 45 mg $mL^{-1}$ concentration were considered suitable for liquid-like inks due to their stable rheological characteristics [62].

As shown in Figure 3, to improve the dispersion and stability of the colloidal MXene suspension, Wu et al. [63] prepared MXenes by the selective etching of $Ti_3C_2T_x$ and then tested four different ligands during the synthesis process, namely (i) sodium ascorbate (SA), (ii) sodium oxalate (SO), (iii) sodium citrate (SC), and (iv) sodium phosphate (SP) wherein it was revealed that ascorbic acid ions could greatly improve the dispersibility, oxidation resistance and that the distance between MXene layers was greatly increased being enhancing the charge/discharge capacity and specific capacitance characteristics. It seems the ascorbic acid may interact with the sub-coordinated Ti atoms of MXene through hydrogen bonds and coordination bonds on the edges of MXene. It was also found that SA-MXene has a higher cycle stability and specific capacity compared to other MXene samples [63].

Like the abovementioned research studies, the majority of studies conducted to date have dispersed MXene nanosheets in aqueous solvents; however, apart from the unique surface hydrophilicity, aqueous solvents present a number of critical issues such as degradation, long-term instability, as well as the increased incompatibility with local polymers in water [64]. One of the most pressing issues in contemporary research has been MXene dispersion in organic solvents. For instance, the tuned microenvironment method (TMM) was used by Zhang et al. [65] to disperse high concentrations of MXene in organic solvents wherein the MXene dispersion is accomplished even without the use of ultrasonic during the process; further, it was found that the insertion of $TBA^+$, adjusting hydrophobicity through surface terminations and functionalizations are the key factors to achieve high-concentration MXene dispersion. Finally, since the size of MXene obtained by TMM was relatively larger than those dispersed in water, it is believed that the MXenes dispersed in organic solvents have higher oxidation resistance [65]. Carey et al. [66] could exfoliate and disperse MXene in non-polar solvents exhibiting high colloidal and oxidation stability; they transformed the lithium cations between the MXene layers by etching to dibenzyl methyl ammonium chloride to create an organophile to evaluate the dispersion and stability of MXene in non-polar liquids. In a non-polar, oxidation-free solvent, an extremely stable colloidal suspension of MXene could be generated and used for industrial settings. Wang et al. [67] suggested an easy method for converting MXene to a novel family of homogenous, solvent-free liquids in another work. Covalent bonding and surface engineering are the two foundations presented in this method. MXene-free liquid has antioxidant stability for 540 days and also has a macroscopic behavior at ambient temperature. Gas adsorption, photoluminescence, composite structures, and magnetic fluids all utilize this novel family of solvent-free liquids. It is believed the preparation of MXene ink still is in its infancy stage, and hence a lot of effort is still demanded to further improve the field.

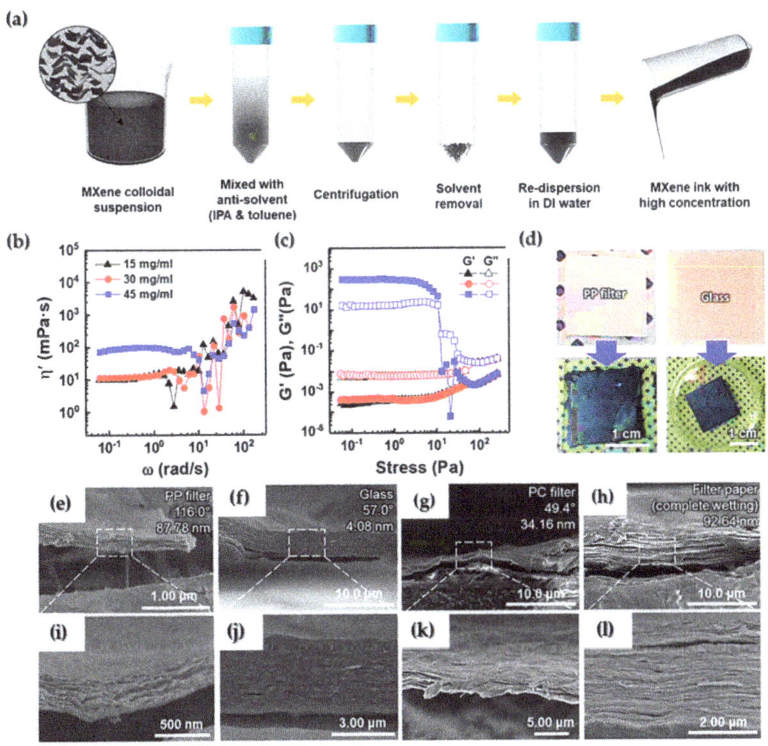


**Figure 2.** (**a**) Schematic illustration of the fabrication process of MXene ink. (**b**) Dynamic shear viscosity ($\eta'$) curves of MXene inks at various concentrations and (**c**) the storage ($G'$) and loss ($G''$) moduli curves of MXene inks as a function of shear stress at different flake concentrations. (**d**) Painted MXene ink on various substrates using a brush. SEM cross−sectional images of painted MXene ink on the (**e**,**i**) PP filter, (**f**,**j**) glass, (**g**,**k**) PC filter, and (**h**,**l**) filter paper. Detailed information (i.e., contact angle and root−mean−square surface roughness) on the bare substrates is included on the upper right in (**e**–**h**), reprinted with permission from [62].

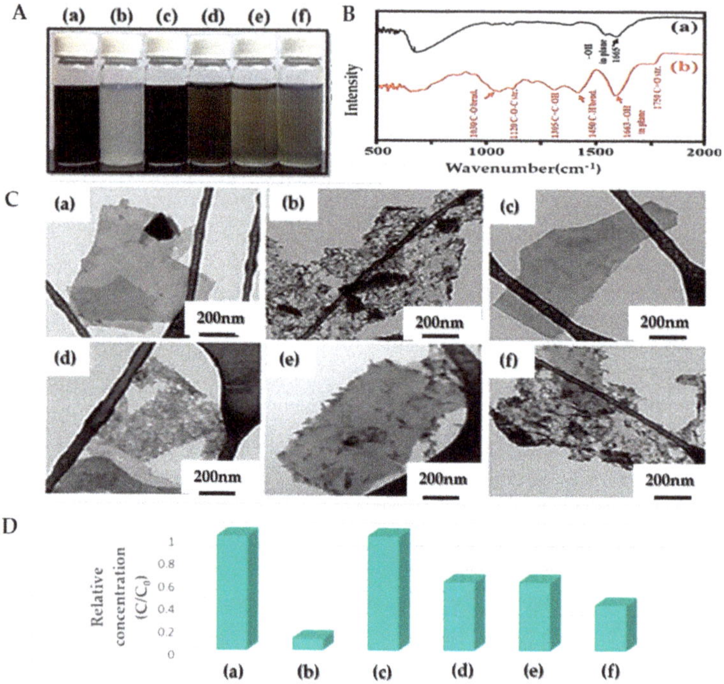


**Figure 3.** (**A**) Photographs, (**B**) FT-IR spectra, (**C**) TEM images, and (**D**) relative concentration (C/C0), (**a**) p-MXene-fresh, (**b**) p-MXene-80, (**c**) SA-MXene-80, (**d**) SC-MXene-80, (**e**) SOMXene-80, and (**f**) SP-MXene-80 suspension with the concentration 0.2 mg $mL^{-1}$ in terms of $Ti_3C_2T_x$ MXene. C0 and C in (**B**) represent the concentration of fresh and stored MXenes, respectively, reprinted with permission from [63].

## 3. Properties of MXene Inks

### *3.1. Chemical Stability*

As discussed earlier, MXene shows excellent dispersibility in various solvents, e.g., water [57,58,68]. Reports show that water-soluble oxygen has a detrimental effect on the chemical degradation of layered flakes, with the high aspect ratio affecting the structural decomposition and the reduced shelf life of MXene inks. In the presence of $TiO_2$ on the $Ti_3C_2T_x$ surface, for example, phase change and structural deformation occur when MXene is exposed to light or high temperatures [57]. Oxidation usually begins at the edges and then progresses to basal surfaces. While it may occur in all environments, those smaller flakes oxidize quicker than the large MXenes; meanwhile, The rate/speed of oxidation depends on the environment as it is maximum in liquids and minimum in solids [69]; therefore, it seems the elimination of the dissolved oxygen in water by saturating it with nitrogen or argon could be an effective way to suppress the oxidation rate. According to Zhang et al. [70], oxidation is very fast in air, and the MXene suspension is degraded by 1.2% after 25 days under argon gas protection. The possible oxidation mechanism is explained as follows [68].

$$Ti_3C_2O_2 + 4H_2O = 3TiO_2 + 2C + 4H_2 \quad (4)$$

To avoid oxidation, MXene-based inks are usually stored in Ar-sealed vials for long-term storage; it is also recommended to store them at low temperatures (T < 4 °C) without exposure to light [68] since long-term storage is still challenging. Storing dried MXene in a vacuum or in an organic solvent could also be an option, as it has been observed that MXene inks remained stable in these conditions for several months [57]. It is noted that when stored in dry conditions, restacking may occur, particularly at elaborated temperatures. Based on research conducted by Zhao et al. [71], it has been revealed that antioxidants such as sodium ascorbate could improve oxidation resistance even in the presence of water and oxygen; sodium ascorbate binds to the positively charged edges of the MXene, resisting against oxidation deterioration. The benefit of antioxidants is that they usually do not change the chemical or colloidal stability or electrical conductivity; for example, in the case of $Ti_3C_2T_x$, after 21 days, the properties were still well preserved. As an alternative strategy, polyionic salts such as polyphosphates or polysilicates can be used against oxidation owing to a few advantages such as easy washing and being cost-effective; they make the scalable production and long-term storage of MXene inks possible [72]. By developing a hydrogen annealing approach to enhance the oxidation resistance of bare MXene, Lee et al. [73] showed that the hydrogen treatment (i) could considerably improve the oxidation stability of MXene films in harsh conditions, i.e., at high temperatures and 100% relative humidity, (ii) hydrogen annealing could also restore the electrical conductivity of the previously oxidized films offering a great potential of MXene in real industrial applications.

### 3.2. *Surface Properties*

Based on preliminary studies [24], it has been predicted that terminated MXenes have negative formation energy with thermodynamic stability. Synthetic $Ti_3C_2T_x$ usually have -OH, -O, or -F surface terminations [74]:

The concentration of functional groups depends on the etching conditions and pH [75,76]. Chemical etching can significantly alter the composition of Tx, and there is still no clear evidence of the identity and coordination of these surface-terminating groups; it is, therefore, important and interesting to study the coordination of $T_x$ [77]. If high-concentration HF is used, more fluorine functional groups and fewer oxygen functional groups are formed; these surface terminations may lead to a hydrophilic surface with a negative zeta potential [75]. MXenes can be chemically modified to control their surface properties and to improve their stability by chemical modification [21], as the surface differences may result in novel chemical, mechanical, optical, and electronic properties [78]. Due to the unique surface properties, it is possible to use a stable suspension of MXene (e.g., $Ti_3C_2T_x$) for the printing processes without additives such as surfactants and/or surface modifiers [79].

After the delamination process, MXene flakes usually have a lateral particle size of a few nano- to micro-scale range which can then be uniformed by centrifugation [80–82]. Surface properties highly depend on the flake size [83]; for example, according to Kylie et al. [84], the electrochemical properties of $Ti_3C_2T_x$ can be improved by controlling the particle size and distribution; smaller flakes usually have less conductivity than large ones attributing to the contact resistance between the flakes due to the defect appeared on the edges.

### 3.3. *Rheological Properties*

#### 3.3.1. Aqueous Single-Layer MXene

Although the first member of the MXene family was discovered about a decade ago, attention on the rheological aspects of this 2D material has only come in the past few years [85]. According to Web of Science, the number of publications on MXene inks is 2, 7, 13, and 23, respectively, in 2017, 2018, 2019, and 2020 while the number of publications on MXene rheology is only 0, 2, 6, and 7, respectively [86]. According to Akuzum et al. [87], aqueous MXene has a viscoelastic nature with shear thinning behavior wherein the viscosity decreased by rotation speed while increased by the addition of higher MXene concentration; a zero-shear viscosity of 5 mPa s at 0.18 mg $mL^{-1}$ was measured. Being an effective parameter in assessing rheological behavior, tan δ is calculated by the following equation:

$$\tan\delta = \frac{G''}{G'} \tag{5}$$

where $\delta$ is the phase difference between the applied stress ($\sigma$) and the corresponding strain ($\varepsilon$), and $G''$ and $G'$, respectively, represent loss modulus (viscous behavior) and elastic modulus (energy storage behavior). To further read about the basics of viscoelasticity, refer to [88]. It was found that for a low weight fraction of aqueous $Ti_3C_2T_x$ (< 0.90 mg $mL^{-1}$), the behavior at low frequencies is mostly elastic with dominant $G'$ values; however, colloidal $Ti_3C_2T_x$ may exhibit a viscous-like behavior with dominant $G''$ values in a wide range of frequencies. The critical volume fraction of MXene nanoflakes ($\varphi^*$) is given by the following relation indicating the strong influence of surface charges and electromagnetic permittivity [89].

$$\varphi^* \propto \frac{h}{d}\left(1 + \frac{2}{d\kappa}\right)^{-3} \tag{6}$$

where $h/d$ is the aspect ratio, and $\kappa^{-1}$ is the Debye screening length.

According to Akuzum et al. [87], $G'$ is frequency dependent in higher weight fraction of MXenes (i.e., $G' \approx 2$ Pa at 3.60 mg $mL^{-1}$), being a suitable condition for those processes with weak gel-like wet spinning (Figure 4). Depending on weight fraction, viscosity, and elastic and loss modulus, $Ti_3C_2T_x$ can be used in a variety of techniques, such as spray coating, wet spinning, and inkjet printing, among others.

It was seen that mono-layer $Ti_3C_2T_x$ could interact with each other even at a long distance via electrostatic forces. Single-layer MXenes usually have higher $\zeta$-potentials than those multi-layers. For instance, it was seen that the rheological behavior might change at a relatively low weight fraction of 0.9 mg $mL^{-1}$ $Ti_3C_2T_x$, wherein the percolation threshold is reached with a gel transition occurring at or above the aforesaid concentration.

Regarding the fluidity of a given ink, both static and dynamic yield stresses should be designed with care; the former represents a value needed for ink to initiate flowing and exist from the printing nozzle, while the latter is required to maintain the fluidity conditions; as a result, the ink should experience a stress of lower than the dynamic stress upon the ink adhesion to a given substrate.

Robot dispensing needs ink with the ability to withstand yield stress, surface tension and the weight of the next layers of printed ink, avoiding any shape distortion or unwanted strain during or after the printing process. To this end, $G'$ must be greater than the sum of the applied forces, including weight and stresses; further, the amount of required yield stress is dependent on how quickly a transition between linear and non-linear viscoelastic behaviors occurs. Flow transition index (FTI), being the ratio of the oscillatory yield stress to the oscillatory stress at which $G' = 0.9G'_{LVR}$, is an important criterion to estimate the required yield stress; the smaller FTI is usually desirable.

When the weight fraction of aqueous monolayer $Ti_3C_2T_x$ is < 10 mg $mL^{-1}$, the storage modulus and the yield stress value are not appropriate for extrusion printing. However, the aqueous ink with a higher concentration than the percolation threshold may form 3D structures by direct ink writing (DIW).

Using a low-waste layer-by-layer extrusion printing technique, Yang et al. [90] could produce 3D freestanding MXene architectures with an ideal rheological property to be used in microsupercapacitors wherein the energy and power densities of 24.4 μWh $cm^{-2}$ and 0.64 mW $cm^{-2}$ were respectively achieved at 4.3 mA $cm^{-2}$; it was seen an increase in the weight fraction of $Ti_3C_2T_x$ MXene might greatly increase the viscosity and elasticity of the given ink system exhibiting the shear thinning of four orders of magnitudes and a quick transition from low (0.01 $s^{-1}$) to high (1000 $s^{-1}$) shear rates when 50 mg $mL^{-1}$ $Ti_3C_2T_x$ (with $G'_{LVR}$ of 36.5 kPa and oscillatory stress value of 206 Pa) is used being promising for the DIW.

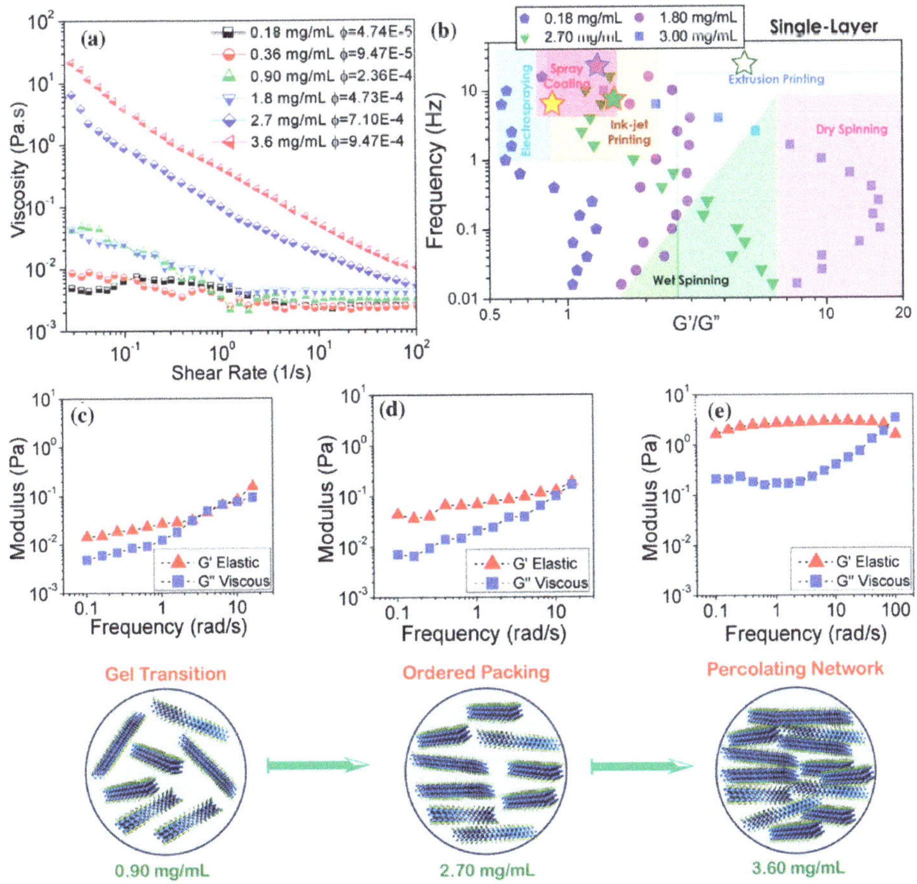


**Figure 4.** Viscoelastic response of 0.18–3.60 mg mL$^{-1}$ (volume fraction, $\phi$, $4.74 \times 10^{-5}$–$9.47 \times 10^{-4}$) mono-layer $Ti_3C_2T_x$ MXene in water. (**a**) Shear rate against viscosity (**b**) Frequency verses $G'/G''$ in different fabrication techniques. The stars show the approximate parameters used in MXene applications obtained from the literature. The viscoelastic behavior of MXene suspensions having (**c**) 0.9 mg mL$^{-1}$, (**d**) 2.7 mg mL$^{-1}$, and (**e**) 3.6 mg mL$^{-1}$ MXene loadings under 0.1% strain amplitude, reprinted with permission from [87].

The effect of aspect ratio and flake size on the rheological behavior of $Ti_3C_2T_x$ MXene flakes in different solvents was studied by Zhang et al. [91], wherein they see both inks with small and large size nanosheets exhibiting a nematic phase behavior; however, those inks prepared with small-size flakes should be much more concentrated than the inks with larger MXene sheets to reach the similar values needed for viscosity and tan $\delta$; they also believe that there is a critical transition concentration value ($C_t$) above which a nematic phase can appear. Finally, 6.3 mg mL$^{-1}$ is a concentration below which liquid-like and above which solid-like behaviors are seen, indicating a sharp transition from the liquid-to-solid-like behavior at this weight fraction. The effect of flake size on the rheological features has later been investigated by Yang et al. [92], confirming the role of flake size in governing the viscoelasticity of an MXene ink system.

In order to tune the rheological properties of MXene inks, other approaches, such as using super absorbent polymer (SAP) beads, have already been examined wherein Orangi et al. [93] achieved a concentration of 290 mg mL$^{-1}$ $Ti_3C_2T_x$ (ca. 28.9 wt.%) with a yield stress of 24 Pa, a flow index of 0.73, and the value of tanδ in a deemed desirable range for extrusion printing. Shen et al. [94] used vacuum filtration to obtain a 3.5 mg mL$^{-1}$ $Ti_3C_2T_x$ ink with decreasing viscosity with similar exponent reported by Orangi et al. 365 Pa 16.3 kPa, and 14 were the values of oscillatory yield stress, small strain storage modulus, and FTI, respectively, which is suitable for DWI. Greeves et al. proposed the following equation with good agreement with the data published in literature to show the effects of flake size (Z), weight fraction (C), and shear rate ($\gamma^{\cdot}$) on the viscosity (η) of $Ti_3C_2T_x$ MXene inks:

$$\eta \approx 1.1(ZC)^{1.5}\gamma^{\cdot -0.9} \tag{7}$$

The discrepancy observed between the results obtained by the above equation and those experimental values reported to date is probably owing to other parameters such as pH, flake size distribution, dispersion quality, defects concentration, the age of colloid, as well as Zeta potential values affecting the final viscosity.

### 3.3.2. Aqueous Multi-Layer MXene

As discussed earlier, inks filled with single-layer MXenes exhibit shear thinning properties as well as viscoelastic behaviors; however, it has been proved that inks with multi-layer $Ti_3C_2T_x$ MXene also have viscoelastic behavior but at much higher loadings than mono-layer MXene inks [87]; a viscosity of 1770 Pa s with shear thinning behavior has been observed at 70 wt.% $Ti_3C_2T_x$ MXene ink. While the loss modulus is greater in lower weight fractions, the value of elastic modulus gradually increases to reach the loss modulus value; then, both loss and elastic values increase by orders of magnitude toward the percolation and gel transition conditions. Like those inks with single-layer MXenes, the rheological behavior of the inks with multi-layer MXene flakes can also be tuned by, for example, the weight fraction. The advantage of multi-layer MXenes over those single-layer flakes is their higher elastic modulus being an asset for printing purposes. However, owing to higher ζ-potentials and much stronger long-distance interactions of mono-layer MXenes, they show viscoelastic behaviors even at low MXene concentrations and moderate strain rates.

By using a screen-printing process to produce conductive paths and micro-supercapacitors, Abdolhosseinzadeh et al. [95] first combined single- and multi-layer $Ti_3C_2T_x$ as well as un-etched $Ti_3AlC_2$ MAX phase to produce an ink with relatively low-shear viscosity of 35 Pa s and an elastic modulus of ca. 370 Pa. Afterwards, they diluted the compound to reach an ink containing 22 wt.% solid additives with the viscosity and the storage modulus of one and two orders of magnitudes, respectively.

### 3.3.3. Non-Aqueous Inks

Apart from water-based inks, many other non-aqueous solvents have also been utilized to produce MXene-based inks, such as dimethyl sulfoxide (DMSO) [96,97].

Vural et al. [97] filled 2.25 mg mL$^{-1}$ $Ti_3C_2T_x$ to DMSO and observed the resultant ink exhibited a Newtonian rheological property having 3.1 mPa s viscosity in a wide range of shear rates of 2–200 s$^{-1}$ wherein the surface tension was 51.5 mN m$^{-1}$ and nozzle aperture 120 μm; although the addition of 0.95 mg mL$^{-1}$ TR42 protein to the ink changed the inverse Ohnesorge number from 27.48 to 21.3, the ink could successfully be printed on a variety of substrate to produce flexible electrodes having excellent electrical conductivities. Zhang et al. not only examined ca. 12 mg mL$^{-1}$ $Ti_3C_2T_x$ in DMSO ink, but they also studied other inks based on water, N-methyl-2-pyrrolidone (NMP), and ethanol wherein viscoelastic behavior was detected in all inks; they also found the values of inverse Ohnesorge number to be much lower for $Ti_3C_2T_x$ than the ones reported by Vural et al., indicating superiority of MXene over TR42 protein. The printed lines made by Zhang et al. [96] exhibited significantly high electrical conductivities of up to 2770 S cm$^{-1}$.

As seen in Figure 5, the rheological properties of different MXene-based inks are similar in terms of viscosity and shear rate behaviors. The MXene-in-water inks seem to have greater viscosity and shear rate values, probably owing to the higher viscosity of shear rate values of pure water than pure DMSO, NMP, and ethanol.

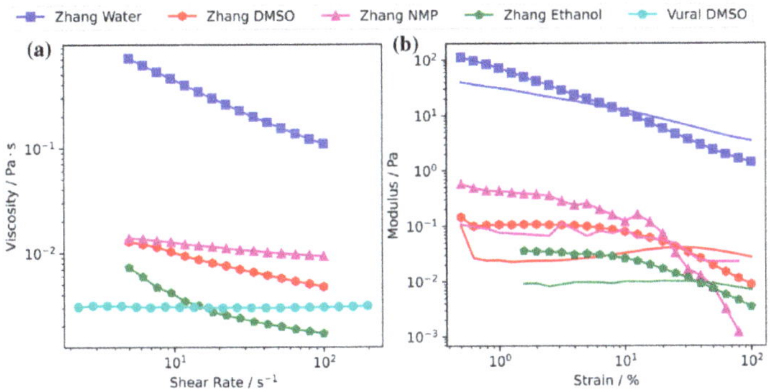


**Figure 5.** Comparison of rheological behavior of $Ti_3C_2T_x$ MXene-based inks reported by Zhang et al. and Vural et al. (**a**) Viscosity vs. shear rate. (**b**) Loss and elastic moduli vs. oscillatory strain amplitude ($G''$ without markers and $G'$ with markers). The amounts of $Ti_3C_2T_x$ MXene in water, ethanol and NMP are respectively 36, 0.7, and 12 mg mL$^{-1}$. 12 mg mL$^{-1}$ and 2.25 mg mL$^{-1}$ are the $Ti_3C_2T_x$ concentrations in DMSO, respectively, used by Zhang et al. and Vural et al.

Apart from the abovementioned solvents, the rheology of MXene can also be studied when added to polymers or other hybrid materials. The number of studies in this area is few, and only a few sporadic papers have studied the rheology of melt polymers filled by MXenes [98]. Gao et al. dealt with the rheological properties of $Ti_3C_2T_x$ MXene/thermoplastic polyurethane nanocomposites at 200 °C; they dispersed 0–1.0 wt% $Ti_3C_2T_x$ nanosheets in thermoplastic polyurethane (TPU) wherein they observed both elastic and loss moduli increased with the frequency and that the highest viscosity was measured in the sample with the highest MXene loadings.

It has been found that chemical modification or hybridization can be considered a rheology tuner. Yu et al. used nitrogen-doped $Ti_3C_2T_x$ (N-$Ti_3C_2T_x$), carbon black, and a polymer in water to produce ink for screen printing; they also used N-$Ti_3C_2T_x$ hybridized with CNTs, activated carbon and graphene oxide in water to produce DIW ink. The former ink had a percolation threshold at about ca. 28 wt%, with no significant changes in $G'$ and $G''$ above the mentioned threshold, indicating gel stability at oscillating frequencies. Further, the N-$Ti_3C_2T_x$/C ink with 34 wt.% solid loadings exhibited a significant viscosity at relatively low shear rates with subsequent high-resolution printed electrodes. The elastic modulus and yield stress value of the N-$Ti_3C_2T_x$/AC/CNT/GO ink with 15 wt.% solid loadings were respectively reported as 2 and 3.5 times higher than the ink without MXene, with 11 wt.% AC/CNT/GO filler.

Although a number of parameters, such as drying time, dot gain, and water pickup, are important, ink rheology and viscosity are the two main parameters that govern printability. Rheology can be defined as how ink flows through the various stages of printing; further, the difference between the term rheology and viscosity is significant. Rheology is the study of fluid over time and the shear and stress gaps, while viscosity is the resistance to liquid flow at a given time. Where its rheology governs how these external forces affect the ink is itself invariant. Figure 6a,b graphically shows the relationship between shear stress and shear strain in a fluid [99,100]. In a Newtonian flow of an ideal fluid, a linear relationship between shear stress and shear strain is observed. The viscosity remains constant for different shear rates. It should be noted that no liquid is completely Newtonian. This issue has been fully discussed elsewhere [44].

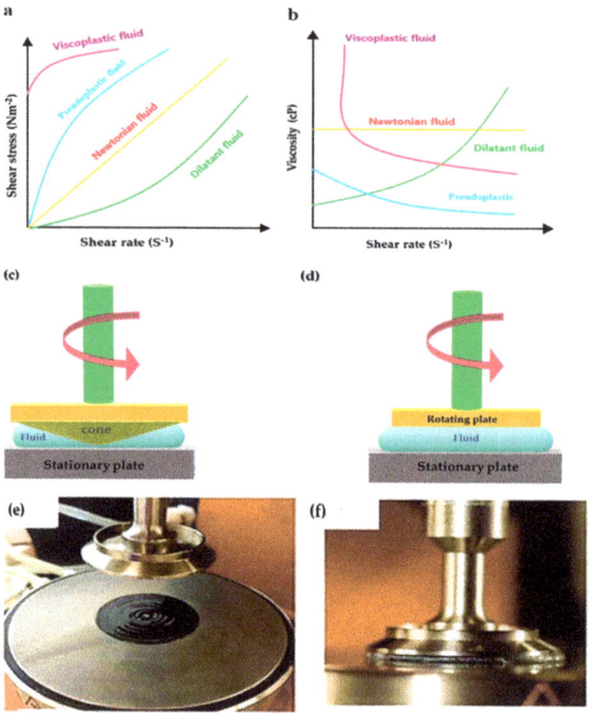


**Figure 6.** (**a**) Shear stress and shear strain flow curves for different fluids and (**b**) viscosity profiles for different fluid systems (**c**) Schematics (not to scale) of CTP and (**d**) PTP viscometers. Photograph of a PTP setup preparing for measurement with (**e**) plate retracted and ink loaded and (**f**) plate down on an ink sample, reprinted with permission from [44].

There are many ways to describe rheology and measure it for inks. Plate-to-plate (PTP) and cone-to-plate (CTP) systems are generally more expensive than conventional methods. Still, beneficial information about the rheological performance of non-Newtonian fluids can be obtained from the results. The principle behind CTP and PTP viscometers is shown in Figure 6c,d [44].

The physical behavior of MXene under applied stresses determines the rheology of MXene ink and is an influential factor in the final print, its quality and performance. Both pseudoplastic and thixotropic properties are descriptive of shear thinners. Shear-thinning fluids are efficient for various printing techniques such as extrusion, screen printing, and spray coating. This type of liquid, when a force is applied, passes through the printer nozzle and after leaving the nozzle, it has a high viscosity, which helps maintain the printing pattern [75,101].

Shear thinning behavior has been observed for MXene inks based on both mono-layered and multi-layered MXenes. The ratio of elastic modulus (G′) to viscous modulus (G″) is another important parameter that determines the rheology of dispersion. Of course, the rheological requirements for different printing methods are different. For example, spray coating requires a high processing rate with a high viscosity modulus; on the other hand, extrusion printing requires a high elastic modulus to maintain the desired print pattern. According to Figure 4a,b and Figure 7, it can be seen that single and multilayered MXenes have shown different behaviors [90].

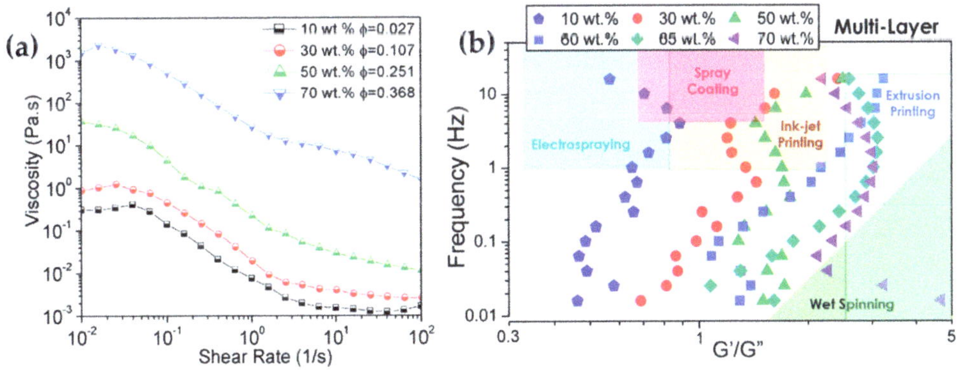


**Figure 7.** (**a**) Measured viscosity versus the shear rate of multi−layer MXene flake suspension (**b**) Frequency dependence of the ratio of the G′ elastic modulus to G″ viscous modulus for multi-layer $Ti_3C_2T_x$ MXene flakes dispersed in water, reprinted with permission from [87].

To match the surface tension between the ink and the substrate, high-quality dispersion of MXene (e.g., $Ti_3C_2T_x$) in different solvents is required. Among the polar solvents suitable for this part, the following can be mentioned [57,65,102]:

1. N-dimethyl formamide (DMF);
2. N-methyl-2-pyrrolidone (NMP);
3. Dimethyl sulfoxide (DMSO);
4. Propylene carbonate (PC);
5. Ethanol.

It should be noted that flakes dispersed in organic polar solvents have better resistance to oxidation than water. The concentration of dispersed $Ti_3C_2T_x$ and the viscosity of the solvents mentioned above have a linear relationship indicating that there is a fluid flow of ink when these solvents are used for various printing techniques [57].

A wide range of rheological properties and different concentrations make it possible to print MXene with different techniques. Due to this reason, MXene inks have been very popular among researchers for various applications, some of which are briefly mentioned in the next section.

## 4. Printing Methods

Although printing has made great strides to date, functional printing and ink development still remain in their infancy. Owing to the difficulty of developing high-performance additive-free 2D inks, great efforts are needed. Among nanomaterials, MXenes have special capabilities offering extraordinary production possibilities [103]. To make MXene films with a thickness of nanometers and micrometers for practical applications, one needs to be able to deposit patterned solid-state dispersions of MXene [75]. There have been several reported methods of printing/coating for achieving this goal [59,96,104–106]. Various printing and coating techniques have differences in fluidic properties, resolution, and scalability [75,103].

### *4.1. Inkjet Printing*

The technology of inkjet printing (IJP) is a digital, non-contact method that is extensively used in research as well as industrial application [103]. Considering that a pattern is provided to the printer as a digital file (which can be modified easily) and the ink consumption is very low (1–2 mL), this is one of the best techniques for parameter optimization and fast prototyping [107]. There are two main types of inkjet printing based on the droplet generation mechanism, i.e., continuous inkjet printing (C-IJP) and drop-on-demand printing (DOD-IJP); C-IJP is mainly used in industry, while DOD-IJP is more appropriate for scientific research in the lab [108].

One of the major challenges in inkjet printing is producing printable inks with suitable physical fluid properties. A good indicator for fluidic properties and ink printability is the inverse Ohnesorge number Z, which is defined as [75]:

$$Z = \frac{\gamma^{1/2}}{\eta} \tag{8}$$

where $\alpha$ is the printer nozzle diameter, $\eta$ is fluidic density, $\gamma$ is the surface tension, and $\eta$ is the fluidic viscosity. Jang et al. recommend an optimal Z range of 4–14 for inkjet-printable ink. Based on the rheological properties of MXene inks, the Z value of (~2.6) for ethanol ink is slightly higher than for DMSO, with a Z value of (~2.5) and NMP ink with a Z of (~2.2). Optimal Z-values for MXene organic inks are ($1 \leq Z \leq 14$) [75].

Wearable electrical biosensors with a great degree of pattern flexibility could be made using the process of inkjet printing, which is a simple and basic route. Inkjet printing with digital patterns eliminates the need for a mask, simplifying the fabrication process. This reduces the amount of material wasted in the process of creating the desired pattern. Owing to their solution-processability, surface chemistry, superior conductivity, and biocompatibility, MXenes sheets have attracted great attention for printing different films such as those used in epidermal applications, out of which only two investigations so far have used MXenes in skin electronics, one to estimate skin bending and another to detect hydrogen peroxide in perspiration [109].

As a biosensing platform, Saleh et al. [110] created an aqueous MXene ink and inkjet-printed MXene sheets on flexible, conductive polymer substrates, as demonstrated in Figure 8. As part of MXene dispersion, a nonionic surfactant saponin is added to the electrodes that can be inkjet printed with water-based MXenes and attached to the skin as a biosensor. An ion-selective layer on printed MXene surfaces can be used to create a $Na^+$ sensor in a sweat-like medium. In addition to a good signal-to-noise ratio, these electrodes are durable, having a shelf life of 50 days in ambient conditions. In addition to paper, glass, and PEDOT: PSS film, the MXene/saponin formulation is printable on a wide range of substrates that have varied mechanical and surface properties. Sticky tapes were used to test for the adhesiveness of the MXene films on these substrates. Paper was less adherent to MXene than it was to glass. This paper shows that MXene printed films may be used for multifunctional biosensors, as $Na^+$ in artificial sweat can be detected when combined with a sodium ($Na^+$) ion-selective membrane ($Na^+$ membrane). Interferon Gamma (a proinflammatory cytokine protein) causes antibodies on films to generate an electrical signal. As a result, by mounting IFN antibodies on the surface of the films, they might be employed as cytokine protein sensors. They've created a platform that shows how printed MXenes could be employed in a variety of sensing applications.

Vural et al. [97] developed 2D MXene inkjet printing, utilizing protein-based binders that can build sequence-controlled assemblies with hydrogen bonding 2D crystals. Cellulose paper, glass, polyethylene terephthalate (PET), and polymethylmethacrylate (PMMA) were all effectively printed with these inks, and they printed MXene patterns on PET films to create deformable light-emitting diode (LED) circuits and EMI shielding materials to demonstrate the potential of these inks in flexible electronics. This technique might be useful in the development of humidity-mediated memory systems and sensors, especially for microfluidic systems.

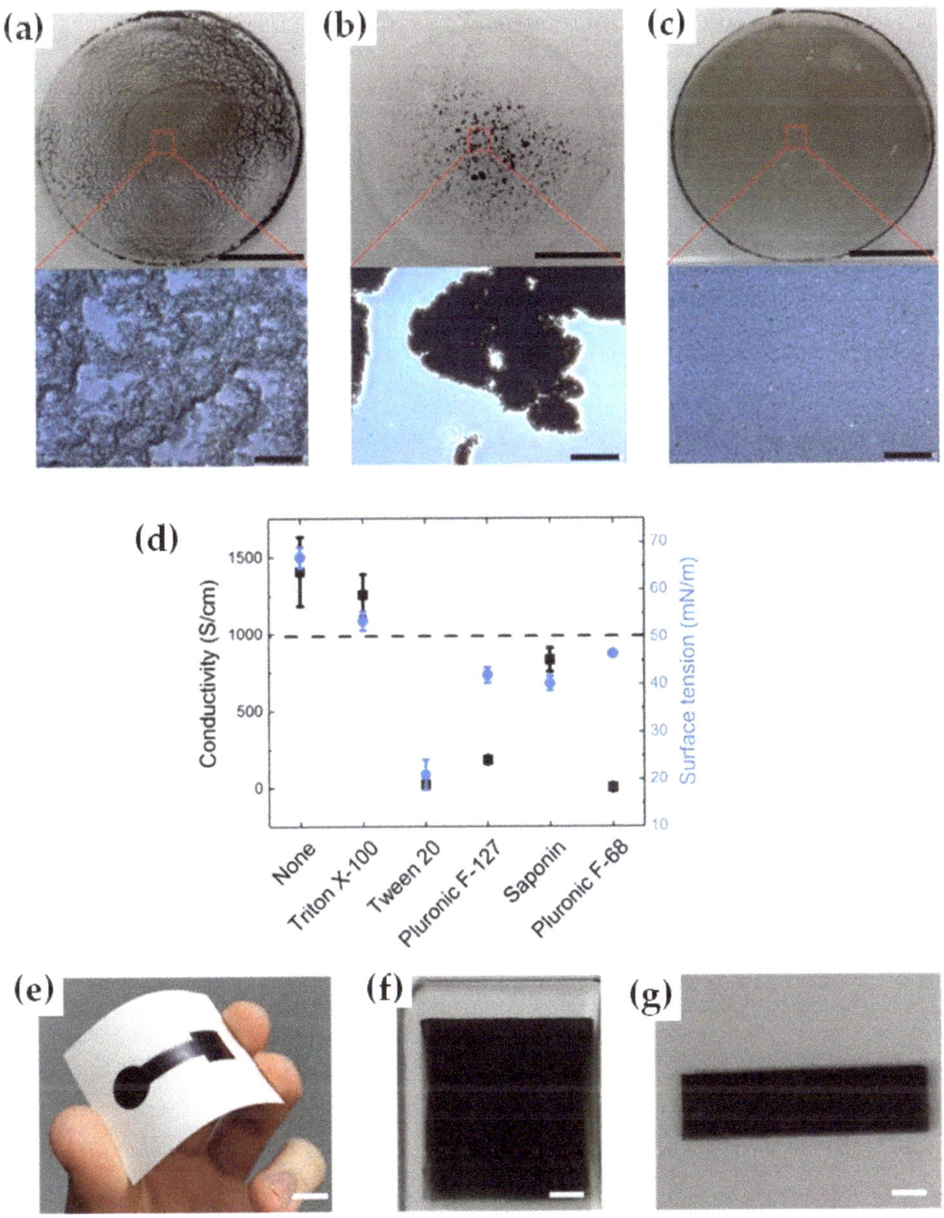


**Figure 8.** Optical photographs of MXene solutions comprising (**a**) an anionic and (**b**) a cationic surfactant, both creating heterogeneous suspensions. (**c**) MXene solution with a nonionic surfactant creates a homogeneous suspension. The bottom images are microscope images of the same suspensions. (**d**) The effect of surfactants on the surface tension of MXene inks and the conductivity of the resulting films (n = 3). Digital photographs of MXene films printed on various substrates: (**e**) tattoo paper, (**f**) glass and (**g**) a PEDOT:PSS coated glass substrate. Scale bars are 5 mm (**a–c** top), 100 µm (**a–c** bottom), 1 cm (**e**), and 2mm (**f**,**g**), reprinted with permission from [110].

MXene inks have rarely been inkjet-printed on textiles, which demand inks with different characteristics than those required for nonporous surfaces [111]. Using a commercial printer, Uzun et al. [111] showed that these inks may be utilized to TIJ print different structures such as conductive patterns and interdigitated devices; they showed how controlled ejection of ink droplets from nozzles onto textile substrates like knit and woven textiles might be used to make wearable micro-supercapacitors (MSCs); this was the first time MXenes have been printed on textiles for textile-based wearable devices. On all surfaces, increasing the $Ti_3C_2T_x$ flake size and concentration resulted in printed lines with reduced resistance. The inkjet-printed circuit on a cotton knit fabric produced with low concentration (9.4 mg $mL^{-1}$) was found to have adequate conductivity to operate three LED lights after only one print pass. When six print passes were conducted, the electrical resistance of the lines ranging from 450 $cm^{-1}$ (S-$Ti_3C_2T_x$ ink at 22.4 $cm^{-1}$) to 55 $cm^{-1}$. Over several cycles, the devices showed outstanding cycling stability and generated reasonably high energy and power densities of 12.36 $\pm$ 4.46 Wh $cm^{-2}$ and 0.16 $\pm$ 0.58 mW $cm^{-2}$, respectively. In this way, the functioning of materials that may be printed for a range of textile-based energy harvesting, storage, and sensing applications was greatly improved [111].

Inkjet printing is a cutting-edge strategy for utilizing the benefits of 2D materials for printed optoelectronic devices with tiny small-footprints, integration, substrate/geometrics compatibility, scalability, and low cost. The selection of ink solutions is crucial. For ink formulation, high-viscosity solutions such as N-methyl-2-pyrrolidone (NMP) and dimethyl sulfoxide (DMSO) can be employed directly. To improve the printed film uniformity, a small amount of polymers or proteins can be added to the solutions that act as "binders". Jiang et al. inkjet-printed MXene nanosheets in laser resonators with both fiber and free-space geometrics, achieving a wide range of spectral band ultrafast laser operations from near-infrared to mid-infrared, with pulse durations up to 100 femtoseconds [112].

Inkjet printing offers the greatest potential for rapidly creating and using novel materials. Wen et al. [109] reported an early example of inkjet-printed MXene transparent films. The key in this study is how to overcome the "coffee ring effect," which complicates the inkjet printing process; it is extremely difficult to manufacture large-area transparent film electrodes due to the fact that the majority of the solute is deposited at the edges of the printed patterns. Furthermore, ink rheological properties and the solvent used in MXene ink affect printing performance. Films made with $Ti_3C_2T_x$ can serve both as transparent conductors and capacitors. They may be used to build symmetric or asymmetric flexible and transparent supercapacitors to increase energy and power densities.

The extensive research done in inkjet printing has provided the technique with many benefits, such as the ability to print complex patterns, the use of different substrates with different properties, and high resolution. All of these point toward the ability to make micrometer-sized devices. However, in addition to all the advantages, some issues are of fundamental importance, especially the preparation of inks with appropriate rheological properties, fluidity, and appropriate physical and morphological characteristics [103].

### 4.2. Screen Printing

The screen printing method (SP) is a fast and efficient method of direct printing of 2D materials [103], and the ink is deposited via a mesh with some open (printing pattern) and blocked (non-printing) areas. Screen printing is basically divided into two methods: flat-bed and rotary [75]. In detail, the flat-bed approach involves pressing ink through a flat, patterned screen onto a substrate and repeating this process to build up layers, while in the rotary process, ink is pressed through an ink cylinder and perforated metal onto the substrate. MXene sediment printing illustrates the feasibility of making wearable smart electronics using waste-free ink formulation for screen printing.

S.Abdolhosseinzadeh et al. [95] demonstrate scalable printing of diverse structures and patterns using additive-free MXene sediment inks with good spatial regularity and high resolution. With the help of patterns with different line and gap sizes, it is possible to create a conductive pattern, letters, and MSCs in a matter of seconds (Figure 9a). To provide energy and power solutions for unique applications, the all-printed MSCs may be linked fast and simply in series or parallel (Figure 9b). All tandem devices have perfect capacitive behavior and little IR drop (Figure 9c,d), making them the optimal approach. Figure 9d,e shows that the tandem device has a very low equivalent series resistance due to its rapid charge/discharge rate, and "yrashed" sediments from MXene may easily power a bright LED (Figure 9f), their high capacitance (158 mF cm$^{-2}$) and energy density (1.64 watts cm$^{-2}$) set them apart from other printed MSCs in terms of energy density and areal capacitance [95].

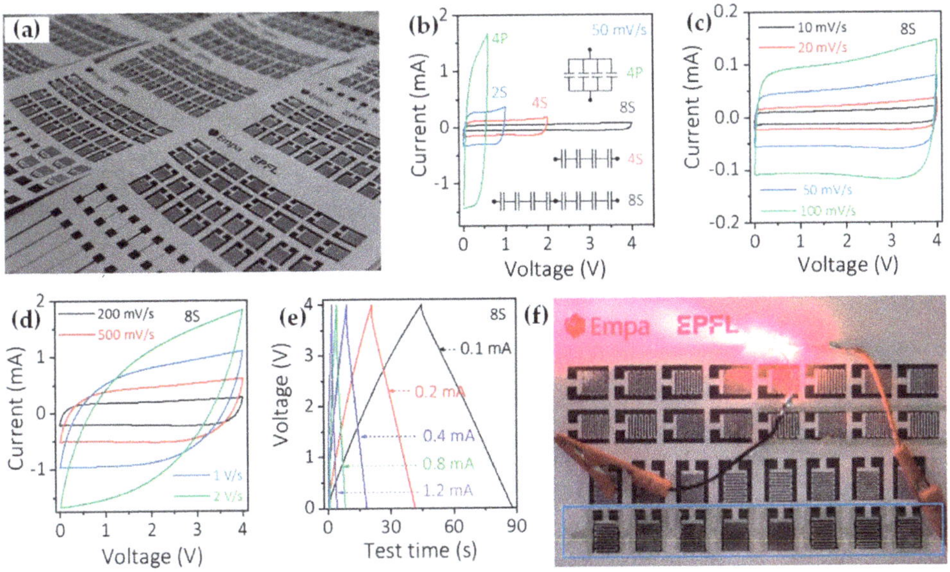


**Figure 9.** Scalable production of micro-supercapacitors based on MXene sediment inks. (**a**) Optical image of a screen-printed MXene-based microsupercapacitor, showing the great promise of scale-up production of high-performance MSCs. (**b**) CV profiles of different tandem devices at 50 mV s-1. (**c**,**d**) CVs of a tandem device with eight MSCs connected in series at different scan rates show a high-rate response. (**e**) GCD profiles of a tandem device with eight MSCs connected in series, indicating symmetric, linear curves at various currents. (**f**) Optical image of the tandem device to power an LED light, demonstrating the feasibility of the as-printed tandem device for practical applications, reprinted with permission from [95].

There are several advantages to the screen-printing method, such as controllable thickness, cost-effectiveness, and mass-production capabilities. Conversely, its major disadvantages are low print resolution, roughness, and thick ink used in this process [75]; all of them indicating that still a great effort should be made in this area.

### 4.3. 3D Printing

Using 3D printing technology, complex-shaped parts can be formed automatically from CAD data (computer-aided design) without any tooling [11]. Extrusion-based additive manufacturing (AM) is a process for fabricating complex structures utilized in a variety of industries, including energy storage. An ink (filament) of a colloidal or gel-type is extruded from a nozzle and layered on a surface to form 3D objects in this printing process [93]. A 3D object is created by layering materials upon each other in additive manufacturing [113]. Unlike previous 2D printing methods, 3D ink consists of polymers that are usually solid-liquid or powder. Comparable to other manual or molding processes, this automated fabrication has some benefits, such as computer-aided design capabilities, fast editing of

CAD models and precise dimensioning [114]. A 3D printing process can be summarized as follows: (1) designing a CAD model, (2) conversion of the model to an STL file, (3) slicing the STL file to several 2D layered cross-sections, (4) 3D printing the prototypes, and then (5) post-processing [115].

In most cases, 3D-printed electrode materials are not electrocatalytic or appropriate for these purposes. Further modification of 3D-printed electrodes can be achieved using atomic layer deposition and electrodeposition. The only downside is that electrodeposition or ALD may not be possible for all materials since not all precursors are available. Electrodeposition and ALD currently cannot be used to deposit MXenes [116–118]. Using highly concentrated MXene ink for extrusion printing at room temperature, Orangi et al. [93] demonstrate that MSCs with a range of topologies and electrode thicknesses can be produced at a large scale. It is possible to print flexible MSCs on polymer and paper substrates using established printing methods. The aforesaid research describes printed solid-state devices exhibiting a very high capacitance area and excellent electrochemical performance.

Regarding 3D bio-printing, Rastin et al. [119] demonstrate that $Ti_3C_2$ MXene nanosheets can be used to build 3D bio-inks despite the fact that most of the current bio-inks have poor electrical conductivity. Studies have shown that hydrogels with electrical conductivity have improved signal transmission between cells. In most cases, these conductive bionics were developed by combining biopolymers with conductive polymers, but studies of 2D conductive materials were limited. It was found that HA/Alg biogels containing $Ti_3C_2$ nanosheets have outstanding printability and good resolution. The conductivity of the mentioned biocomposite ink could reach 5500$\pm$ 85 μs cm$^{-1}$ and 7200 $\pm$ 126 μs cm$^{-1}$, respectively, when 1 mg mL$^{-1}$ and 5 mg mL$^{-1}$ of $Ti_3C_2$ were respectively added to the ink, surprisingly higher than those values obtained from the neat ink, being promising for many biomedical applications such as neural tissue engineering and the like [119].

3D printing is also an optimal solution for integrating functional materials into multidimensional architectures. Cao et al. [120] successfully developed TOCNFs/$Ti_3C$ smart fibers and textiles with woodpile and fishing net structures with enhanced electrical, photonic and mechanical properties (Figure 10); this type of hybrid inks is promising in versatile applications such as smart textiles, electronic skins, wearable sensors, soft robotics, and human-machine interaction [120].

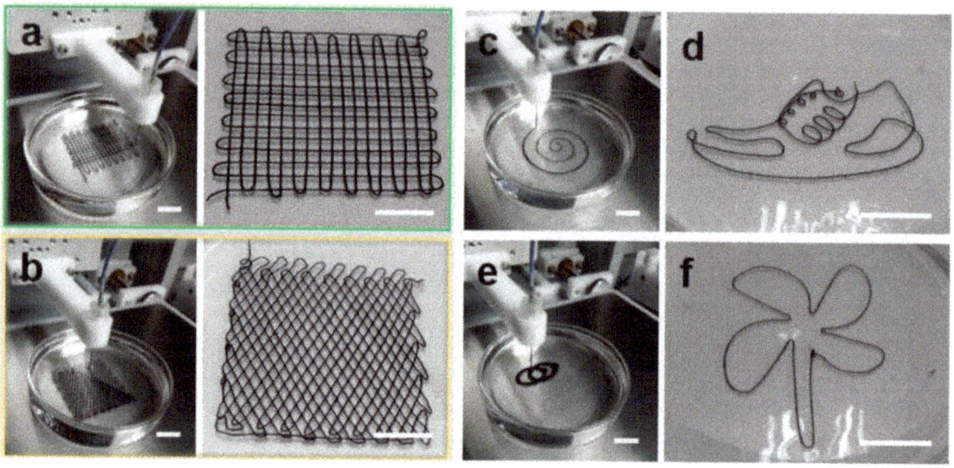


**Figure 10.** Optical images of the 3D-printed TOCNFs/$Ti_3C_2$ fabrics with (**a**) woodpile and (**b**) fishing net structures. (**c**–**f**) Programmable printed TOCNFs/$Ti_3C_2$ fibers with designed geometric structures, reprinted with permission from [120].

Huang et al. [121] used an in-situ ice template to produce MXene/CNT aerogel electrodes through a 3D printing approach. During the surfactant/MXene assembly process, shear stress is applied to align the flakes vertically. Three-dimensionally printed all-MXene micro-supercapacitors (MSCs) can provide an ultrahigh capacitance of 2.0 F $cm^{-2}$ at 1.2 $mAcm^{-2}$ and retain a record-high energy density (0.1 mwh $cm^{-2}$ at 0.38 mW $cm^{-2}$) [121]. 3D printing ink relies on extrusion-based technology to ensure smooth filament extrusion with good accuracy and shape retention. It is possible to create oriented microstructures even in high concentrations of MXene, and this 3D printing strategy may be applicable for additive manufacturing of MXene micro-supercapacitors as shown in Figure 11a, wherein MXene slurry forms a viscous, thixotropic ink with $3.3 \times 10^3$ Pa s and without aggregates (Figure 11c). A scatter plot displaying the storage modulus and loss modulus of MXene ink as a function of shear strain is shown in Figure 11d. Plateaus of $G'$ are orders of magnitude greater than those of $G$, which show primarily solid-like behavior. Extrusion will continue through micron-sized nozzles at relatively low pressures using this feature [121].

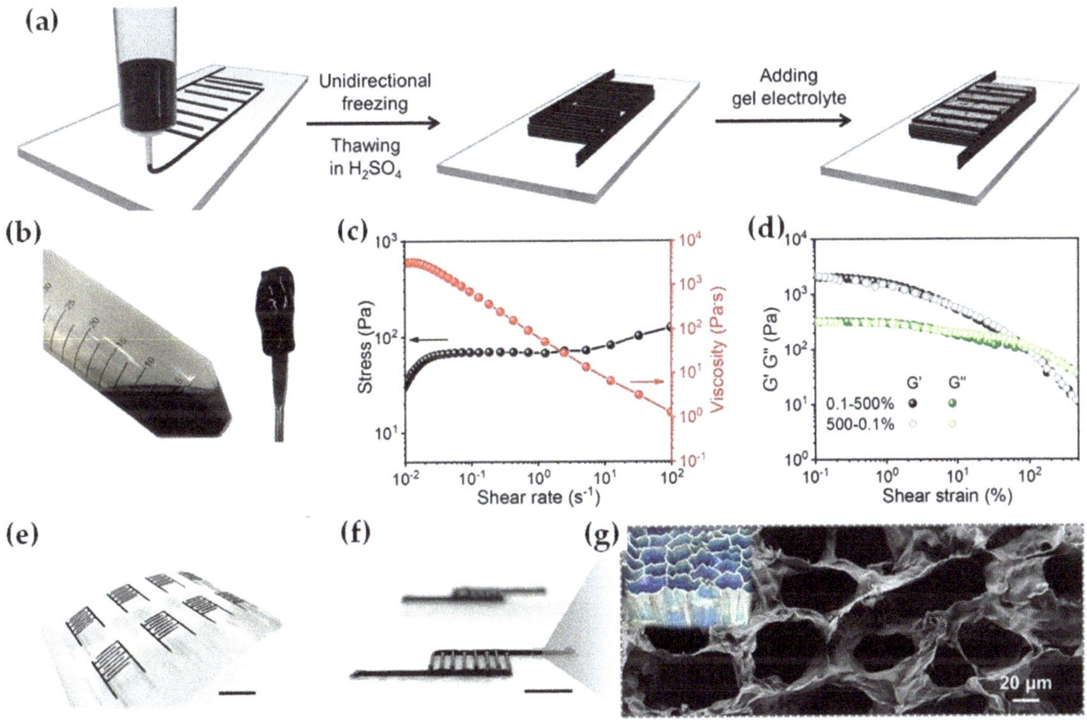


**Figure 11.** (**a**) The fabrication process of 3D-printing all-MXene MSC via MSES. (**b**) The digital photographs of MXene slurry. (**c**) Shear-thinning behavior of the MXene inks (stress and viscosity against different shear rates). (**d**) The oscillatory measurements (1 Hz) of the MXene ink, which sweeps from 0.1% to 500% and back to 0.1% strain. (**e**,**f**) Photographs of 3D-printed MXene MSC, scale bar is 1 cm. (**g**) The top-view SEM image of MXene hydrogel MSC, reprinted with permission from [121].

In a recent study on 3D printing MXene-based hybrid ink, Jambhulkar et al. [122] explained how a hybrid process is able to obtain nano-aligned and micropatterned MXene nanoflakes with hierarchical 2D morphologies. The developed 3D printing process included a surface topography design by a microcontinuous liquid interface production (μCLIP) as well as a direct nanoparticle-assembling through capillarity-driven DIW method wherein the final hierarchical product exhibited anisotropic conductivity values with a wide piezoresistive sensing range, and enhanced mechanical durability. Using this hybrid 3D printing technology, nanoparticles can be patterned and assembled for broad applications in a fast and scalable manner, indicating both manufacturing feasibility and device functionality [122].

In all, 3D printing has many advantages, including environmental friendliness, economic efficiency, controllable thickness and geometry, and flexibility of the printed pattern; however, it is limited by the materials that can be used for printing, being the most critical challenge of this method [75].

### 4.4. Stamping

There are several methods of non-pattern printing, including screen printing, spray coating, and stamping. Stamping is typically used to create effective deposition, but the adhesion between the ink, donor, and substrate requires to be precisely optimized. High printing speeds and scalability are the advantages of stamping, while the high cost of setting up and prototyping gravure rollers and plates, as well as the large volume requirements for inks, are the major disadvantages of this technique [75]. Stamping is a good option for producing flexible, scalable, and effective devices. Inks used for stamping should have characteristics such as high electrical conductivity, a long service life, excellent mechanical properties, good strength to withstand deformation, as well as excellent physical properties such as rheology, viscosity, surface tension, and adhesion [123,124].

This type of manufacturing method is divided into four categories depending on the use of a template: gravure, letterpress, foam-assisted stamping, and stamping. A stamped pattern's printing parameters might vary based on the type of stamping technique utilized. It should be pointed out that the ink is cheap, and the stamp may be recyclable. Stamping is usually cost-effective and does not require expensive technologies such as laser scribing or photolithography. Since this printing technology can manufacture flexible devices with easy procedures, it provides a new resource for MSC device production [123].

Zhang et al. [104] developed a new stamping strategy, as shown in Figure 12, to 3D print a viscous water-based MXene ink to quickly prepare high-performance coplanar MXene MSC; using continuous thin-film nanostructures, these MSCs no longer have metal current collectors and may achieve a rapid electron transfer, thereby obtaining a high-speed response in solid-state devices. In addition, functionalized MXene imprinted on the hydrophilic porous paper also improves the accessibility of the MXene sheet to electrolyte ions. The combination of the pseudocapacitive behavior, high density and significant thickness of the $Ti_3C_2T_x$ film may lead to a significant increase in the surface capacitance of each electrode. For example, the interdigitated MSC $Ti_3C_2T_x$ has areal capacitances of 61 mF $cm^{-2}$ and 50 mF $cm^{-2}$, respectively, under 25 μA $cm^{-2}$ and 800 μA $cm^{-2}$. It was also presented that the production of MXene-MSCs can be increased by designing pads and cylindrical stamps, followed by a cold rolling method. Based on this method and the use of cold rolling and pressing, it is possible to produce dozens of MSCs with a high capacity of 56.8 $mFcm^{-2}$ at 10 mV $S^{-1}$ in a very short time. It is thought that by further optimizing the synthesis of MXenes, adjusting their particle size and chemical composition, and controlling surface functional groups, an MSC with greater surface/volume capacity and faster solid processing speed could be developed, opening up many exciting possibilities for the production of high-performance energy storage devices [104]. In all, rare studies on MXene stamping are available in the literature, and hence there still exist a lot of questions unanswered regarding the process parameters, materials, characterization and potential applications.

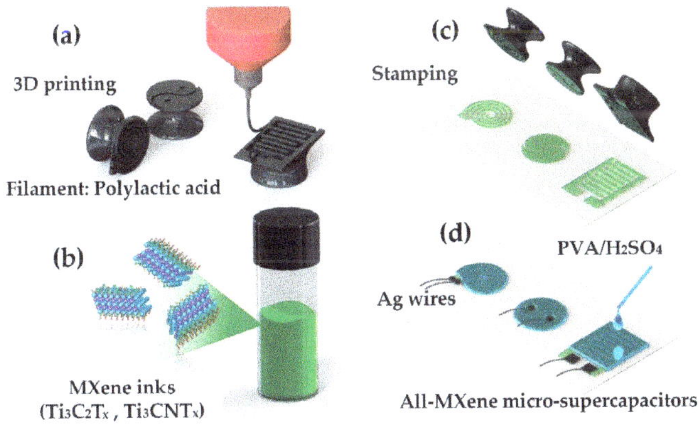


**Figure 12.** Fabrication of all-MXene-based MSCs using the stamping strategy. (**a**) stamps are first 3D printed, (**b**) MXene ink is controlled and prepared, (**c**) MXene ink is firmly pressed and stamped onto a rough hydrophilic substrate, and (**d**) upon attaching Ag wires and casting the electrolyte, the device is left to dry in air to form solid MXene-based MSC, reprinted with permission from [104].

### 4.5. Patterned Coating

In spite of the advantages of printing in terms of resolution, scalability, and pattern complexity, equipment and ink properties often pose a challenge, limiting the number of materials that can be printed, not to mention even more complicated techniques to form printable inks from raw materials [75]. In this regard, traditional coating methods, e.g., drop-casting [125–127], spin coating [128–130], spray coating [131–134], vacuum filtration [125,135], etc. are ideal for testing materials and prototyping devices, as they have less stringent equipment and ink requirements. Because of the surface functional groups on MXene nanosheets, they can be processed in solution, which can be sprayed, spun, or dip-coated in order to make highly conductive thin films [136]. When compared to printing methods, patterned coatings are more cost-effective; however, the pattern complexity, tunable thicknesses, and surface roughness are much more challenging to manipulate with accuracy.

Hu et al. [137] described an approach to creating coplanar and adjustable interdigital electrodes by patterning completely delaminated few-layered MXene flakes onto printed paper, as presented in Figure 13. A solid-state MSC's volumetric capacitance may increase by at least 460 percent with increasing thickness of the MXene active layer, compared to advanced carbon-based planar symmetric MSCs (0.1–6 mF $cm^2$). Devices with volumetric energy densities of 5.48 mWh $cm^3$ or more are equivalent to or better than MSCs produced from other types of material, such as carbon or metallic oxides or paper-based polymers. High conductivity, fast intercalation, and hydrophilic surfaces are only a few of the great characteristics of MXene-based planar solid-state symmetric MSCs that exhibit good electrochemical performance. A major distinction is that this manufacturing procedure does not use any binder, conducting additives, or organic cosolvents. A unique, layered porous electrode structure has been created by Hu et al. [137] by aligning MXene flakes along the c-axis with no, or very minimal, binder/conductive additives. The electrode structure was highly electrically conductive due to the short ion diffusion distance between the electrode materials and electrolyte. Furthermore, by designing the electrode stacks so that the ions discharge rapidly, the resistance could be further decreased, allowing for the rapid diffusion of electrolytes. In light of these results, MXene-based symmetric MSCs may be suitable for use as miniaturized electronic devices [137].

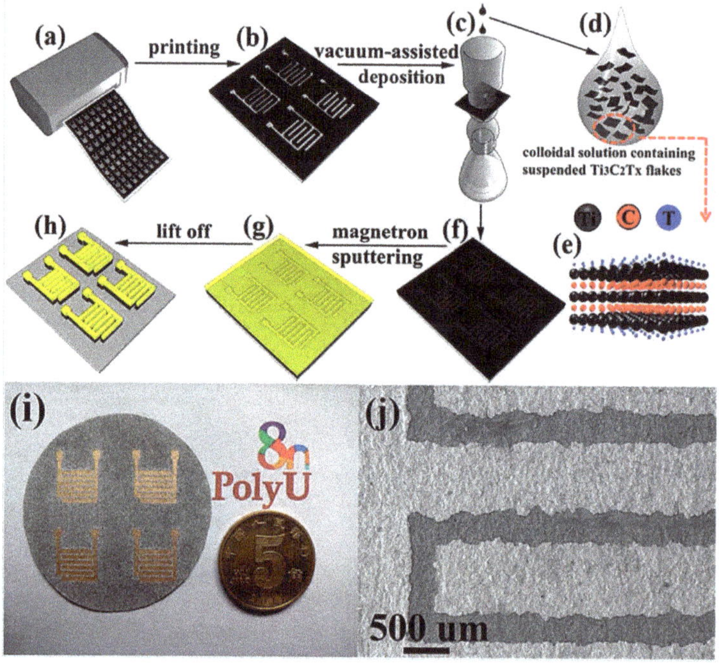


**Figure 13.** Flowchart of the easily manipulated protocol for patterning of fully delaminated few-layered MXene ($Ti_3C_2T_x$) flakes on paper for planar symmetric MSCs: (**a**) a common laser printer for printing an interdigital circuit template designed with drawing software; (**b**) paper substrate with the laser-printed template; (**c**) vacuum-assisted deposition of fully delaminated few-layered MXene flakes on the printed paper to fabricate the electroactive layer at the bottom; (**d**) schematic of colloidal solution containing fully delaminated suspended fewlayered MXene flakes; (**e**) schematic of the crystal structure of $Ti_3C_2T_x$ flakes (MXene); "T" symbolizes the surface atoms and atomic group, such as OH, F, and O; (**f**) the bottom binder/conductive-additive free electroactive layer based on few-layered $Ti_3C_2T_x$ flakes; (**g**) magnetron sputtering of a top Au film that served as a current collector on the electroactive layer at the bottom; (**h**) final lift-off to form the MXene-based planar MSCs; (**i**) a photo of the finished MXene-based planar MSCs; (**j**) the corresponding SEM photo of the fabricated interdigital electrodes of the devices, reprinted with permission from [137].

MXene films with high transmittance, conductivity and excellent mechanical stability are good candidates for transparent flexible electrodes for high-performance organic photovoltaic (OPV) [136]. A major draw of organic photovoltaics (OPVs) is their high flexibility, lightweight, low cost, and ease of printing for long-term renewable energy production. As a result of the electrical characteristics of MXenes, Qin et al. [136] have developed vertically stacked $Ti_3C_2T_x$ MXene-based solid-state photovoltaic supercapacitors. A series of highly conducting, transparent and flexible films were produced, consisting of nanoflakes aligned parallel to the substrates, by spin-casting colloidal solutions of $Ti_3C_2T_x$ nanosheets. On any substrate, spin-coating may produce smoother, more flexible transparent electrodes for OPVs by using tiny MXene flakes (Figure 14a). $Ti_3C_2T_x$ spin-coated films displayed apparent arcs of diffracted intensity in the two-dimensional GIWAXS patterns (Figure 14b), indicating the development of regular orientations, being a substantial difference between the in-plane and out-of-plane diffraction signals, which indicates that flakes are aligned parallel to the substrate plane in the presence of shear force. In order to gain insight into the characteristics of $Ti_3C_2T_x$ optoelectronic film, transmittance spectra with film thickness were measured (Figure 14d), wherein the transmittance reduced with increasing film thickness as well as a large visible peak. Spin-coated $Ti_3C_2T_x$ films were measured for their transmittance at 550 nm and conductivity, and the results are shown in Figure 14e [136].

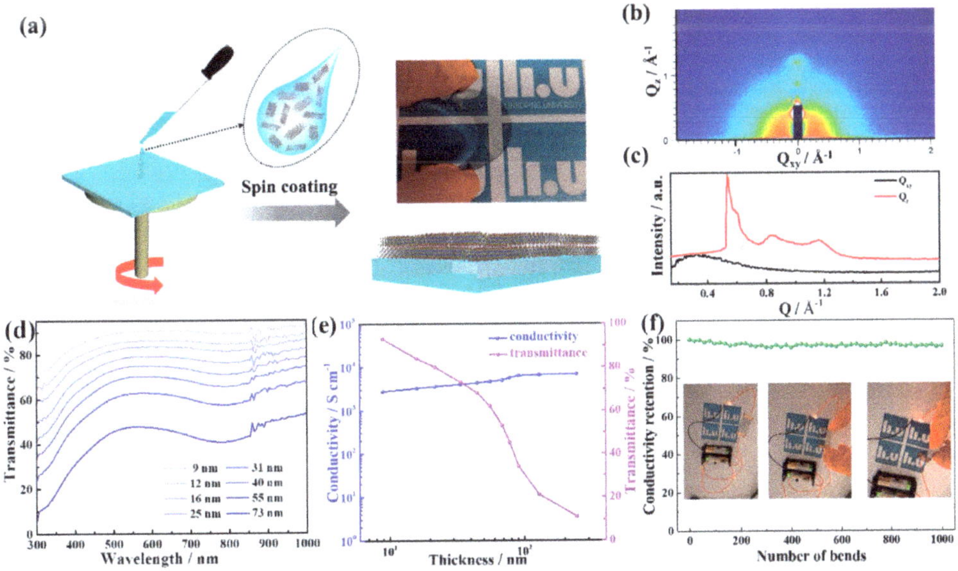

**Figure 14.** Optoelectronic properties of $Ti_3C_2T_x$ films. (**a**) Schematic of the preparation of a transparent flexible electrode. (**b**) Two-dimensional GIWAXS pattern of $Ti_3C_2T_x$ film prepared by the spin-coating method. (**c**) Out-of-plane and in-plane line-cut profiles from the GIWAXS pattern. (**d**) Transmittance spectra of $Ti_3C_2T_x$ films of various thicknesses. (**e**) Variations in conductivity and transmittance of $Ti_3C_2T_x$ electrodes as a function of thickness. (**f**) Conductivity retention with the number of bends for flexible $Ti_3C_2T_x$ electrodes on PET substrates. The inset image shows the conductivity of the $Ti_3C_2T_x$ electrodes in bent and twisted states, reprinted with permission from [136].

Sarycheva et al. [133] used 2D titanium carbide MXenein wireless communication devices due to their unique properties. They produced a clear MXene antenna through one step of spray coating. According to the results, a thickness of 100 nm and a reflection coefficient of less than 10 dB was obtained. It is also possible to achieve a reflection coefficient of -65 dB by increasing the thickness of the antenna to millimeter dimensions. In the study, titanium carbide MXene is found to operate below the skin depth of copper and other metals, presenting an opportunity to produce transparent antennas. Due to their high conductivity and water dispersibility, MXenes have been found to be an ideal material for the industrial production of various portable, flexible, and wearable electronic devices [133].

As opposed to other printing processes, they are less in demand and also less expensive. The disadvantage of this method is that there is no control over the thickness and roughness of the surface of the design, which is a major drawback. In addition, complex patterns will not be easy to achieve, unlike other methods [75].

## 5. Summary, Pitfalls, and Perspectives

MXenes have recently emerged as a revolutionary class of material displaying exceptional tailored-made properties. The onward journey and remarkable rise are establishing MXene-based materials as multifaceted playgrounds for technology-oriented explorations and are offering a tool-box for the ad hoc tailoring of advanced materials capable of effectively addressing current and future societal challenges. Unexpected applications have witnessed tremendous growth owing to the material's unique chemical and physical properties, including, among others, optical, electrical, mechanical and thermal characteristics. Attaining an in-depth and critical understanding of the broadest arsenal of such unique and new properties, as well as the synergistic effects of the assorted characteristics, will play a pivotal role in new discoveries in both research and industrial sectors. The present paper aims to keep the readers updated with the recent developments in MXene-based inks, including their formulations, preparations and applications in a variety of fields ranging from energy and storage areas to biomaterials. Unlike other 2D sheets like graphene or dichalcogenides being well-implemented in real industrial applications such as batteries and electrodes, among others, the implementation of MXene inks in real industrial applications is still in its infancy stage and hence still, much attention should be made to advance the field and mature at different aspects ranging from fundamental sciences and manufacturing expenses to those practical challenges. It is also emphasized that unlike many other types of 2D materials, MXene has special traits such as mechano-ceramic behavior, hydrophilicity as well as rich surface chemistry, making them a promising additive in many liquid-based hybrid materials and structures. First of all, the basics regarding the MAX phases and the process of turning MAX to MXene nanoflakes have been proposed. The composition of MXene inks with the preparation of MXene-based inks has then been introduced. Due to their vital importance, the stability of MXene nanosheets in different media has been discussed, and then considerable attention is paid to the fluidity and rheological properties and features of the MXenes, being essential parameters dictating the final quality and performance of the MXene ink. Further, different printing techniques ranging from inkjet printing and stamping to 3D printing, are elaborated to discuss the current state of the art in this ever-growing field. Finally, despite the brilliant progress made, there are still multiple longstanding hurdles that need to be overcome, so it is believed that still, innumerable questions are to be answered to fill the existing voids. For instance, MXene synthesis with controllable surface terminations is one of the major challenges ahead. The uniformity of these functional groups effectively improves the rheological properties and colloidal stability of MXene inks and also changes the final print quality. Regarding computational study, rare studies are available in this field, and much more attention is needed to explore ink behavior in different situations. Further, few types of MXene (almost always $Ti_3C_2T_x$) have been used in inks and subsequent printing techniques, and it is strongly thought other different MXenes also have different features when hybridized in varied inks. Most of the current research has been undertaken in the field of energy storage and electrodes for capacitors; however, other 2D nanomaterials have been used in a variety of research and industrial sectors with promising results. It is believed a great variety of transition metals with various surface terminations makes it possible to achieve new compounds for specific and unique applications. Printing of heterogeneous structures, including MXene and other two-dimensional nanomaterials, could also be considered in the future. Finally, as one of the most challenging issues, MXene inks have relatively low stability and oxidation resistance in different materials. In order to commercialize MXene-based inks, there is a long way to go for scalability, longer shelf life, higher performance and functionality.

**Author Contributions:** Z.A., Y.Z. and M.M. contributed equally. All authors contribute to the conception of the project, data collection and analysis, and the writing of the manuscript. All authors have read and agreed to the published version of the manuscript.

**Funding:** No funding has been provided in the present study.

**Institutional Review Board Statement:** Not applicable.

**Informed Consent Statement:** Not applicable.

**Data Availability Statement:** All data is contained within the article.

**Conflicts of Interest:** The authors declare no conflict of interest.

## Leferences


. Aleeva, Y.; Pignataro, B. Recent advances in upscalable wet methods and ink formulations for printed electronics. *J. Mater. Chem. C* **2014**, *2*, 6436–6453. [CrossRef]

. Altin, S.; Bulut, F.; Yasar, S. The production of a low cost printing device for energy storage systems and the application for supercapacitors. *J. Energy Storage* **2019**, *25*, 100882. [CrossRef]

. Chen, L.; Zhou, Y.; Wu, M.; Hong, M. Remote-mode microsphere nano-imaging: New boundaries for optical microscopes. *Opto-Electron. Adv.* **2018**, *1*, 170001. [CrossRef]

. Li, W.; Li, Y.; Su, M.; An, B.; Liu, J.; Su, D.; Li, L.; Li, F.; Song, Y. Printing assembly and structural regulation of graphene towards three-dimensional flexible micro-supercapacitors. *J. Mater. Chem. A* **2017**, *5*, 16281–16288. [CrossRef]

. Beidaghi, M.; Gogotsi, Y. Capacitive energy storage in micro-scale devices: Recent advances in design and fabrication of micro-supercapacitors. *Energy Environ. Sci.* **2014**, *7*, 867–884. [CrossRef]

. Zhang, C.J.; Nicolosi, V. Graphene and MXene-based transparent conductive electrodes and supercapacitors. *Energy Storage Mater.* **2019**, *16*, 102–125. [CrossRef]

'. Li, J.; Ye, F.; Vaziri, S.; Muhammed, M.; Lemme, M.C.; Östling, M. Efficient inkjet printing of graphene. *Adv. Mater.* **2013**, *25*, 3985–3992. [CrossRef]

. Secor, E.B.; Ahn, B.Y.; Gao, T.Z.; Lewis, J.A.; Hersam, M.C. Rapid and versatile photonic annealing of graphene inks for flexible printed electronics. *Adv. Mater.* **2015**, *27*, 6683–6688. [CrossRef]

. Li, J.; Sollami Delekta, S.; Zhang, P.; Yang, S.; Lohe, M.R.; Zhuang, X.; Feng, X.; Ostling, M. Scalable fabrication and integration of graphene microsupercapacitors through full inkjet printing. *ACS Nano* **2017**, *11*, 8249–8256. [CrossRef]

0. Jun, H.Y.; Ryu, S.O.; Kim, S.H.; Kim, J.Y.; Chang, C.H.; Ryu, S.O.; Choi, C.H. Inkjet Printing of Few-Layer Enriched Black Phosphorus Nanosheets for Electronic Devices. *Adv. Electron. Mater.* **2021**, *7*, 2100577. [CrossRef]

1. Hu, G.; Kang, J.; Ng, L.W.; Zhu, X.; Howe, R.C.; Jones, C.G.; Hersam, M.C.; Hasan, T. Functional inks and printing of two-dimensional materials. *Chem. Soc. Rev.* **2018**, *47*, 3265–3300. [CrossRef] [PubMed]

2. McManus, D.; Vranic, S.; Withers, F.; Sanchez-Romaguera, V.; Macucci, M.; Yang, H.; Sorrentino, R.; Parvez, K.; Son, S.-K.; Iannaccone, G. Water-based and biocompatible 2D crystal inks for all-inkjet-printed heterostructures. *Nat. Nanotechnol.* **2017**, *12*, 343–350. [CrossRef] [PubMed]

3. Wang, Z.; Liang, X.; Zhao, T.; Hu, Y.; Zhu, P.; Sun, R. Facile synthesis of monodisperse silver nanoparticles for screen printing conductive inks. *J. Mater. Sci. Mater. Electron.* **2017**, *28*, 16939–16947. [CrossRef]

4. Liu, F.; Qiu, X.; Xu, J.; Huang, J.; Chen, D.; Chen, G. High conductivity and transparency of graphene-based conductive ink: Prepared from a multi-component synergistic stabilization method. *Prog. Org. Coat.* **2019**, *133*, 125–130. [CrossRef]

5. Aguilar-Banegas, A.D.; Reyes-Cruz, F.D.; Vargas-Pineda, J.A.; Ortega-Jimenez, C.H. Literature review of gallium: Conductive ink alternative? *Mater. Sci. Forum* **2020**, *975*, 139–144. [CrossRef]

6. Pinilla, S.; Coelho, J.; Li, K.; Liu, J.; Nicolosi, V. Two-dimensional material inks. *Nat. Rev. Mater.* **2022**, *7*, 717–735. [CrossRef]

7. Gogotsi, Y.; Anasori, B. The Rise of MXenes. *ACS Nano* **2019**, *13*, 8491–8494. [CrossRef]

8. Naguib, M.; Come, J.; Dyatkin, B.; Presser, V.; Taberna, P.-L.; Simon, P.; Barsoum, M.W.; Gogotsi, Y. MXene: A promising transition metal carbide anode for lithium-ion batteries. *Electrochem. Commun.* **2012**, *16*, 61–64. [CrossRef]

9. Iqbal, A.; Sambyal, P.; Koo, C.M. 2D MXenes for Electromagnetic Shielding: A Review. *Adv. Funct. Mater.* **2020**, *30*, 2000883. [CrossRef]

20. Malaki, M.; Varma, R.S. Mechanotribological Aspects of MXene-Reinforced Nanocomposites. *Adv. Mater.* **2020**, *32*, 2003154. [CrossRef]

21. Soleymaniha, M.; Shahbazi, M.A.; Rafieerad, A.R.; Maleki, A.; Amiri, A. Promoting Role of MXene Nanosheets in Biomedical Sciences: Therapeutic and Biosensing Innovations. *Adv. Healthc. Mater.* **2018**, *8*, 1801137. [CrossRef]

22. Chen, J.; Huang, Q.; Huang, H.; Mao, L.; Liu, M.; Zhang, X.; Wei, Y. Recent progress and advances in the environmental applications of MXene related materials. *Nanoscale* **2020**, *12*, 3574–3592. [CrossRef] [PubMed]

23. Shi, Z.; Khaledialidusti, R.; Malaki, M.; Zhang, H. MXene-based materials for solar cell applications. *Nanomaterials* **2021**, *11*, 3170. [CrossRef] [PubMed]

24. Naguib, M.; Kurtoglu, M.; Presser, V.; Lu, J.; Niu, J., Heon, M.; Hultman, L.; Gogotsi, Y.; Barsoum, M.W. Two-dimensional nanocrystals produced by exfoliation of $Ti_3AlC_2$. *Adv. Mater.* **2011**, *23*, 4248–4253. [CrossRef]

25. Zhang, C.; Ma, Y.; Zhang, X.; Abdolhosseinzadeh, S.; Sheng, H.; Lan, W.; Pakdel, A.; Heier, J.; Nüesch, F. Two-dimensional transition metal carbides and nitrides (MXenes): Synthesis, properties, and electrochemical energy storage applications. *Energy Environ. Mater.* **2020**, *3*, 29–55. [CrossRef]

26. Anasori, B.; Lukatskaya, M.; Gogotsi, Y. 2D metal carbides and nitrides (MXenes) for energy storage. *Nat. Rev. Mater.* **2017**, *2*, 16098. [CrossRef]

27. Zhang, S.; Zhuo, H.; Li, S.; Bao, Z.; Deng, S.; Zhuang, G.; Zhong, X.; Wei, Z.; Yao, Z.; Wang, J.-G. Effects of surface functionalization of MXene-based nanocatalysts on hydrogen evolution reaction performance. *Catal. Today* **2021**, *368*, 187–195. [CrossRef]

28. Malaki, M.; Maleki, A.; Varma, R.S. MXenes and ultrasonication. *J. Mater. Chem. A* **2019**, *7*, 10843–10857. [CrossRef]

29. Venkateshalu, S.; Grace, A.N. MXenes-a new class of 2D layered materials: Synthesis, properties, applications as supercapacitor electrode and beyond. *Appl. Mater. Today* **2020**, *18*, 100509. [CrossRef]

30. Naguib, M.; Unocic, R.R.; Armstrong, B.L.; Nanda, J. Large-scale delamination of multi-layers transition metal carbides an carbonitrides "MXenes". *Dalton Trans.* **2015**, *44*, 9353–9358. [CrossRef]
31. Anasori, B.; Gogotsi, Y. *2D Metal Carbides and Nitrides (MXenes)*; Springer: Berlin/Heidelberg, Germany, 2019.
32. Wyatt, B.C.; Rosenkranz, A.; Anasori, B. 2D MXenes: Tunable Mechanical and Tribological Properties. *Adv. Mater.* **202** *33*, 2007973. [CrossRef]
33. Hart, J.L.; Hantanasirisakul, K.; Lang, A.C.; Anasori, B.; Pinto, D.; Pivak, Y.; van Omme, J.T.; May, S.J.; Gogotsi, Y.; Taheri, M.L Control of MXenes' electronic properties through termination and intercalation. *Nat. Commun.* **2019**, *10*, 522. [CrossRef] [PubMed
34. Anichini, C.; Czepa, W.; Pakulski, D.; Aliprandi, A.; Ciesielski, A.; Samorì, P. Chemical sensing with 2D materials. *Chem. Soc. Re* **2018**, *47*, 4860–4908. [CrossRef] [PubMed]
35. Ghidiu, M.; Lukatskaya, M.R.; Zhao, M.-Q.; Gogotsi, Y.; Barsoum, M.W. Conductive two-dimensional titanium carbide 'clay' wit high volumetric capacitance. *Nature* **2014**, *516*, 78. [CrossRef]
36. Shen, C.; Wang, L.; Zhou, A.; Wang, B.; Wang, X.; Lian, W.; Hu, Q.; Qin, G.; Liu, X. Synthesis and Electrochemical Properties c Two-Dimensional RGO/Ti3C2T x Nanocomposites. *Nanomaterials* **2018**, *8*, 80. [CrossRef] [PubMed]
37. Liu, F.; Zhou, A.; Chen, J.; Jia, J.; Zhou, W.; Wang, L.; Hu, Q. Preparation of Ti3C2 and Ti2C MXenes by fluoride salts etching an methane adsorptive properties. *Appl. Surf. Sci.* **2017**, *416*, 781–789. [CrossRef]
38. Halim, J.; Lukatskaya, M.R.; Cook, K.M.; Lu, J.; Smith, C.R.; Näslund, L.-Å.; May, S.J.; Hultman, L.; Gogotsi, Y.; Eklund, Transparent conductive two-dimensional titanium carbide epitaxial thin films. *Chem. Mater.* **2014**, *26*, 2374–2381. [CrossRef [PubMed]
39. Lee, H.-H.; Chou, K.-S.; Huang, K.-C. Inkjet printing of nanosized silver colloids. *Nanotechnology* **2005**, *16*, 2436. [CrossRef [PubMed]
40. Sirringhaus, H.; Kawase, T.; Friend, R.; Shimoda, T.; Inbasekaran, M.; Wu, W.; Woo, E. High-resolution inkjet printing o all-polymer transistor circuits. *Science* **2000**, *290*, 2123–2126. [CrossRef]
41. De Gans, B.J.; Duineveld, P.C.; Schubert, U.S. Inkjet printing of polymers: State of the art and future developments. *Adv. Mate* **2004**, *16*, 203–213. [CrossRef]
42. Liu, Z.; Zhang, Z.-B.; Chen, Q.; Zheng, L.-R.; Zhang, S.-L. Solution-processable nanotube/polymer composite for high performance TFTs. *IEEE Electron Device Lett.* **2011**, *32*, 1299–1301. [CrossRef]
43. Liu, Z.; Li, H.; Qiu, Z.; Zhang, S.L.; Zhang, Z.B. SMALL-hysteresis thin-film transistors achieved by facile dip-coating o nanotube/polymer composite. *Adv. Mater.* **2012**, *24*, 3633–3638. [CrossRef] [PubMed]
44. Ng, L.W.; Hu, G.; Howe, R.C.; Zhu, X.; Yang, Z.; Jones, C.G.; Hasan, T. 2D ink design. In *Printing of Graphene and Related 2L Materials*; Springer: Berlin/Heidelberg, Germany, 2019; pp. 103–134.
45. Secor, E.B.; Lim, S.; Zhang, H.; Frisbie, C.D.; Francis, L.F.; Hersam, M.C. Gravure printing of graphene for large-area flexibl electronics. *Adv. Mater.* **2014**, *26*, 4533–4538. [CrossRef]
46. Dua, V.; Surwade, S.P.; Ammu, S.; Agnihotra, S.R.; Jain, S.; Roberts, K.E.; Park, S.; Ruoff, R.S.; Manohar, S.K. All-organic vapo sensor using inkjet-printed reduced graphene oxide. *Angew. Chem. Int. Ed.* **2010**, *49*, 2154–2157. [CrossRef] [PubMed]
47. Finn, D.J.; Lotya, M.; Cunningham, G.; Smith, R.J.; McCloskey, D.; Donegan, J.F.; Coleman, J.N. Inkjet deposition of liquid exfoliated graphene and MoS 2 nanosheets for printed device applications. *J. Mater. Chem. C* **2014**, *2*, 925–932. [CrossRef]
48. Kelly, A.G.; Finn, D.; Harvey, A.; Hallam, T.; Coleman, J.N. All-printed capacitors from graphene-BN-graphene nanoshee heterostructures. *Appl. Phys. Lett.* **2016**, *109*, 023107. [CrossRef]
49. Hu, G.; Albrow-Owen, T.; Jin, X.; Ali, A.; Hu, Y.; Howe, R.C.; Shehzad, K.; Yang, Z.; Zhu, X.; Woodward, R.I. Black phosphorus ink formulation for inkjet printing of optoelectronics and photonics. *Nat. Commun.* **2017**, *8*, 278. [CrossRef]
50. Howe, R.C.; Hu, G.; Yang, Z.; Hasan, T. Functional inks of graphene, metal dichalcogenides and black phosphorus for photonics and (opto) electronics. *Proc. SPIE* **2015**, *9953*, 81–97.
51. Smith, R.J.; King, P.J.; Lotya, M.; Wirtz, C.; Khan, U.; De, S.; O'Neill, A.; Duesberg, G.S.; Grunlan, J.C.; Moriarty, G. Large-scale exfoliation of inorganic layered compounds in aqueous surfactant solutions. *Adv. Mater.* **2011**, *23*, 3944–3948. [CrossRef]
52. Sun, Y.; Rogers, J.A. Inorganic semiconductors for flexible electronics. *Adv. Mater.* **2007**, *19*, 1897–1916. [CrossRef]
53. Son, D.; Lee, J.; Qiao, S.; Ghaffari, R.; Kim, J.; Lee, J.E.; Song, C.; Kim, S.J.; Lee, D.J.; Jun, S.W. Multifunctional wearable devices for diagnosis and therapy of movement disorders. *Nat. Nanotechnol.* **2014**, *9*, 397–404. [CrossRef] [PubMed]
54. Goosey, M. Printed Circuits Handbook. *Circuit World* **2010**, *36*, 150–160. [CrossRef]
55. Shahzad, F.; Alhabeb, M.; Hatter, C.B.; Anasori, B.; Hong, S.M.; Koo, C.M.; Gogotsi, Y. Electromagnetic interference shielding with 2D transition metal carbides (MXenes). *Science* **2016**, *353*, 1137–1140. [CrossRef] [PubMed]
56. Levi, M.D.; Lukatskaya, M.R.; Sigalov, S.; Beidaghi, M.; Shpigel, N.; Daikhin, L.; Aurbach, D.; Barsoum, M.W.; Gogotsi, Y Solving the capacitive paradox of 2D MXene using electrochemical quartz-crystal admittance and in situ electronic conductance measurements. *Adv. Energy Mater.* **2015**, *5*, 1400815. [CrossRef]
57. Maleski, K.; Mochalin, V.N.; Gogotsi, Y. Dispersions of two-dimensional titanium carbide MXene in organic solvents. *Chem. Mater.* **2017**, *29*, 1632–1640. [CrossRef]
58. Alhabeb, M.; Maleski, K.; Anasori, B.; Lelyukh, P.; Clark, L.; Sin, S.; Gogotsi, Y. Guidelines for Synthesis and Processing of Two-Dimensional Titanium Carbide ($Ti_3C_2T_x$ MXene). *Chem. Mater.* **2017**, *29*, 7633–7644. [CrossRef]

59. Peng, Y.-Y.; Akuzum, B.; Kurra, N.; Zhao, M.-Q.; Alhabeb, M.; Anasori, B.; Kumbur, E.C.; Alshareef, H.N.; Ger, M.-D.; Gogotsi, Y. All-MXene (2D titanium carbide) solid-state microsupercapacitors for on-chip energy storage. *Energy Environ. Sci.* **2016**, *9*, 2847–2854. [CrossRef]
60. Anasori, B.; Gogotsi, Y. MXenes: Trends, growth, and future directions. *Graphene 2D Mater.* **2022**, *7*, 75–79. [CrossRef]
61. Kurra, N.; Ahmed, B.; Gogotsi, Y.; Alshareef, H.N. MXene-on-paper coplanar microsupercapacitors. *Adv. Energy Mater.* **2016**, *6*, 1601372. [CrossRef]
62. Seok, S.-H.; Choo, S.; Kwak, J.; Ju, H.; Han, J.-H.; Kang, W.-S.; Lee, J.; Kim, S.-Y.; Lee, J.; Wang, J. Synthesis of high quality 2D carbide MXene flakes using a highly purified MAX precursor for ink applications. *Nanoscale Adv.* **2021**, *3*, 517–527. [CrossRef]
63. Wu, C.-W.; Unnikrishnan, B.; Chen, I.-W.P.; Harroun, S.G.; Chang, H.-T.; Huang, C.-C. Excellent oxidation resistive MXene aqueous ink for micro-supercapacitor application. *Energy Storage Mater.* **2020**, *25*, 563–571. [CrossRef]
64. Lim, S.; Park, H.; Yang, J.; Kwak, C.; Lee, J. Stable colloidal dispersion of octylated Ti3C2-MXenes in a nonpolar solvent. *Colloids Surf. A Physicochem. Eng. Asp.* **2019**, *579*, 123648. [CrossRef]
65. Zhang, Q.; Lai, H.; Fan, R.; Ji, P.; Fu, X.; Li, H. High concentration of Ti3C2T x MXene in organic solvent. *ACS Nano* **2021**, *15*, 5249–5262. [CrossRef]
66. Carey, M.; Hinton, Z.; Natu, V.; Pai, R.; Sokol, M.; Alvarez, N.J.; Kalra, V.; Barsoum, M.W. Dispersion and stabilization of alkylated 2D MXene in nonpolar solvents and their pseudocapacitive behavior. *Cell Rep. Phys. Sci.* **2020**, *1*, 100042. [CrossRef]
67. Wang, D.; Xin, Y.; Wang, Y.; Li, X.; Wu, H.; Zhang, W.; Yao, D.; Wang, H.; Zheng, Y.; He, Z. A general way to transform Ti3C2Tx MXene into solvent-free fluids for filler phase applications. *Chem. Eng. J.* **2021**, *409*, 128082. [CrossRef]
68. Li, X.; Huang, Z.; Zhi, C. Environmental stability of MXenes as energy storage materials. *Front. Mater.* **2019**, *6*, 312. [CrossRef]
69. Habib, T.; Zhao, X.; Shah, S.A.; Chen, Y.; Sun, W.; An, H.; Lutkenhaus, J.L.; Radovic, M.; Green, M.J. Oxidation stability of Ti3C2Tx MXene nanosheets in solvents and composite films. *npj 2D Mater. Appl.* **2019**, *3*, 8. [CrossRef]
70. Zhang, C.J.; Pinilla, S.; McEvoy, N.; Cullen, C.P.; Anasori, B.; Long, E.; Park, S.-H.; Seral-Ascaso, A.s.; Shmeliov, A.; Krishnan, D. Oxidation stability of colloidal two-dimensional titanium carbides (MXenes). *Chem. Mater.* **2017**, *29*, 4848–4856. [CrossRef]
71. Zhao, X.; Vashisth, A.; Prehn, E.; Sun, W.; Shah, S.A.; Habib, T.; Chen, Y.; Tan, Z.; Lutkenhaus, J.L.; Radovic, M. Antioxidants unlock shelf-stable Ti3C2Tx (MXene) nanosheet dispersions. *Matter* **2019**, *1*, 513–526. [CrossRef]
72. Natu, V.; Hart, J.L.; Sokol, M.; Chiang, H.; Taheri, M.L.; Barsoum, M.W. Edge capping of 2D-MXene sheets with polyanionic salts to mitigate oxidation in aqueous colloidal suspensions. *Angew. Chem.* **2019**, *131*, 12785–12790. [CrossRef]
73. Lee, Y.; Kim, S.J.; Kim, Y.-J.; Lim, Y.; Chae, Y.; Lee, B.-J.; Kim, Y.-T.; Han, H.; Gogotsi, Y.; Ahn, C.W. Oxidation-resistant titanium carbide MXene films. *J. Mater. Chem. A* **2020**, *8*, 573–581. [CrossRef]
74. Ibragimova, R.; Erhart, P.; Rinke, P.; Komsa, H.-P. Surface functionalization of 2D MXenes: Trends in distribution, composition, and electronic properties. *J. Phys. Chem. Lett.* **2021**, *12*, 2377–2384. [CrossRef]
75. Zhang, Y.Z.; Wang, Y.; Jiang, Q.; El-Demellawi, J.K.; Kim, H.; Alshareef, H.N. MXene printing and patterned coating for device applications. *Adv. Mater.* **2020**, *32*, 1908486. [CrossRef] [PubMed]
76. Ibragimova, R.; Puska, M.J.; Komsa, H.-P. pH-dependent distribution of functional groups on titanium-based MXenes. *ACS Nano* **2019**, *13*, 9171–9181. [CrossRef]
77. Persson, I.; Näslund, L.-Å.; Halim, J.; Barsoum, M.W.; Darakchieva, V.; Palisaitis, J.; Rosen, J.; Persson, P.O.Å. On the organization and thermal behavior of functional groups on Ti3C2 MXene surfaces in vacuum. *2D Mater.* **2017**, *5*, 015002. [CrossRef]
78. Ji, J.; Zhao, L.; Shen, Y.; Liu, S.; Zhang, Y. Covalent stabilization and functionalization of MXene via silylation reactions with improved surface properties. *FlatChem* **2019**, *17*, 100128. [CrossRef]
79. Kang, J.; Sangwan, V.K.; Wood, J.D.; Hersam, M.C. Solution-based processing of monodisperse two-dimensional nanomaterials. *Acc. Chem. Res.* **2017**, *50*, 943–951. [CrossRef]
80. Qian, A.; Seo, J.Y.; Shi, H.; Lee, J.Y.; Chung, C.H. Surface functional groups and electrochemical behavior in dimethyl sulfoxide-delaminated Ti3C2Tx MXene. *ChemSusChem* **2018**, *11*, 3719–3723. [CrossRef]
81. Hu, M.; Hu, T.; Li, Z.; Yang, Y.; Cheng, R.; Yang, J.; Cui, C.; Wang, X. Surface functional groups and interlayer water determine the electrochemical capacitance of Ti3C2 T x MXene. *ACS Nano* **2018**, *12*, 3578–3586. [CrossRef]
82. Shekhirev, M.; Busa, J.; Shuck, C.E.; Torres, A.; Bagheri, S.; Sinitskii, A.; Gogotsi, Y. Ultralarge flakes of $Ti_3C_2T_x$ MXene via soft delamination. *ACS Nano* **2022**, *16*, 13695–13703. [CrossRef]
83. Zhang, J.; Kong, N.; Uzun, S.; Levitt, A.; Seyedin, S.; Lynch, P.A.; Qin, S.; Han, M.; Yang, W.; Liu, J. Scalable Manufacturing of Free-Standing, Strong $Ti_3C_2T_x$ MXene Films with Outstanding Conductivity. *Adv. Mater.* **2020**, *32*, 2001093. [CrossRef] [PubMed]
84. Kayali, E.; VahidMohammadi, A.; Orangi, J.; Beidaghi, M. Controlling the dimensions of 2D MXenes for ultrahigh-rate pseudocapacitive energy storage. *ACS Appl. Mater. Interfaces* **2018**, *10*, 25949–25954. [CrossRef] [PubMed]
85. Lim, K.R.G.; Shekhirev, M.; Wyatt, B.C.; Anasori, B.; Gogotsi, Y.; Seh, Z.W. Fundamentals of MXene synthesis. *Nat. Synth.* **2022**, *1*, 601–614. [CrossRef]
86. Greaves, M.; Mende, M.; Wang, J.; Yang, W.; Barg, S. Investigating the rheology of 2D titanium carbide (MXene) dispersions for colloidal processing: Progress and challenges. *J. Mater. Res.* **2021**, *36*, 4578–4600. [CrossRef]
87. Akuzum, B.; Maleski, K.; Anasori, B.; Lelyukh, P.; Alvarez, N.J.; Kumbur, E.C.; Gogotsi, Y. Rheological Characteristics of 2D Titanium Carbide (MXene) Dispersions: A Guide for Processing MXenes. *ACS Nano* **2018**, *12*, 2685–2694. [CrossRef] [PubMed]
88. Phan-Thien, N.; Mai-Duy, N. *Understanding Viscoelasticity: An Introduction to Rheology*; Springer: Berlin/Heidelberg, Germany, 2013.

89. Jogun, S.; Zukoski, C. Rheology and microstructure of dense suspensions of plate-shaped colloidal particles. *J. Rheol.* **1999**, *43*, 847–871. [CrossRef]
90. Yang, W.; Yang, J.; Byun, J.J.; Moissinac, F.P.; Xu, J.; Haigh, S.J.; Domingos, M.; Bissett, M.A.; Dryfe, R.A.; Barg, S. 3D printing of freestanding MXene architectures for current-collector-free supercapacitors. *Adv. Mater.* **2019**, *31*, 1902725. [CrossRef]
91. Zhang, J.; Uzun, S.; Seyedin, S.; Lynch, P.A.; Akuzum, B.; Wang, Z.; Qin, S.; Alhabeb, M.; Shuck, C.E.; Lei, W. Additive-free MXene liquid crystals and fibers. *ACS Cent. Sci.* **2020**, *6*, 254–265. [CrossRef]
92. Yang, W.; Byun, J.J.; Yang, J.; Moissinac, F.P.; Peng, Y.; Tontini, G.; Dryfe, R.A.; Barg, S. Freeze-assisted tape casting of vertically aligned MXene films for high rate performance supercapacitors. *Energy Environ. Mater.* **2020**, *3*, 380–388. [CrossRef]
93. Orangi, J.; Hamade, F.; Davis, V.A.; Beidaghi, M. 3D printing of additive-free 2D $Ti_3C_2T_x$ (MXene) ink for fabrication of micro-supercapacitors with ultra-high energy densities. *ACS Nano* **2019**, *14*, 640–650. [CrossRef] [PubMed]
94. Shen, K.; Li, B.; Yang, S. 3D printing dendrite-free lithium anodes based on the nucleated MXene arrays. *Energy Storage Mater.* **2020**, *24*, 670–675. [CrossRef]
95. Abdolhosseinzadeh, S.; Schneider, R.; Verma, A.; Heier, J.; Nüesch, F.; Zhang, C. Turning trash into treasure: Additive free MXene sediment inks for screen-printed micro-supercapacitors. *Adv. Mater.* **2020**, *32*, 2000716. [CrossRef] [PubMed]
96. Zhang, C.J.; McKeon, L.; Kremer, M.P.; Park, S.-H.; Ronan, O.; Seral-Ascaso, A.; Barwich, S.; Coileáin, C.Ó.; McEvoy, N.; Nerl, H.C. Additive-free MXene inks and direct printing of micro-supercapacitors. *Nat. Commun.* **2019**, *10*, 1795. [CrossRef] [PubMed]
97. Vural, M.; Pena-Francesch, A.; Bars-Pomes, J.; Jung, H.; Gudapati, H.; Hatter, C.B.; Allen, B.D.; Anasori, B.; Ozbolat, I.T.; Gogotsi, Y. Inkjet printing of self-assembled 2D titanium carbide and protein electrodes for stimuli-responsive electromagnetic shielding. *Adv. Funct. Mater.* **2018**, *28*, 1801972. [CrossRef]
98. Gao, Q.; Feng, M.; Li, E.; Liu, C.; Shen, C.; Liu, X. Mechanical, thermal, and rheological properties of Ti3C2Tx MXene/thermoplastic polyurethane nanocomposites. *Macromol. Mater. Eng.* **2020**, *305*, 2000343. [CrossRef]
99. Hubbe, M.A. Flocculation and redispersion of cellulosic fiber suspensions: A review of effects of hydrodynamic shear and polyelectrolytes. *BioResources* **2007**, *2*, 296–331.
100. Watanabe, K.; Amari, T. Rheological properties of coatings during drying processes. *J. Appl. Polym. Sci.* **1986**, *32*, 3435–3443. [CrossRef]
101. Fan, Z.; Wei, C.; Yu, L.; Xia, Z.; Cai, J.; Tian, Z.; Zou, G.; Dou, S.X.; Sun, J. 3D printing of porous nitrogen-doped Ti3C2 MXene scaffolds for high-performance sodium-ion hybrid capacitors. *ACS Nano* **2020**, *14*, 867–876. [CrossRef]
102. Xu, G.; Wang, X.; Gong, S.; Wei, S.; Liu, J.; Xu, Y. Solvent-regulated preparation of well-intercalated Ti3C2Tx MXene nanosheets and application for highly effective electromagnetic wave absorption. *Nanotechnology* **2018**, *29*, 355201. [CrossRef]
103. Abdolhosseinzadeh, S.; Jiang, X.; Zhang, H.; Qiu, J.; Zhang, C.J. Perspectives on solution processing of two-dimensional MXenes. *Mater. Today* **2021**, *48*, 214–240. [CrossRef]
104. Zhang, C.; Kremer, M.P.; Seral-Ascaso, A.; Park, S.H.; McEvoy, N.; Anasori, B.; Gogotsi, Y.; Nicolosi, V. Stamping of flexible, coplanar micro-supercapacitors using MXene inks. *Adv. Funct. Mater.* **2018**, *28*, 1705506. [CrossRef]
105. Quain, E.; Mathis, T.S.; Kurra, N.; Maleski, K.; Van Aken, K.L.; Alhabeb, M.; Alshareef, H.N.; Gogotsi, Y. Direct writing of additive-free MXene-in-Water ink for electronics and energy storage. *Adv. Mater. Technol.* **2019**, *4*, 1800256. [CrossRef]
106. Yu, L.; Fan, Z.; Shao, Y.; Tian, Z.; Sun, J.; Liu, Z. Versatile N-doped MXene ink for printed electrochemical energy storage application. *Adv. Energy Mater.* **2019**, *9*, 1901839. [CrossRef]
107. Dai, J.; Ogbeide, O.; Macadam, N.; Sun, Q.; Yu, W.; Li, Y.; Su, B.-L.; Hasan, T.; Huang, X.; Huang, W. Printed gas sensors. *Chem. Soc. Rev.* **2020**, *49*, 1756–1789. [CrossRef]
108. Derby, B. Inkjet Printing of Functional and Structural Materials: Fluid Property Requirements, Feature Stability, and Resolution. *Annu. Rev. Mater. Res.* **2010**, *40*, 395–414. [CrossRef]
109. Wen, D.; Wang, X.; Liu, L.; Hu, C.; Sun, C.; Wu, Y.; Zhao, Y.; Zhang, J.; Liu, X.; Ying, G. Inkjet Printing Transparent and Conductive MXene ($Ti_3C_2T_x$) Films: A Strategy for Flexible Energy Storage Devices. *ACS Appl. Mater. Interfaces* **2021**, *13*, 17766–17780. [CrossRef]
110. Saleh, A.; Wustoni, S.; Bihar, E.; El-Demellawi, J.K.; Zhang, Y.; Hama, A.; Druet, V.; Yudhanto, A.; Lubineau, G.; Alshareef, H.N.; et al. Inkjet-printed $Ti_3C_2T_x$ MXene electrodes for multimodal cutaneous biosensing. *J. Phys. Mater.* **2020**, *3*, 044004. [CrossRef]
111. Uzun, S.; Schelling, M.; Hantanasirisakul, K.; Mathis, T.S.; Askeland, R.; Dion, G.; Gogotsi, Y. Additive-Free Aqueous MXene Inks for Thermal Inkjet Printing on Textiles. *Small* **2021**, *17*, 2006376. [CrossRef]
112. Jiang, X.; Li, W.; Hai, T.; Yue, R.; Chen, Z.; Lao, C.; Ge, Y.; Xie, G.; Wen, Q.; Zhang, H. Inkjet-printed MXene micro-scale devices for integrated broadband ultrafast photonics. *npj 2D Mater. Appl.* **2019**, *3*, 34. [CrossRef]
113. Schubert, C.; Van Langeveld, M.C.; Donoso, L.A. Innovations in 3D printing: A 3D overview from optics to organs. *Br. J. Ophthalmol.* **2014**, *98*, 159–161. [CrossRef] [PubMed]
114. Hassan, K.; Nine, M.J.; Tung, T.T.; Stanley, N.; Yap, P.L.; Rastin, H.; Yu, L.; Losic, D. Functional inks and extrusion-based 3D printing of 2D materials: A review of current research and applications. *Nanoscale* **2020**, *12*, 19007–19042. [CrossRef] [PubMed]
115. Fu, K.; Wang, Y.; Yan, C.; Yao, Y.; Chen, Y.; Dai, J.; Lacey, S.; Wang, Y.; Wan, J.; Li, T. Graphene oxide-based electrode inks for 3D-printed lithium-ion batteries. *Adv. Mater.* **2016**, *28*, 2587–2594. [CrossRef] [PubMed]
116. Redondo, E.; Pumera, M. MXene-functionalised 3D-printed electrodes for electrochemical capacitors. *Electrochem. Commun.* **2021**, *124*, 106920. [CrossRef]

17. Foo, C.Y.; Lim, H.N.; Mahdi, M.A.; Wahid, M.H.; Huang, N.M. Three-dimensional printed electrode and its novel applications in electronic devices. *Sci. Rep.* **2018**, *8*, 7399. [CrossRef] [PubMed]
18. Iffelsberger, C.; Ng, S.; Pumera, M. Catalyst coating of 3D printed structures via electrochemical deposition: Case of the transition metal chalcogenide MoSx for hydrogen evolution reaction. *Appl. Mater. Today* **2020**, *20*, 100654. [CrossRef]
19. Rastin, H.; Zhang, B.; Mazinani, A.; Hassan, K.; Bi, J.; Tung, T.T.; Losic, D. 3D bioprinting of cell-laden electroconductive MXene nanocomposite bioinks. *Nanoscale* **2020**, *12*, 16069–16080. [CrossRef]
20. Cao, W.-T.; Ma, C.; Mao, D.-S.; Zhang, J.; Ma, M.-G.; Chen, F. MXene-Reinforced Cellulose Nanofibril Inks for 3D-Printed Smart Fibres and Textiles. *Adv. Funct. Mater.* **2019**, *29*, 1905898. [CrossRef]
21. Huang, X.; Huang, J.; Yang, D.; Wu, P. A Multi-Scale Structural Engineering Strategy for High-Performance MXene Hydrogel Supercapacitor Electrode. *Adv. Sci.* **2021**, *8*, 2101664. [CrossRef]
22. Jambhulkar, S.; Liu, S.; Vala, P.; Xu, W.; Ravichandran, D.; Zhu, Y.; Bi, K.; Nian, Q.; Chen, X.; Song, K. Aligned Ti3C2T x MXene for 3D Micropatterning via Additive Manufacturing. *ACS Nano* **2021**, *15*, 12057–12068. [CrossRef]
23. Li, F.; Li, Y.; Qu, J.; Wang, J.; Bandari, V.K.; Zhu, F.; Schmidt, O.G. Recent developments of stamped planar micro-supercapacitors: Materials, fabrication and perspectives. *Nano Mater. Sci.* **2021**, *3*, 154–169. [CrossRef]
24. Sundriyal, P.; Bhattacharya, S. Scalable micro-fabrication of flexible, solid-state, inexpensive, and high-performance planar micro-supercapacitors through inkjet printing. *ACS Appl. Energy Mater.* **2019**, *2*, 1876–1890. [CrossRef]
25. Lin, Z.; Barbara, D.; Taberna, P.-L.; Van Aken, K.L.; Anasori, B.; Gogotsi, Y.; Simon, P. Capacitance of $Ti_3C_2T_x$ MXene in ionic liquid electrolyte. *J. Power Sources* **2016**, *326*, 575–579. [CrossRef]
26. Chen, W.; Luo, M.; Yang, K.; Liu, C.; Zhang, D.; Zhou, X. MXene loaded onto clean wiper by a dot-matrix drop-casting method as a free-standing electrode for stretchable and flexible supercapacitors. *Chem. Eng. J.* **2021**, *423*, 130242. [CrossRef]
27. Yu, M.; Feng, X. Scalable manufacturing of MXene films: Moving toward industrialization. *Matter* **2020**, *3*, 335–336. [CrossRef]
28. Huang, H.; He, J.; Wang, Z.; Zhang, H.; Jin, L.; Chen, N.; Xie, Y.; Chu, X.; Gu, B.; Deng, W.; et al. Scalable, and low-cost treating-cutting-coating manufacture platform for MXene-based on-chip micro-supercapacitors. *Nano Energy* **2020**, *69*, 104431. [CrossRef]
29. Montazeri, K.; Currie, M.; Verger, L.; Dianat, P.; Barsoum, M.W.; Nabet, B. Beyond Gold: Spin-Coated Ti3C2-Based MXene Photodetectors. *Adv. Mater.* **2019**, *31*, 1903271. [CrossRef]
30. Zhang, M.; Wang, Y.; Gao, F.; Wang, Y.; Shen, X.; He, N.; Zhu, J.; Chen, Y.; Wan, X.; Lian, X. Formation of new MXene film using spinning coating method with DMSO solution and its application in advanced memristive device. *Ceram. Int.* **2019**, *45*, 19467–19472. [CrossRef]
31. Hantanasirisakul, K.; Zhao, M.Q.; Urbankowski, P.; Halim, J.; Anasori, B.; Kota, S.; Ren, C.E.; Barsoum, M.W.; Gogotsi, Y. Fabrication of Ti3C2Tx MXene transparent thin films with tunable optoelectronic properties. *Adv. Electron. Mater.* **2016**, *2*, 1600050. [CrossRef]
32. Wang, Z.; Kim, H.; Alshareef, H.N. Oxide thin-film electronics using all-MXene electrical contacts. *Adv. Mater.* **2018**, *30*, 1706656. [CrossRef] [PubMed]
33. Sarycheva, A.; Polemi, A.; Liu, Y.; Dandekar, K.; Anasori, B.; Gogotsi, Y. 2D titanium carbide (MXene) for wireless communication. *Sci. Adv.* **2018**, *4*, eaau0920. [CrossRef]
34. Couly, C.; Alhabeb, M.; Van Aken, K.L.; Kurra, N.; Gomes, L.; Navarro-Suárez, A.M.; Anasori, B.; Alshareef, H.N.; Gogotsi, Y. Asymmetric flexible MXene-reduced graphene oxide micro-supercapacitor. *Adv. Electron. Mater.* **2018**, *4*, 1700339. [CrossRef]
35. Zhou, H.; Wang, Y.; Wang, F.; Deng, H.; Song, Y.; Li, C.; Ling, Z. Water permeability in MXene membranes: Process matters. *Chin. Chem. Lett.* **2020**, *31*, 1665–1669. [CrossRef]
36. Qin, L.; Jiang, J.; Tao, Q.; Wang, C.; Persson, I.; Fahlman, M.; Persson, P.O.; Hou, L.; Rosen, J.; Zhang, F. A flexible semitransparent photovoltaic supercapacitor based on water-processed MXene electrodes. *J. Mater. Chem. A* **2020**, *8*, 5467–5475. [CrossRef]
37. Hu, H.; Hua, T. An easily manipulated protocol for patterning of MXenes on paper for planar micro-supercapacitors. *J. Mater. Chem. A* **2017**, *5*, 19639–19648. [CrossRef]

MDPI AG
Grosspeteranlage 5
4052 Basel
Switzerland
Tel.: +41 61 683 77 34

*Nanomaterials* Editorial Office
E-mail: nanomaterials@mdpi.com
www.mdpi.com/journal/nanomaterials

www.ingramcontent.com/pod-product-compliance
Lightning Source LLC
Chambersburg PA
CBHW062043220926
51850CB00010B/2649
* 9 7 8 3 7 2 5 8 5 4 8 7 5 *